Transforming Agriculture with AI, IoT, and Blockchain

Htet Ne Oo • Binay Kumar Singh
Anuj Kumar
Editors

Transforming Agriculture with AI, IoT, and Blockchain

Smart Solutions for a Digital Future

Springer

Editors
Htet Ne Oo
Department of Computer Science
and Engineering
Chitkara University Institute of Engineering
and Technology (CUIET), Chitkara
University,
Rajpura, Punjab, India

Binay Kumar Singh
Indian Institute of Wheat and Barley
Research
Karnal, Haryana, India

Anuj Kumar
Indian Institute of Wheat and Barley
Research
Karnal, Haryana, India

ISBN 978-981-92-1490-7 ISBN 978-981-92-1491-4 (eBook)
https://doi.org/10.1007/978-981-92-1491-4

This Springer imprint is published by the registered company Springer Nature Singapore Pte Ltd.
The registered company address is: 152 Beach Road, #21-01/04 Gateway East, Singapore 189721, Singapore

Foreword

Agriculture is changing rapidly due to climate change, resource scarcity, environmental degradation, and the need for sustainable food systems to support a growing population. Traditional methods are no longer sufficient for these complex challenges. Integrating advanced digital technologies such as deep learning, the Internet of Things (IoT), and blockchain provides a promising path to intelligent, adaptive, and transparent agricultural systems. The volume *Transforming Agriculture with AI, IoT, and Blockchain—Smart Solutions for a Digital Future* is a timely and comprehensive resource that links advanced computational innovations with practical agricultural applications.

A key strength of this volume is its systems-oriented approach, presenting digital technologies as interconnected drivers of next-generation agriculture. Editors **Drs. Htet Ne Oo**, **Binay Kumar Singh**, and **Anuj Kumar** have curated a well-organized collection integrating theoretical foundations, technological architectures, and global case studies. The chapters provide a deep understanding of artificial intelligence for predictive analytics and decision support; IoT-enabled precision farming and real-time monitoring; and blockchain frameworks for data integrity, traceability, and decentralized transactions. The volume also covers data governance, cybersecurity, ethical algorithm deployment, and digital inclusion, ensuring technological progress supports equity, transparency, and sustainability.

The inclusion of diverse case studies and interdisciplinary perspectives enhances the practical relevance and global applicability of this work. It offers an overview of adoption pathways, scalability challenges, and impact assessment across different contexts. By combining scientific rigor with actionable knowledge, this volume sets a high standard for interdisciplinary scholarship. It is an essential reference for researchers, policymakers, industry professionals, and development practitioners committed to advancing smart, inclusive, and sustainable agricultural transformation.

Director, ICAR-Indian Institute
of Wheat and Barley Research
Karnal, Haryana, India

Ratan Tiwari

Preface

Agriculture is undergoing transformation as digital technologies rapidly integrate with biological and ecological systems. This shift is driven by mounting challenges, including climate variability, natural resource degradation, labor shortages, and the rising demand for food and nutritional security. These factors require a move from traditional, experience-based methods to intelligent, data-driven, and adaptive agricultural systems. *Transforming Agriculture with AI, IoT, and Blockchain—Smart Solutions for a Digital Future* has been developed to address a notable gap in the literature, where these technologies are frequently examined in isolation rather than as interconnected elements within a unified agricultural innovation ecosystem. By employing a systems-level perspective, this volume explores the scientific foundations, technological architectures, and integrated applications of deep learning, IoT, and blockchain throughout the agricultural value chain. It also addresses critical issues such as data privacy, cybersecurity, algorithmic accountability, and the potential for digital exclusion.

This book is designed for a diverse, interdisciplinary audience, including researchers, academics, data scientists, engineers, policymakers, development practitioners, extension professionals, and postgraduate students seeking advanced knowledge of smart agricultural systems. The volume is structured into 17 chapters, progressing from foundational concepts and global perspectives to domain-specific applications of deep learning, IoT, and blockchain. Subsequent chapters present integrative frameworks, case studies from both developed and developing regions, and analyses of socioeconomic impacts, policy considerations, and future innovation pathways. As an edited collection, it features contributions from experts across disciplines and regions, ensuring a balanced integration of theoretical rigor, empirical evidence, and practical relevance. We, the editors, gratefully acknowledge the contributions of all authors, the support of their institutions, and the guidance of the publisher and reviewers. We hope this volume will meaningfully advance sustainable, inclusive, and technology-enabled agricultural transformation.

Rajpura, Punjab, India — Htet Ne Oo
Karnal, Haryana, India — Binay Kumar Singh
Karnal, Haryana, India — Anuj Kumar

Contents

1 **Introduction to Next-Generation Agriculture** 1
Heena Wadhwa, Htet Ne Oo, Astha Gupta, and Ochin Sharma

2 **Deep Learning and Artificial Intelligence Applications in Agriculture** 15
Ramya Sakthivel and Kalimuthu Krishnan

3 **Agri-Tech Ecosystems: The Sustainable Agriculture by Smart Digital Innovations and Global Drivers** 53
Srikanta Kumar Mohapatra, Jayashree Mohanty, Sumant Kumar Mohapatra, and Htet Ne Oo

4 **AI and Deep Learning for Crop and Soil Health Monitoring** 65
Rajwinder Kaur, Saravjeet Singh, and Rajender Kumar

5 **Predictive Analysis and Decision Support Systems in Agriculture** ... 85
Nilotpal Das, Atin Kumar, and Sharad Sachan

6 **Intelligent Automation and Robotics for Smart Agriculture: Transforming Farming Through Technological Synergy** 101
Anuradha Reddy, G. S. Pradeep Ghantasala, A. V. Lakshmi Prasuna, Shaik Sharmila, Aswathi Sheelan, and M. Arvindhan

7 **IoT for Precision Farming and Resource Optimization** 113
Prachi Arora and Sandeep Mahato

8 **Sensor Networks and Real-Time Farm Data Acquisition** 135
Ruchika Bindal, Mandeep Kaur, and Righa Tandon

9 **Smart Greenhouses and Controlled Environment Agriculture** 153
G. S. Pradeep Ghantasala, Pellakuri Vidyullatha, and Iacovos Ioannou

10 **Blockchain for Agri-Supply Chain Traceability** 171
Shruti Mishra and Diksha Bali

11 Smart Contracts and Decentralized Marketplaces in Agriculture 187
Soumya Mohapatra, P. Sruthi, Ramadas Sendhil, and Anuj Kumar

12 Smart Sensing and AI for Monitoring Rancidity in PUFA-Rich Agricultural Products 201
Anupma Sharma, Ritesh Kumar, Rishemjit Kaur, Sudeshna Bagchi, Saurav Kumar, and Amol P. Bhondekar

13 Socioeconomic Impact of Smart Agriculture 223
Debi Kalyan Jayasingh and Abhijeet Satapathy

14 Transformative Egg Industry Revolution: A Comprehensive Framework for AI, IoT, Machine Learning, and Blockchain Integration 241
Saurav Kumar, Binay Kumar Singh, Vikash K. Chaudhary, Akshita Sharma, Anupma Sharma, and Nandkishore Thombare

15 Policy Frameworks for Capacity Building and Education in Agriculture 257
Astha Gupta, Pardeep Singh Tiwana, and Gagandeep Kaur

16 Case Studies of Transforming Agriculture in Developing and Developed Nations 277
Htet Ne Oo, Heena Wadhwa, Srikanta Kumar Mohapatra, Binay Kumar Singh, and Saurav Kumar

Editors and Contributors

About the Editors

Htet Ne Oo is an accomplished academician and researcher with over 15 years of experience in Computer Science and Information Technology, covering teaching, research, and international academic coordination. She currently serves as a Professor in the Department of Computer Science and Engineering at Chitkara University, Punjab, India. She holds a PhD in Information Technology from the University of Technology, Yatanarpon Cyber City, Myanmar. She has contributed to several international and national research projects in collaboration with organizations such as the United Nations' World Food Programme (WFP), the Food and Agriculture Organization (FAO), and the ASEAN Cyber University Project with Busan Digital University, Republic of Korea. Her research focuses on artificial intelligence (AI), machine learning (ML), GIS, and remote sensing, with applications in urban analytics, emergency response systems, and sustainable development. She has made notable contributions to smart agriculture through patents on AI-driven livestock monitoring, smart scarecrow systems, and integrated crop–livestock solutions. She has supervised postgraduate and doctoral research, published widely in reputed journals and conferences, and holds professional certifications, including the Wipro Faculty Certification in Full Stack Java. She has held key administrative roles, including Head of International Relations, contributing to global academic partnerships.

Binay Kumar Singh is Principal Scientist at ICAR–Indian Institute of Wheat and Barley Research (IIWBR), Karnal, India. He has over 18 years of experience in genomics and molecular breeding for stress resilience in major crops, including rice, wheat, and mustard. He holds a PhD and MSc in Molecular Biology and Biotechnology from ICAR–Indian Agricultural Research Institute (IARI), New Delhi, and a BSc in Agriculture from Banaras Hindu University, Varanasi. He has served as Head of the Division of Crop Science at ICAR–Research Complex for North Eastern Hill Region, Meghalaya, and held scientific positions at the Directorate of Rapeseed-Mustard Research, Bharatpur, and ICAR–Indian Institute of Agricultural Biotechnology (IIAB), Ranchi. He has developed seven improved crop varieties and registered three genetic stocks. He has supervised numerous postgraduate and doctoral students and established over 20 institutional collaborations, including partnerships with start-ups. He has published more than 80 research papers, as well as books, book chapters, and technical bulletins, and has reviewed and edited manuscripts for high-impact journals.

Anuj Kumar is Principal Scientist at ICAR–Indian Institute of Wheat and Barley Research (IIWBR), Karnal, India, with over 25 years of experience in agricultural research, education, and extension. He earned his PhD in Agricultural Extension from ICAR–National Dairy Research Institute (NDRI), Karnal. His work focuses on translating scientific innovations into field-level impact, enhancing farmers' livelihoods, and promoting sustainable, climate-resilient agriculture. He has been involved in national and international collaborations with leading organizations such as ACIAR, ICARDA, and ICRISAT. His key interests include capacity building of farmers, rural youth, and farm women; conservation agriculture; and resource-efficient and natural farming systems. As a visiting faculty member at the Extension Education Institute, Nilokheri, Haryana, and an author of numerous scientific and extension publications, he bridges the gap among research, policy, and practice.

Contributors

Prachi Arora Indian Institute of Information Technology, Una, Himachal Pradesh, India

M. Arvindhan American School of Management, American University of Science, Hồ Chí Minh, Vietnam

Sudeshna Bagchi CSIR-Central Scientific Instruments Organization, Chandigarh, Punjab, India

Diksha Bali ICAR-National Institute of Agricultural Economics and Policy Research, New Delhi, India

Amol P. Bhondekar CSIR-Central Scientific Instruments Organization, Chandigarh, Punjab, India

Ruchika Bindal Chitkara University Institute of Engineering and Technology, Chitkara University, Rajpura, Punjab, India

Vikash K. Chaudhary Divyansh Agro, Ranchi, Jharkhand, India

Nilotpal Das School of Agriculture, Uttaranchal University, Dehradun, Uttarakhand, India

Astha Gupta Chitkara University Institute of Engineering and Technology, Chitkara University, Rajpura, Punjab, India

Iacovos Ioannou Department of Computer Science, Philips University, Nicosia, Cyprus

Debi Kalyan Jayasingh Odisha University of Agriculture and Technology, Bhubaneswar, Odisha, India

Gagandeep Kaur CGC, College of Engineering, Mohali, Punjab, India

Mandeep Kaur Chitkara University Institute of Engineering and Technology, Chitkara University, Rajpura, Punjab, India

Rajwinder Kaur Chitkara University Institute of Engineering and Technology, Chitkara University, Rajpura, Punjab, India

Rishemjit Kaur CSIR-Central Scientific Instruments Organization, Chandigarh, Punjab, India

Kalimuthu Krishnan Department of Electronics and Communication Engineering, College of Engineering and Technology, SRM Institute of Science and Technology, Kattankulathur, Tamil Nadu, India

Anuj Kumar Indian Institute of Wheat and Barley Research, Karnal, Haryana, India

Atin Kumar School of Agriculture, Uttaranchal University, Dehradun, Uttarakhand, India

Rajender Kumar Chitkara University Institute of Engineering and Technology, Chitkara University, Rajpura, Punjab, India

Ritesh Kumar CSIR-Central Scientific Instruments Organization, Chandigarh, Punjab, India

Saurav Kumar CSIR-Central Scientific Instruments Organization, Chandigarh, Punjab, India

Academy of Scientific and Innovative Research, Ghaziabad, Uttar Pradesh, India

A. V. Lakshmi Prasuna Department of Information Technology, Mahatma Gandhi Institute of Technology, Hyderabad, Telangana, India

Sandeep Mahato Chaibasa Engineering College, Kelende, Jharkhand, India

Shruti Mishra ICAR-National Institute of Agricultural Economics and Policy Research, New Delhi, India

Jayashree Mohanty Department of Computer Science and Engineering, Chandigarh University, Mohali, Punjab, India

Soumya Mohapatra ICAR-Central Institute of Post-Harvest Engineering & Technology, Ludhiana, Punjab, India

Srikanta Kumar Mohapatra Chitkara University Institute of Engineering and Technology, Chitkara University, Rajpura, Punjab, India

Sumant Kumar Mohapatra Department of CSE (AI&ML), School of Engineering, Dayananda Sagar University, Bengaluru, Karnataka, India

Htet Ne Oo Department of Computer Science and Engineering, Chitkara University Institute of Engineering and Technology (CUIET), Chitkara University, Rajpura, Punjab, India

G. S. Pradeep Ghantasala Department of Computer Science and Engineering, Alliance School of Advanced Computing, Alliance University, Bengaluru, Karnataka, India

Anuradha Reddy Sridevi Women's Engineering College, Hyderabad, Telangana, India

Sharad Sachan Department of Agricultural Economics and Extension, School of Agriculture, Lovely Professional University, Phagwara, Punjab, India

Ramya Sakthivel Department of Electronics and Communication Engineering, College of Engineering and Technology, SRM Institute of Science and Technology, Kattankulathur, Tamil Nadu, India

Abhijeet Satapathy Birsa Agricultural University, Ranchi, Jharkhand, India

Ramadas Sendhil Pondicherry University, Kalapet, Puducherry, India

Akshita Sharma CSIR-Central Scientific Instruments Organization, Chandigarh, Punjab, India

Anupma Sharma CSIR-Central Scientific Instruments Organization, Chandigarh, Punjab, India

Academy of Scientific and Innovative Research, Ghaziabad, Uttar Pradesh, India

Ochin Sharma Chitkara University Institute of Engineering and Technology, Chitkara University, Rajpura, Punjab, India

Shaik Sharmila Sridevi Women's Engineering College, Hyderabad, Telangana, India

Aswathi Sheelan Sridevi Women's Engineering College, Hyderabad, Telangana, India

Binay Kumar Singh Indian Institute of Wheat and Barley Research, Karnal, Haryana, India

Saravjeet Singh Chitkara University Institute of Engineering and Technology, Chitkara University, Rajpura, Punjab, India

P. Sruthi ICAR-National Dairy Research Institute, Karnal, Haryana, India

Righa Tandon Chitkara University Institute of Engineering and Technology, Chitkara University, Rajpura, Punjab, India

Nandkishore Thombare ICAR-National Institute of Secondary Agriculture, Ranchi, Jharkhand, India

Pardeep Singh Tiwana CGC, College of Engineering, Mohali, Punjab, India

Pellakuri Vidyullatha Department of Computer Science and Engineering, Koneru Lakshmaiah Education Foundation, Vaddeswaram, Guntur, Andhra Pradesh, India

Heena Wadhwa Chitkara University Institute of Engineering and Technology, Chitkara University, Rajpura, Punjab, India

Introduction to Next-Generation Agriculture

1

Heena Wadhwa, Htet Ne Oo, Astha Gupta,
and Ochin Sharma

Abstract

Modern agriculture extends beyond seeds and soil. It now incorporates sensors and smart devices that monitor plant health, soil quality, and weather conditions. These interconnected technologies enable real-time tracking of crop yields, thereby enhancing resource efficiency and productivity. Digital innovations such as the Internet of Things (IoT) and deep learning (DL) facilitate disease detection and yield prediction through advanced data-driven models. Additionally, blockchain technology contributes to a more transparent and secure agricultural supply chain, fostering a smart agricultural ecosystem that addresses resource availability and promotes sustainable farming systems globally. This chapter examines the historical evolution of agricultural practices, tracing the progression from traditional methods to contemporary technology-driven approaches.

Keywords

Digital agriculture · Ecosystem · Sustainable farming · Crop monitoring

H. Wadhwa (✉) · A. Gupta · O. Sharma
Chitkara University Institute of Engineering and Technology, Chitkara University, Rajpura, Punjab, India

H. N. Oo
Department of Computer Science and Engineering Chitkara University Institute of Engineering and Technology (CUIET), Chitkara University, Rajpura, Punjab, India

H. N. Oo et al. (eds.), *Transforming Agriculture with AI, IoT, and Blockchain*,
https://doi.org/10.1007/978-981-92-1491-4_1

1.1 Introduction

Accurately predicting crop yields is extremely challenging due to the complex interplay of factors such as pest outbreaks, soil erosion, changing weather patterns, and water scarcity. Traditional statistical methods that rely on historical data often fail to account for these dynamic conditions, resulting in less precise and sometimes unreliable estimates (Mohan et al. 2025). The impact of climate change, including extreme heat and unpredictable weather, further complicates crop yield estimation. Traditional farming methods are insufficient to address these evolving challenges. In response, smart agriculture techniques that integrate modern technologies have emerged to enhance sustainability, productivity, efficiency, and transparency. Despite these advancements, crop yield forecasting remains complicated by the multifaceted effects of pests, weather variability, soil degradation, and water shortages. Previous forecasting approaches primarily rely on historical data and statistics, leading to high failure rates due to their inability to adapt to rapidly changing conditions, resulting in low accuracy (Mohan et al. 2025). Factors such as extreme heat, prolonged droughts, and unpredictable weather, all driven by climate change, continue to affect crop production annually. Additionally, labor shortages and resource scarcity further complicate crop yield prediction.

Farming is undergoing a significant transformation through the adoption of smart agriculture, which takes advantage of modern technologies to enhance sustainability, productivity, and transparency. Furthermore, advanced computational technologies such as artificial intelligence (AI) and deep learning (DL) facilitate improved decision-making in crop yield prediction, crop growth monitoring, and pest management. The Internet of Things (IoT) enables precision farming through automation, real-time sensing, and data analytics to optimize resource efficiency. Smart greenhouses with controlled environments support climate-resilient food production and facilitate year-round cultivation. Additionally, blockchain technology enhances efficiency and trust by introducing traceability and transparency throughout the agricultural supply chain (Assimakopoulos et al. 2025). This chapter examines the historical evolution of agricultural practices and the transition from traditional to technology-driven agriculture.

1.2 Historical Evolution of Agricultural Practices

Agriculture represents one of the oldest ecosystems, having undergone significant transformations over time. The agricultural system continues to adapt to emerging challenges, technological advancements, and evolving local priorities (Suhail 2022). Early agricultural societies developed advanced techniques such as terrace farming, crop rotation, and organic fertilization, which laid the foundation for future agricultural progress (Suhail 2022; Liu et al. 2022). Numerous ancient civilizations, including those in Mesopotamia and China, pioneered the development of irrigation, shifting agriculture, and domestication. These methods established sustainable

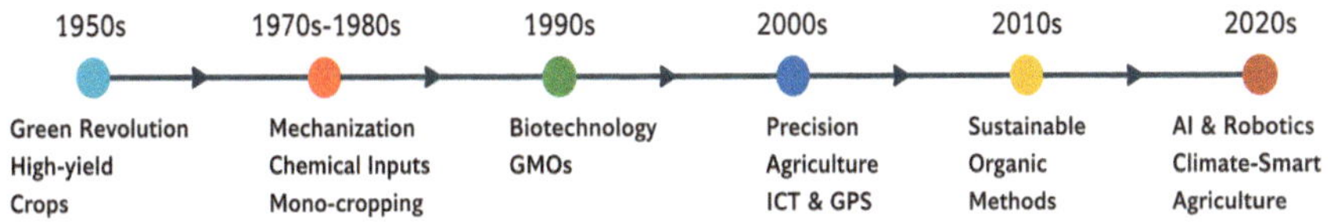

Fig. 1.1 Agricultural evolution timeline

systems that have influenced land management practices for decades (Angelakis et al. 2012).

Figure 1.1 illustrates the historical evolution of agricultural practices from the 1950s onward, highlighting significant transitions and key innovations of the era. Dr. Norman Borlaug, who developed high-yield wheat varieties, led the global Green Revolution during the 1940s and 1950s, particularly in Mexico (Hedden 2003). High-yielding varieties, rapid crop growth, irrigation, crop intensification, and improved farm management collectively contributed to the first Green Revolution, increasing productivity per unit of land (Thenkabail 2010). Peng et al. (2022) demonstrated improvements in agricultural production resulting from mechanization, examining the impact of mechanization levels on agricultural output and revenue using a threshold effect model. Genetically modified crops, also referred to as biotech crops, have been developed through genetic engineering techniques, and more than 30 nations have approved their cultivation (Bekele-Alemu et al. 2025). In the 2000s, precision agriculture and Information and Communication Technology (ICT) enabled real-time data collection and crop record tracking. These technologies facilitate monitoring of soil and water conditions using sensors and drones and transmit data from farms to cloud platforms. Rovira-Más et al. (2008) discussed the effectiveness of precision nutrient management and the accessibility of technologies for assessing soil and climate conditions. Ongoing evaluation and collaboration between farmers and researchers are essential for assessing the effectiveness of these strategies.

Patle et al. (2020) described the impact of organic farming on enhancing the health of agro-ecosystems. Public awareness of healthy eating has increased in response to the rising incidence of health problems and concerns about toxic substances, leading to greater interest in organic foods. It is widely acknowledged that organically grown foods are healthier than those produced through conventional farming methods. Several researchers have investigated the application of AI and DL in agriculture (Benos et al. 2021). Benos et al. (2021) identified the effects of sensors and machine learning (ML) algorithms on crop and livestock management. Their work demonstrates that remote sensing and satellite sensors are used to collect the extensive data required for ML analytics, as illustrated in Fig. 1.2.

Kumari et al. (2025) highlighted the need to increase agricultural productivity while reducing environmental impact. Their approach integrates climate-oriented sustainable agriculture with modern technologies, including sensors, IoT, AI, ML, and remote sensing.

Fig. 1.2 Overview of agricultural technology

1.3 Transition from Traditional to Technology-Driven Agriculture

Agriculture has undergone a significant transformation over time. While it previously relied on manual labor and seasonal cycles, the sector is now adopting modern, technology-based methods to enhance productivity and promote long-term sustainability. Analyzing the strengths, limitations, and challenges of both traditional and modern farming practices informs policy development, supports sustainable growth, and improves livelihoods in rural communities (Hussein et al. 2025).

1.3.1 Historical Overview and Evolution

1.3.1.1 Traditional Agriculture

Traditional farming primarily involves cultivating crops to meet household needs, utilizing basic tools and minimal financial resources. Farmers typically rely on personal experience and seasonal variations to guide their decisions, rather than employing mechanized equipment or contemporary agricultural techniques.

1.3.1.2 Technology-Oriented Agriculture

Modern agriculture employs sensors, machinery, and computer systems to manage and monitor crops with greater precision. These methods collectively constitute smart farming. Smart farming integrates connected devices, data tracking systems, and automation to enhance operational efficiency and improve agricultural

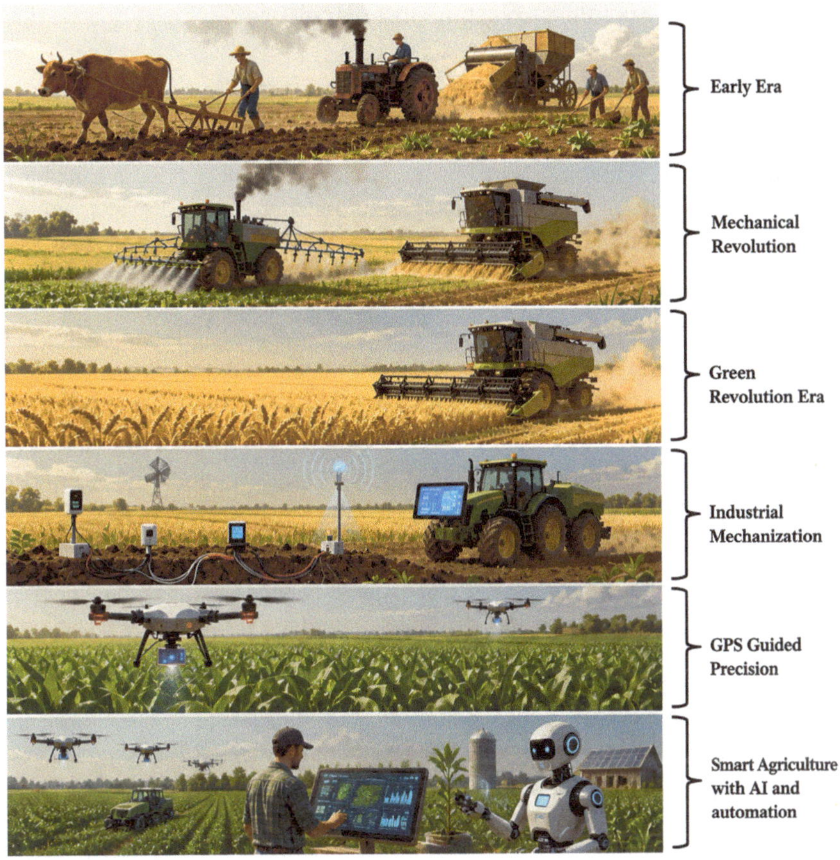

Fig. 1.3 Historical progression of agricultural techniques and farming methods

outcomes (Yang et al. 2021). Over time, farming techniques have evolved with the introduction of new tools and technologies, as depicted in Fig. 1.3. The initial stage illustrates the use of hand tools, representing early agriculture that depended primarily on human and animal labor. The subsequent stage demonstrates the mechanical revolution, during which steam-powered tractors and threshing machines replaced manual labor with mechanical energy. The third stage highlights the adoption of tractors and chemical fertilizers, which significantly increased crop yields but also raised concerns regarding soil health and environmental sustainability. The fourth stage marks the emergence of large-scale machinery, which further improved efficiency and enabled mass production, though often at the expense of soil quality and biodiversity. The final stage depicts contemporary farming practices utilizing advanced sensors and data-driven irrigation decisions. These technologies enhance crop production efficiency and support optimized fertilization. This stage exemplifies the era of digital and connected agriculture, where field data is collected and analyzed through cloud or fog computing systems.

1.3.2 Technological Innovations in Agriculture

1.3.2.1 IoT and Sensor Networks

Real-time monitoring of soil conditions, crop health, soil moisture, and weather helps farmers manage resources better and respond quickly to emerging issues. The use of connected devices, or IoT devices, is growing worldwide. These technologies provide farmers in rural areas with important information to support better agricultural decisions (Kumar et al. 2024).

1.3.2.2 AI and ML

AI enables the identification of patterns, prediction of pest infestations, weather fluctuations, and crop yields, as well as the automation of tasks such as irrigation and machinery operation. ML facilitates improved crop planning by delivering accurate forecasts and enhancing farm management practices (Abbasi et al. 2022; Dhanasekar et al. 2023).

1.3.2.3 Big Data Analytics and Cloud Platforms

Data collected from sensors, drones, and supply chains are analyzed to improve agricultural efficiency, minimize waste, and inform better planning and policy decisions. Online tools such as SmartFarm and Agrivi integrate information from multiple sources, enabling both small- and large-scale farmers to manage their operations more effectively (Dhanasekar et al. 2023). Robots and drones are increasingly being deployed for planting, weeding, spraying, and harvesting, enabling faster work and continuous operations. Research indicates that these technologies can increase crop yields by up to 30% and reduce material waste by nearly 60% compared to traditional farming methods (Gamage et al. 2024).

1.4 Major Challenges in Contemporary Agriculture

Agriculture plays a critical role in providing food, supporting economic development, and sustaining rural communities. Despite its significance, the sector faces numerous challenges that threaten its long-term growth and stability. Climate change, for example, alters rainfall patterns and disrupts established farming seasons, increasing food production unpredictability and reducing crop yields (Dar and Laxmipathi Gowda 2013; Nautiyal et al. 2025). Production is further constrained by limited access to seeds, fertilizers, water, and high-quality soil. Additionally, urban migration has led to a declining agricultural workforce, as fewer individuals choose farming as a profession. The aging farming population and lack of new personnel hinder the adoption of modern technologies and exacerbate labor shortages (Hussein et al. 2025).

Ahmed and Shakoor (2025) state that farming, the environment, and the global food supply are significantly affected by climate change, particularly in vulnerable regions. Habib-ur-Rahman et al. (2022) analyzed the declining interest in agricultural labor and its consequences for farming. The adoption of technology and

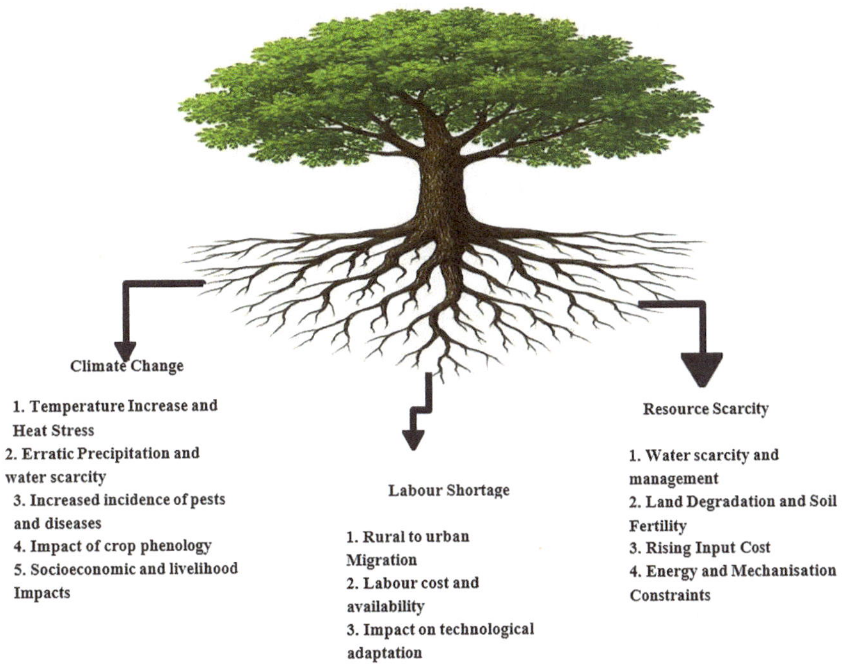

Fig. 1.4 Declining agricultural productivity

machinery is identified as a potential solution to address labor shortages. These authors also provided recommendations to enhance agricultural practices. Rising greenhouse gas concentrations in the atmosphere are widely recognized as having profound effects on Earth's climate, weather patterns, and global temperatures (Dar and Laxmipathi Gowda 2013). Additional factors contributing to reduced agricultural productivity, as illustrated in Fig. 1.4, include labor shortages and insufficient resources. To alleviate the pressures of climate change, high food prices, and environmental challenges on rural populations in developing countries, the development of innovative technical and policy solutions is essential.

1.4.1 Climate Change Impacts

Climate change poses a significant challenge to modern agriculture, affecting all stages of global food production. Elevated atmospheric CO_2, unpredictable rainfall, extreme weather events, and rising temperatures disrupt agricultural development, compromise crop health, and reduce yields. These environmental changes also disturb the natural ecological balance (Tchonkouang et al. 2024). Smallholder farmers and regions dependent on rainfall are particularly vulnerable due to limited resources and adaptive capacity (Touch et al. 2024). Rising temperatures place additional stress on crops, shorten optimal growing periods, and diminish grain yields (Muralikrishnan et al. 2021; Aijaz et al. 2025). Altered rainfall patterns and increased drought frequency further disrupt rainfed agriculture and irrigation strategies (Ringler et al. 2022). Crop modeling studies indicate that reduced growing periods, insufficient rainfall, and extreme temperatures may lead to decreased harvests in many vulnerable regions (Syed et al. 2021; Awuni et al. 2023).

1.4.2 Resource Constraints

Contemporary agriculture faces challenges in achieving sustainable food production due to the depletion of essential resources. As soil nutrient levels decline, increased fertilizer application becomes necessary, resulting in higher costs for farmers and potential environmental harm. Although environmentally sustainable soil management practices offer potential solutions, their adoption remains limited among resource-constrained farmers (Luna Juncal et al. 2023).

1.4.3 Labor Shortages

The scarcity of agricultural laborers results in delays in essential tasks such as planting, harvesting, pest control, and food storage (Xiong et al. 2025). These delays contribute to financial losses and reduced yields for farmers. Furthermore, labor shortages hinder the adoption of advanced machinery and environmentally sustainable practices, limiting farmers' ability to adapt to climate variability and market fluctuations. These challenges intensify the strain on traditional agricultural methods, making it increasingly difficult for farmers to maintain stable incomes. Addressing these issues requires the implementation of sustainable agricultural practices, increased technological adoption to conserve water and other resources, and enhanced government support, particularly in response to ongoing labor shortages.

1.5 Role of Innovation in Ensuring Food Security and Sustainability

With the continued growth of the global population and the increasing challenges posed by climate change, the adoption of modern technological, biological, and digital methods is essential for sustainable food production, improved nutrition, and enhanced supply chain efficiency (Wang et al. 2025). Advancing food security and sustainability requires developing and implementing improved methods, tools, and systems that increase food production, strengthen agricultural resilience and equity, and safeguard the environment. Currently, a wide range of tools and sensors are transforming agricultural practices through precision agriculture, as illustrated in Fig. 1.5.

Precision agriculture significantly contributes to sustainable food production by enhancing crop yields, minimizing environmental impact, and improving resource efficiency (Hiywotu 2025; El Bilali and Allahyari 2018). AI and ML are employed

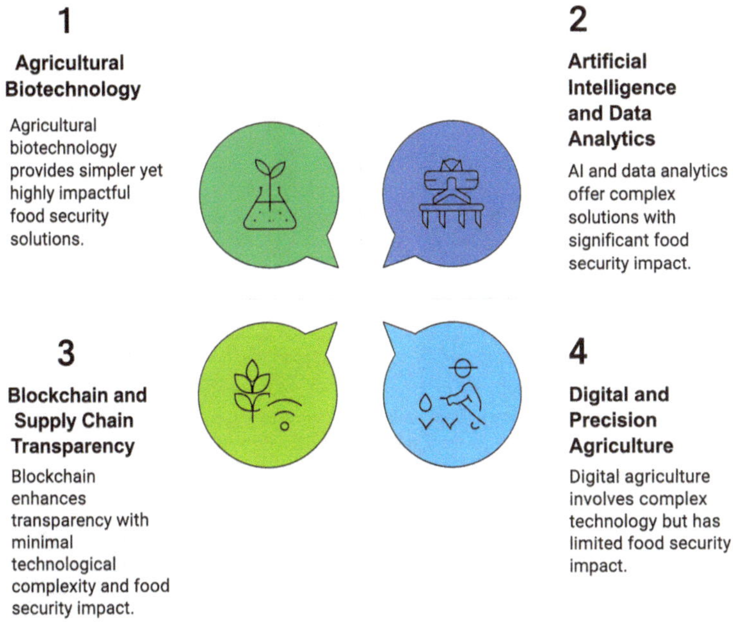

Fig. 1.5 Innovation in food security

to forecast crop yields, predict pest outbreaks, and optimize irrigation management. Blockchain technology enhances trust and transparency across food supply chains. It also supports sustainability certifications and ethical sourcing practices, thereby increasing consumer confidence and promoting equitable access to food. Collectively, these technologies help maintain stable food supply chains and strengthen climate resilience by enabling timely, informed decision-making, thereby reducing potential losses (Namkhah et al. 2023). Figure 1.6 illustrates the integration of resources through a cloud platform. IoT sensors, including soil, weather, and moisture sensors, collect data on crops, soil conditions, and current weather, as shown in Fig. 1.6a. This data is transmitted to the cloud platform for processing, as depicted in Fig. 1.6b. Based on this framework, resources such as water, fertilizer, and energy can be effectively managed. ML algorithms on the cloud platform analyze current conditions and make decisions regarding resource allocation.

1.6 Applications of Smart Agriculture

DL, IoT, and cloud computing are three digital advancements that are fundamentally transforming agriculture. These technologies address challenges related to environmental sustainability, operational efficiency, food security, and climate resilience (Durai and Shamili 2022). ML algorithms and cloud computing techniques

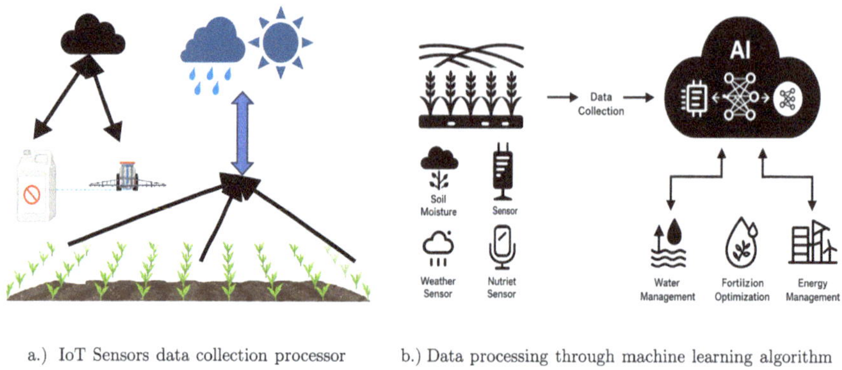

Fig. 1.6 Representation of the resource management system using cloud platform. (**a**) IoT sensor data collection processor, (**b**) Data processing through ML algorithm

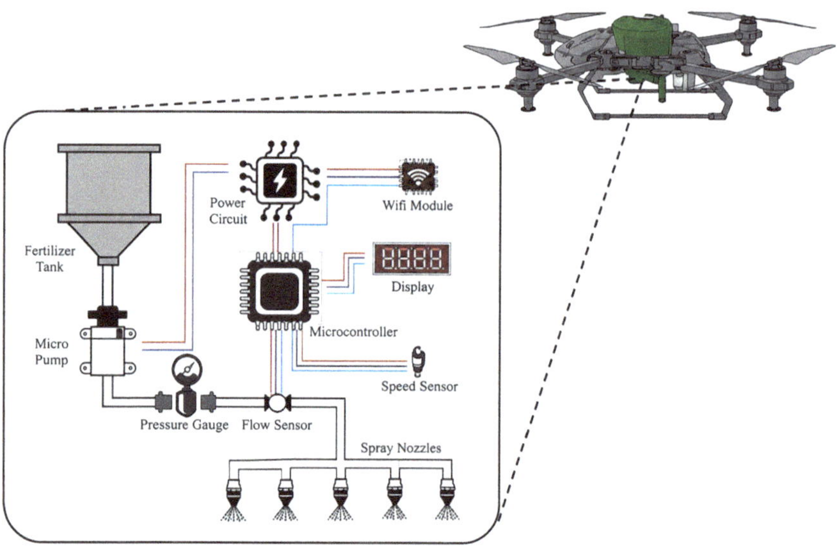

Fig. 1.7 UAV-VFSS structure (Hedden 2003)

facilitate the automatic extraction of patterns from complex, multimodal datasets, including images, sensor data, time series, and textual information, which are otherwise difficult to analyze manually (Dhanya et al. 2021). Additionally, the fog computing paradigm enhances smart agriculture by enabling faster data processing, reducing latency, and supporting real-time decision making at the field level (Wadhwa and Aron 2022).

Ghanimi et al. (2024) introduced innovative concepts, including Smart Farming (SF) and Precision Agriculture (PA). Figure 1.7 presents a conceptual model of the

UAV-VFSS system, illustrating its complex design and functionality. The researchers integrated smart agriculture with IoT multisensor systems to develop a prototype for SF fertilization. The deployed sensors collected environmental and soil parameter data using IoT technology, as depicted in Fig. 1.6. Following data collection and processing, the system determines precise fertilizer requirements, which may vary across different agricultural zones. A field experiment conducted in Maharashtra, India, validated the functionality of the Smart Fertilizing Using IoT Multisensor and Variable-Rate Sprayer Integrated UAV system, demonstrating significant potential to transform and modernize smart fertilization practices.

1.7 Conclusion and Future Directions

Agriculture plays a critical role in providing food, supporting economic growth, and sustaining rural communities. Modern smart farming, which incorporates technologies such as DL, IoT, and blockchain, aims to address contemporary challenges in food sustainability. The continued development of these technologies will require collaborative efforts across multiple disciplines, strengthened global policy frameworks, and equitable access for all stakeholders. Advanced tools in smart farming enable real-time monitoring of agricultural activities. Future progress in smart agriculture will depend on increased interdisciplinary collaboration among researchers, technologists, and farming communities. Enhancing access to affordable digital tools and reinforcing international policy frameworks will be essential to ensure the inclusive and ethical adoption of emerging technologies. Future research should prioritize developing scalable IoT infrastructure, energy-efficient AI models, and secure, interoperable data platforms that perform effectively across diverse agricultural environments. The adoption of advanced technologies such as edge computing, digital twins, robotics, and autonomous decision-making systems will accelerate the transition toward resilient, climate-smart, and sustainable agricultural practices.

References

Abbasi R, Martinez P, Ahmad R (2022) The digitization of agricultural industry—a systematic literature review on agriculture 4.0. Smart Agric Technol 2:100042. https://doi.org/10.1016/j.atech.2022.100042

Ahmed N, Shakoor N (2025) Advancing agriculture through IoT, big data, and AI: a review of smart technologies enabling sustainability. Smart Agric Technol 10:100848. https://doi.org/10.1016/j.atech.2025.100848

Aijaz N, Lan H, Raza T et al (2025) Artificial intelligence in agriculture: advancing crop productivity and sustainability. J Agric Food Res 20:101762. https://doi.org/10.1016/j.jafr.2025.101762

Angelakis AN, Mays LW, Koutsoyiannis D et al (2012) Evolution of water supply through the millennia, vol 11. IWA Publishing. https://doi.org/10.2166/9781780401041

Assimakopoulos F, Vassilakis C, Margaris D et al (2025) AI and related technologies in the fields of smart agriculture: a review. Information 16(2):100. https://doi.org/10.3390/info16020100

Awuni S, Adarkwah F, Ofori BD et al (2023) Managing the challenges of climate change mitigation and adaptation strategies in Ghana. Heliyon 9(5):e15491. https://doi.org/10.1016/j.heliyon.2023.e15491

Bekele-Alemu A, Dessalegn-Hora O, Safawo-Jarso T et al (2025) Rethinking progress: harmonizing the discourse on genetically modified crops. Front Plant Sci 16:1547928. https://doi.org/10.3389/fpls.2025.1547928
Benos L, Tagarakis AC, Dolias G et al (2021) Machine learning in agriculture: a comprehensive updated review. Sensors 21(11):3758. https://doi.org/10.3390/s21113758
Dar WD, Laxmipathi Gowda CL (2013) Declining agricultural productivity and global food security. J Crop Improv 27(2):242–254. https://doi.org/10.1080/15427528.2011.653097
Dhanasekar S, Abarna VK, Gayathri V et al (2023) An efficient smart agriculture system based on the internet of things using aeroponics method. In: 2023 9th Int Conf Adv Comput Commun Syst (ICACCS). IEEE, pp 1386–1391
Dhanya V, Subeesh A, Kushwaha N et al (2021) Deep learning based computer vision approaches for smart agricultural applications. Artif Intell Agric 6:211–229. https://doi.org/10.1016/j.aiia.2022.09.007
Durai SKS, Shamili MD (2022) Smart farming using machine learning and deep learning techniques. Decis Anal J 3:100041. https://doi.org/10.1016/j.dajour.2022.100041
El Bilali H, Allahyari MS (2018) Transition towards sustainability in agriculture and food systems: role of information and communication technologies. Inf Process Agric 5(4):456–464. https://doi.org/10.1016/j.inpa.2018.06.006
Gamage A, Gangahagedara R, Subasinghe S et al (2024) Advancing sustainability: the impact of emerging technologies in agriculture. Curr Plant Biol 40:100420. https://doi.org/10.1016/j.cpb.2024.100420
Ghanimi HM, Suguna R, Jeyaraj JPG et al (2024) Smart fertilizing using IoT multi-sensor and variable rate sprayer integrated UAV. Scalable Comput Pract Exp 25(5):3766–3777. https://doi.org/10.12694/scpe.v25i5.3132
Habib-ur-Rahman M, Ahmad A, Raza A et al (2022) Impact of climate change on agricultural production: issues, challenges, and opportunities in Asia. Front Plant Sci 13:925548. https://doi.org/10.3389/fpls.2022.925548
Hedden P (2003) The genes of the green revolution. Trends Genet 19(1):5–9. https://doi.org/10.1016/s0168-9525(02)00009-4
Hiywotu AM (2025) Advancing sustainable agriculture for goal 2: zero hunger—a comprehensive overview of practices, policies, and technologies. Agroecol Sustain Food Syst 49(7):1027–1055. https://doi.org/10.1080/21683565.2025.2451344
Hussein JB, Workneh TS, Kassim A et al (2025) A review on the impact of big data analytics in transforming agricultural practices, food processing, and preservation strategies. Appl Food Res 5(2):101234. https://doi.org/10.1016/j.afres.2025.101234
Kumar V, Sharma KV, Kedam N et al (2024) A comprehensive review on smart and sustainable agriculture using IoT technologies. Smart Agric Technol 8:100487. https://doi.org/10.1016/j.atech.2024.100487
Kumari K, Nafchi AM, Mirzaee S et al (2025) AI-driven future farming: achieving climate-smart and sustainable agriculture. AgriEngineering 7(3):89. https://doi.org/10.3390/agriengineering7030089
Liu J, Xu Q, Zhou T (2022) Role of mechanization: the impact of the cropland use scale on fertilizer reduction. Front Environ Sci 10:1053715. https://doi.org/10.3389/fenvs.2022.1053715
Luna Juncal MJ, Masino P, Bertone E et al (2023) Towards nutrient neutrality: a review of agricultural runoff mitigation strategies and the development of a decision-making framework. Sci Total Environ 874:162408. https://doi.org/10.1016/j.scitotenv.2023.162408
Mohan RN, Rayanoothala PS, Sree RP (2025) Next-gen agriculture: integrating AI and XAI for precision crop yield predictions. Front Plant Sci 15:1451607. https://doi.org/10.3389/fpls.2024.1451607
Muralikrishnan L, Padaria RN, Choudhary AK et al (2021) Climate change-induced drought impacts, adaptation and mitigation measures in semi-arid pastoral and agricultural watersheds. Sustainability 14(1):6. https://doi.org/10.3390/su14010006

Namkhah Z, Fatemi SF, Mansoori A et al (2023) Advancing sustainability in the food and nutrition system: a review of artificial intelligence applications. Front Nutr 10:1295241. https://doi.org/10.3389/fnut.2023.1295241

Nautiyal M, Joshi S, Hussain I et al (2025) Revolutionizing agriculture: a comprehensive review on artificial intelligence applications in enhancing properties of agricultural produce. Food Chem X 29:102748. https://doi.org/10.1016/j.fochx.2025.102748

Patle GT, Kharpude SN, Dabral PP et al (2020) Impact of organic farming on sustainable agriculture system and marketing potential: a review. Int J Environ Clim Change 10(11):100–120. https://doi.org/10.9734/ijecc/2020/v10i1130270

Peng J, Zhao Z, Liu D (2022) Impact of agricultural mechanization on agricultural production, income, and mechanism: evidence from Hubei Province, China. Front Environ Sci 10:838686. https://doi.org/10.3389/fenvs.2022.838686

Ringler C, Agbonlahor M, Barron J et al (2022) The role of water in transforming food systems. Glob Food Sec 33:100639. https://doi.org/10.1016/j.gfs.2022.100639

Rovira-Más F, Zhang Q, Reid JF (2008) Stereo vision three-dimensional terrain maps for precision agriculture. Comput Electron Agric 60(2):133–143. https://doi.org/10.1016/j.compag.2007.07.007

Suhail I (2022) Historical evolution of agricultural practices in sustainable agriculture. In: Glick TF, Livesey S, Williams F (eds) Sustainable agriculture: a modern agronomist's perspective. KD Publications, pp 176–179

Syed A, Raza T, Bhatti TT et al (2021) Climate impacts on the agricultural sector of Pakistan: risks and solutions. Environ Chall 6:100433. https://doi.org/10.1016/j.envc.2021.100433

Tchonkouang RD, Onyeaka H, Nkoutchou H (2024) Assessing the vulnerability of food supply chains to climate change-induced disruptions. Sci Total Environ 920:171047. https://doi.org/10.1016/j.scitotenv.2024.171047

Thenkabail PS (2010) Global croplands and their importance for water and food security in the twenty-first century: towards an ever green revolution that combines a second green revolution with a blue revolution. Remote Sens 2(9):2305–2312. https://doi.org/10.3390/rs2092305

Touch V, Tan DK, Cook BR et al (2024) Smallholder farmers' challenges and opportunities: implications for agricultural production, environment and food security. J Environ Manag 370:122536. https://doi.org/10.1016/j.jenvman.2024.122536

Wadhwa H, Aron R (2022) Resource utilization for IoT oriented framework using zero hour policy. Wirel Pers Commun 122:2285–2308. https://doi.org/10.1007/s11277-021-08993-0

Wang W, Fu Z, Liu C et al (2025) A study of the impact of technological innovation in food production on the resilience of the food industry. J Innov Knowl 10(4):100756. https://doi.org/10.1016/j.jik.2025.100756

Xiong W, Wu DD, Yeung JHY (2025) Semiconductor supply chain resilience and disruption: insights, mitigation, and future directions. Int J Prod Res 63(9):3442–3465. https://doi.org/10.1080/00207543.2024.2387074

Yang X, Shu L, Chen J et al (2021) A survey on smart agriculture: development modes, technologies, and security and privacy challenges. IEEE/CAA J Autom Sin 8(2):273–302. https://doi.org/10.1109/JAS.2020.1003536

Deep Learning and Artificial Intelligence Applications in Agriculture

2

Ramya Sakthivel and Kalimuthu Krishnan

Abstract

The global agriculture industry faces mounting challenges, including population growth, climate change, resource limitations, and labor shortages. As a result, the adoption of sustainable and precise agricultural practices has become essential. Artificial Intelligence (AI) and Deep Learning (DL) serve as foundational technologies for data-driven farming systems. This chapter provides an overview of various DL architectures, such as Convolutional Neural Networks (CNNs) for image analysis, Recurrent Neural Networks (RNNs) for temporal data, and both Generative Adversarial Networks (GANs) and transformer-based models for extracting hierarchical features from multimodal data. Key applications of AI and DL in agriculture include disease and pest recognition and prediction, intelligent weed management with reduced herbicide use, livestock health and behavior monitoring, and the deployment of robotics for harvesting and pruning. This chapter also examines challenges related to model generalization, data scarcity, ethical considerations, and deployment constraints in AI-enabled agriculture. Additionally, it highlights future research directions, including explainable AI, cross-domain adaptability, and the development of lightweight architectures for agricultural information systems to support real-time applications and scalability in response to increasing user demand and environmental concerns. Optimization strategies such as transfer learning, federated learning, and edge-cloud computing are also discussed.

R. Sakthivel (✉) · K. Krishnan
Department of Electronics and Communication Engineering, College of Engineering and Technology, SRM Institute of Science and Technology, Kattankulathur, Tamil Nadu, India
e-mail: rs7035@srmist.edu.in

H. N. Oo et al. (eds.), *Transforming Agriculture with AI, IoT, and Blockchain*,
https://doi.org/10.1007/978-981-92-1491-4_2

Keywords

Computer vision · Crop monitoring · Plant disease detection · Yield prediction · Convolutional neural networks (CNNs) · Remote sensing · Agricultural robotics

2.1 Introduction

2.1.1 The Agricultural Revolution in the Digital Age

Agriculture has experienced multiple transformative revolutions throughout human history, including the domestication of plants and animals, mechanization, and the Green Revolution's introduction of high-yield crop varieties. Currently, the sector is on the verge of a new revolution driven by Artificial Intelligence (AI) and Deep Learning (DL) technologies. This digital transformation is expected to address critical challenges in modern agriculture, such as increasing food production to meet the needs of a global population projected to reach 9.7 billion by 2050, adapting to climate change, reducing environmental impacts, and optimizing resource utilization (Rikendra and Chauhan 2025).

Traditional agricultural practices often depend on manual inspection, experience-based decision-making, and reactive management strategies. Although these methods have value, they are increasingly insufficient for addressing the complexity and scale of contemporary agricultural operations. DL and AI technologies provide data-driven, precise, and scalable solutions that process large volumes of information from diverse sources, such as satellite imagery, drone surveillance, ground sensors, weather data, and historical records, to deliver actionable insights in real time (Akkem et al. 2023). Figure 2.1 illustrates the application of AI and DL in agriculture.

Why is AI important in Indian agriculture?

- The low efficiency of irrigation, estimated at approximately 38%, contributes to both groundwater depletion and environmental pollution.
- Large-scale postharvest losses are prevalent, particularly in perishable commodities such as fruits and vegetables. Additionally, significant losses occur during food grain procurement, storage, and distribution.
- The agricultural labor force has declined over recent years, and the sector is failing to attract younger generations.
- Climate change, manifested through increased temperatures and altered rainfall patterns, reduces agricultural productivity.
- Farmers increasingly face both biotic and abiotic stresses.
- Land degradation remains a persistent challenge.

The agricultural sector faces several challenges, including a rising population, increasing demand for food grains, inefficient use of inputs such as fertilizers, greater incidence of pesticide resistance, and elevated pesticide residues in food. The integration of AI in agriculture offers potential solutions by relieving labor shortages, attracting younger generations, reducing physical drudgery, and

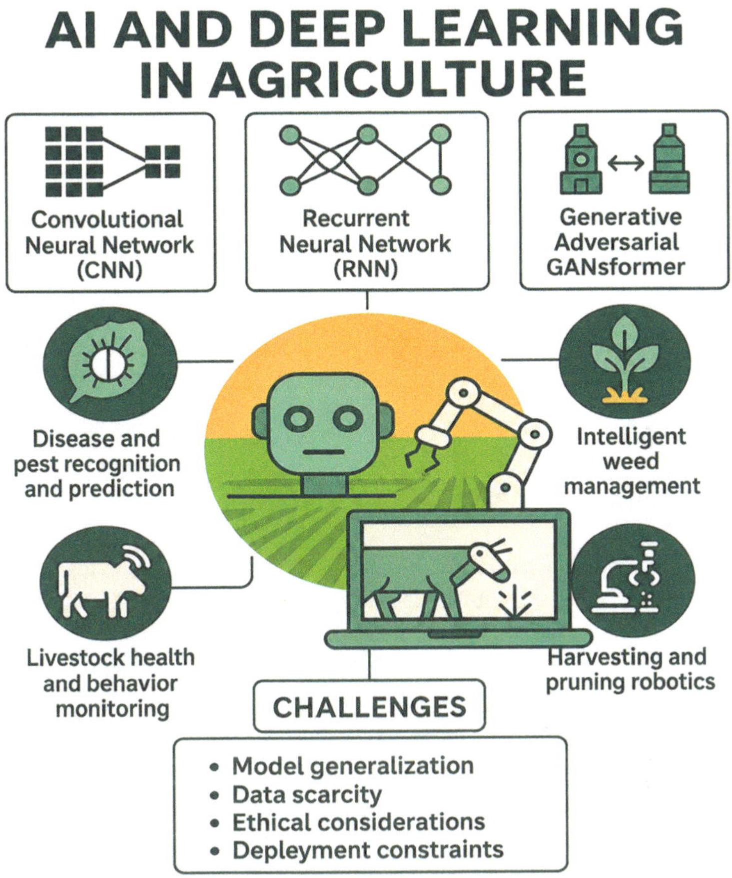

Fig. 2.1 Applications of AI and DL in agriculture

conserving diminishing natural resources. To address these challenges, agriculture must adopt intelligent approaches that enhance production through labor- and resource-saving technologies and practices. The application of AI enables smart farming, and there is growing interest among farmers to utilize this technology to overcome persistent agricultural challenges.

2.1.2 Defining DL and AI in an Agricultural Context

AI encompasses a wide array of computational methods that enable machines to perform tasks traditionally associated with human intelligence, such as pattern recognition, decision-making, and prediction. DL, a subset of machine learning (ML), utilizes artificial neural networks with multiple layers to autonomously learn hierarchical data representations, thereby reducing the need for manual feature engineering (Guntuka 2024).

Within agriculture, these technologies are applied through several principal capabilities:

- *Computer vision*: Automated analysis of images and video enables identification of plants, detection of diseases, fruit counting, assessment of crop health, and monitoring of growth stages.
- *Predictive analytics*: These methods facilitate yield forecasting, disease outbreak prediction, and anticipation of optimal harvest periods using both historical and real-time data.
- *Decision support systems*: These systems deliver actionable recommendations to farmers regarding irrigation, fertilization, pest control, and resource allocation.
- *Autonomous systems*: These systems enable robots and drones to perform tasks including selective weeding, targeted spraying, and automated harvesting.

2.1.3 The Promise of Precision Agriculture

Precision agriculture is a management strategy that utilizes information technology to deliver precise inputs for optimal crop health and productivity. AI and DL constitute the computational foundation of precision agriculture, enabling the following advancements:

- *Resource optimization*: The targeted application of water, fertilizers, and pesticides minimizes waste and environmental impact while sustaining or enhancing crop yields.
- *Early problem detection*: AI systems can identify diseases, pests, and nutrient deficiencies prior to the manifestation of visible symptoms.
- *Scalability*: Large-scale agricultural operations can be monitored efficiently, overcoming the limitations of manual inspection.
- *Data-driven decisions*: Transitioning from intuition-based to evidence-based farming practices.

Recent studies highlight the measurable advantages of integrating AI into precision agriculture. For example, smart irrigation systems that combine AI with Internet of Things (IoT) sensors have reduced water usage by 20% and increased crop yields by 15% in field trials (Singh et al. 2025). In addition, AI-powered disease detection systems implemented in India have achieved 90% accuracy in identifying crop diseases, which facilitates timely intervention (Singh et al. 2025).

2.2 Major Application Areas of AI in Agriculture

DL and AI have been applied in agriculture across multiple domains, each targeting specific challenges and opportunities in contemporary farming. The following section reviews five primary application areas where these technologies have had a substantial impact.

2.2.1 Crop Monitoring and Growth Analysis

Crop monitoring refers to the systematic observation of crop development during the growing season to evaluate health, growth stages, and overall vigor. Conventional monitoring methods demand significant manual labor and yield only limited, time-specific assessments. In contrast, AI-powered monitoring systems enable continuous, comprehensive, and objective evaluations.

2.2.1.1 Remote Sensing and Image Analysis

DL models analyze imagery from diverse sources, including satellites, Unmanned Aerial Vehicles (UAVs), and ground-based cameras, to monitor crop development across spatial and temporal scales (Han and Wang 2025). Convolutional Neural Networks (CNNs) are particularly effective at extracting visual features that correspond to key agronomic parameters, including:

- *Phenological stages*: Identification of germination, vegetative growth, flowering, and maturation phases.
- *Canopy coverage*: Measurement of the proportion of ground covered by vegetation.
- *Plant stress*: Detection of early indicators of water stress, nutrient deficiency, or environmental damage.
- *Growth rate*: Quantification of temporal changes in plant size and biomass.

Multispectral and hyperspectral imaging provide information beyond the visible spectrum. The near-infrared (NIR) and red-edge bands are especially useful for evaluating chlorophyll content and photosynthetic activity. DL models trained on multiband images can detect subtle physiological changes that are not visible to human observers or standard RGB cameras.

2.2.1.2 Phenotyping and Trait Analysis

The integration of high-throughput phenotyping platforms with DL facilitates automated, large-scale measurement of plant traits. This approach is especially valuable for breeding programs and agricultural research. Computer vision systems are capable of automatically quantifying the following traits:

- Leaf area and morphological characteristics
- Plant height and structural architecture
- Root system structure, assessed through specialized imaging techniques
- Flowering patterns and fruit set dynamics
- Color and surface texture parameters

Generative models, including Generative Adversarial Networks (GANs) and Variational Autoencoders (VAEs), are increasingly utilized for phenotype monitoring and representation learning. These models assist researchers in elucidating genetic variation and plant responses to environmental factors (Guntuka 2024).

2.2.2 Disease and Pest Detection

Plant diseases and pest infestations pose significant risks to agricultural productivity, resulting in estimated annual losses of 20–40% of global crop yields. Timely and precise detection is essential for effective management and reduction of crop damage.

2.2.2.1 Visual Disease Diagnosis

Image-based disease detection is among the most effective applications of DL in agriculture. CNNs trained on labeled images of healthy and diseased plants are capable of classifying disease types with high accuracy. For instance, the DeepCrop system, which employs a ResNet-50 architecture and transfer learning, achieved 98.98% accuracy on the PlantVillage dataset, comprising images of diseased and healthy leaves from various crop species (Islam et al. 2023).

Such systems generally adhere to a standardized processing pipeline:

- *Image acquisition*: Images of potentially diseased plants are captured using smartphones or dedicated cameras.
- *Preprocessing*: Images are normalized, resized, and augmented to enhance model robustness.
- *Feature extraction*: Deep CNN layers autonomously learn discriminative features relevant to disease identification.
- *Classification*: The final layers predict the specific disease type or confirm the healthy status of the plant.
- *Recommendation*: The system provides treatment suggestions based on the identified disease.

Web and mobile applications make these capabilities accessible to farmers worldwide. The DeepCrop system, for example, has been deployed as a web application allowing farmers to upload leaf images and receive immediate diagnostic results (Islam et al. 2023).

2.2.2.2 Multicrop Disease Systems

Advanced systems can process multiple crop types concurrently. For example, a hybrid CNN-LSTM-Attention architecture developed for multicrop disease classification in maize, potato, rice, sugarcane, and wheat achieved an overall accuracy of 86.9% (Rikendra and Chauhan 2025). These results indicate the potential for unified platforms to support diverse agricultural operations without requiring separate models for each crop type.

2.2.2.3 Pest Detection and Monitoring

In addition to disease management, AI systems are being developed for pest detection and population monitoring. Computer vision models can identify specific insect species in field images or in trap monitoring systems, enabling targeted interventions and reducing reliance on broad-spectrum pesticides. The integration of these models with pheromone traps and automated counting systems provides real-time pest population data to support decision-making.

2.2.3 Yield Prediction and Forecasting

Accurate yield prediction plays a critical role in agricultural planning, market forecasting, supply chain management, and food security assessment. AI-based yield prediction systems utilize diverse data sources to estimate productivity prior to harvest.

2.2.3.1 Multimodal Fusion Approaches

Contemporary yield prediction systems utilize multimodal fusion, integrating multiple data sources to enhance predictive accuracy.

- *Visual features*: These are extracted from satellite, drone, or ground imagery using CNNs.
- *Disease severity scores*: These are computed using disease detection models.
- *Environmental variables*: These include weather data, temperature, precipitation, and solar radiation.
- *Soil characteristics*: These encompass nutrient levels, moisture content, and pH.
- *Management practices*: These include planting dates, fertilization schedules, and irrigation regimes.

A hybrid pipeline that integrates CNN feature extraction, disease severity estimates, and environmental data into an XGBoost ensemble model achieved an average validation accuracy of 93.84% for yield estimation across five major crops (Rikendra and Chauhan 2025). These results indicate that fusion approaches consistently outperform single-modality models.

2.2.3.2 Temporal Modeling

Crop yield is fundamentally a temporal process that depends on conditions throughout the growing season. Recurrent Neural Networks (RNNs), particularly Long Short-Term Memory (LSTM) networks, are well-suited to modeling these temporal dependencies. By processing sequences of multidate imagery or sensor readings, LSTM networks capture growth dynamics and the cumulative effects of environmental conditions (Rikendra and Chauhan 2025).

Attention mechanisms further improve temporal modeling by enabling models to focus on critical growth stages. For example, conditions during flowering and grain-filling periods often exert a disproportionate influence on final yield, and attention-weighted models can automatically learn to emphasize these phases.

2.2.3.3 Performance Benchmarks

A comprehensive review of DL methods for crop yield prediction and bloom recognition reported an average accuracy of 92.51% across the surveyed studies. Consistent results across various research groups and crop types suggest that

AI-based yield prediction has developed into a reliable technology suitable for operational deployment.

2.2.4 Weed Management and Selective Control

Weeds compete with crops for essential resources such as nutrients, water, and light, which leads to reduced yields and diminished crop quality. Conventional weed control methods primarily depend on herbicides, which raises significant environmental and health concerns. AI-powered weed management enables precise targeting, thereby reducing chemical use while maintaining effective weed control.

2.2.4.1 Weed Detection and Classification

Computer vision systems differentiate crops from weeds and classify weed species by analyzing leaf shape, color, texture, and growth patterns. These capabilities facilitate the following applications:

- *Precision spraying*: Targeted application of herbicides exclusively to areas where weeds are detected
- *Mechanical weeding*: Directing autonomous robots to physically extract weeds from crop fields
- *Weed mapping*: Generating spatial distribution maps to inform weed management strategies

DL models trained on labeled datasets of crop and weed images demonstrate high accuracy under real-time field conditions. However, distinguishing young crop plants from visually similar weed species remains challenging, necessitating that models extract subtle discriminative features.

2.2.4.2 Robotic Weed Control

Autonomous weeding robots combine computer vision with mechanical or thermal weed control mechanisms. These systems autonomously navigate agricultural fields, identify weeds, and implement targeted control measures. DL algorithms enable these robots to perceive and operate safely and effectively within complex, unstructured agricultural environments.

2.2.5 Soil Analysis and Management

Soil health is essential for sustainable agriculture; however, conventional soil testing methods are labor-intensive and yield only point-in-time data. AI systems facilitate continuous soil monitoring and predictive management.

2.2.5.1 Sensor-Based Soil Monitoring

IoT sensor networks deployed in agricultural fields continuously measure soil moisture, temperature, electrical conductivity (as an indicator of salinity), and nutrient concentrations. DL models analyze these sensor data streams to:

- Predict irrigation requirements and optimize water distribution
- Detect nutrient deficiencies prior to the manifestation of visible symptoms
- Identify soil compaction or drainage issues
- Forecast the risk of soil-borne diseases

Integration of sensor data with DL models has enabled precision irrigation frameworks that significantly reduce water consumption while maintaining or improving yields.

2.2.5.2 Spectral Soil Analysis

Hyperspectral imaging and spectroscopy yield comprehensive data regarding soil composition. DL models trained on spectral signatures are capable of estimating the following parameters:

- Organic matter content
- Nutrient concentrations (nitrogen, phosphorus, and potassium)
- Soil texture and structure
- Contamination levels

When combined with AI-based interpretation, these noninvasive optical methods provide rapid, cost-effective alternatives to conventional laboratory analyses.

2.3 Key DL Architectures in Agriculture

The effectiveness of AI applications in agriculture largely depends on selecting appropriate DL architectures for specific tasks. This section reviews the principal architectures utilized and evaluates their respective advantages across various agricultural applications. Figure 2.2 presents the overall workflow for AI- and DL-based agricultural systems.

2.3.1 Convolutional Neural Networks (CNNs)

CNNs continue to serve as the primary architecture for image-based agricultural tasks. These networks are specifically designed to process grid-structured data, such as images, by applying convolutional filters that detect local patterns and progressively construct complex feature representations (Guntuka 2024).

Fig. 2.2 Overall workflow adopted for AI and DL-based agricultural systems

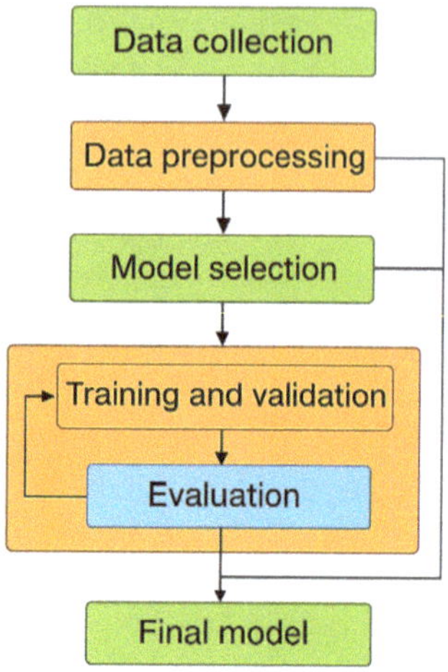

2.3.1.1 Architecture Fundamentals

A typical CNN comprises the following components:

- *Convolutional layers*: These layers apply learnable filters to extract features such as edges, textures, and patterns.
- *Pooling layers*: These layers reduce the spatial dimensions of feature maps while preserving essential information.
- *Activation functions*: These functions introduce nonlinearity into the network, with the rectified linear unit (ReLU) being the most commonly used.
- *Fully connected layers*: These layers integrate extracted features to perform final classification or regression tasks.

2.3.1.2 Popular CNN Architectures in Agriculture

Several CNN architectures have demonstrated significant effectiveness in agricultural applications.

- *ResNet (Residual Networks)*: ResNet architectures, particularly ResNet-50 and ResNet-101, are extensively utilized for disease detection and crop classification. The inclusion of residual connections, also known as skip connections, facilitates the training of deep networks without performance degradation and supports the extraction of highly discriminative features. The DeepCrop system achieved 98.98% accuracy in plant disease classification using ResNet-50 (Islam et al. 2023).
- *VGG Networks*: VGG architectures, characterized by their simple and uniform structure, are frequently employed in transfer learning applications. Pretrained VGG models are commonly used as feature extractors for a range of agricultural tasks.
- *EfficientNet*: EfficientNet architectures are designed to optimize the balance between accuracy and computational efficiency, which makes them suitable for deployment on resource-constrained devices, including edge computers and mobile phones used in field environments.

- *U-Net and related segmentation networks*: These are well-suited for tasks that require pixel-level classification, such as identifying diseased regions within leaves or counting individual fruits. These architectures enable precise spatial localization.

2.3.1.3 Transfer Learning with CNNs

Agricultural datasets are typically smaller than general image datasets such as ImageNet. Transfer learning mitigates this limitation by initializing models with weights pretrained on large-scale datasets and subsequently fine-tuning them using agricultural data. This method:

- Reduces both training time and the amount of data required
- Enhances generalization performance
- Facilitates high accuracy even when only modest-sized agricultural datasets are available

Empirical studies consistently demonstrate that transfer learning with ImageNet-pretrained models outperforms training from scratch for agricultural image classification tasks (Islam et al. 2023).

2.3.2 Recurrent Neural Networks and Temporal Models

Although CNNs are effective for spatial pattern recognition, numerous agricultural processes exhibit temporal dynamics. RNNs and their advanced variants are designed to model temporal dependencies in sequential data.

2.3.2.1 LSTM and GRU Networks

Long Short-Term Memory (LSTM) networks and Gated Recurrent Units (GRUs) address the vanishing gradient problem that constrains standard RNNs, thereby facilitating the modeling of long-term dependencies. In agricultural applications, these architectures are utilized for the following purposes:

- *Time-series yield prediction*: Analyzing sequences of satellite imagery or sensor data collected throughout the growing season
- *Growth modeling*: Estimating future crop growth stages using historical development patterns
- *Weather-dependent forecasting*: Integrating temporal weather patterns into crop yield and growth predictions

A hybrid CNN-LSTM architecture for multicrop disease classification and yield prediction illustrates the effectiveness of integrating spatial (CNN) and temporal (LSTM) processing (Rikendra and Chauhan 2025). The CNN extracts visual features from images, while the LSTM models the progression of disease severity and environmental conditions over time, thereby influencing final yields.

2.3.2.2 Sequence-to-Sequence Models

Tasks that involve predicting future sequences, such as forecasting crop growth trajectories, can be effectively addressed using sequence-to-sequence (Seq2Seq) encoder-decoder models. In these models, the encoder processes historical data, and the decoder generates future predictions.

2.3.3 Attention Mechanisms and Transformers

Attention mechanisms and transformer architectures represent recent advancements in DL, providing significant benefits for agricultural applications that require global context and multimodal data integration.

2.3.3.1 Attention Mechanisms

Attention mechanisms enable models to selectively focus on the most relevant components of the input data. In agricultural applications, attention can be categorized as follows:

- Spatial attention highlights significant regions within images, such as diseased areas present in agricultural fields.
- Temporal attention emphasizes key growth stages or significant agricultural events.
- Channel attention identifies and selects the most informative spectral bands within multispectral imagery.

The integration of attention modules with CNN-LSTM pipelines has enhanced disease classification accuracy in multicrop systems (Rikendra and Chauhan 2025). Attention mechanisms enable models to autonomously identify the features and temporal intervals that are most predictive of agricultural outcomes.

2.3.3.2 Vision Transformers (ViTs)

Vision Transformers (ViTs) utilize the transformer architecture, initially designed for natural language processing, in the context of image analysis. ViTs segment images into patches and process these as sequences, enabling the capture of long-range dependencies that CNNs may overlook. Although ViTs are computationally intensive, they demonstrate significant potential in the following applications:

- Large-scale crop monitoring using satellite imagery
- Complex scene understanding within diverse agricultural environments
- Multimodal fusion of image, textual, and sensor data

2.3.3.3 Spatio-temporal Transformers

Recent transformer architectures tailored for video and spatio-temporal data have demonstrated significant potential in agricultural monitoring. These models can process image sequences, such as daily satellite observations, while preserving both spatial and temporal context.

2.3.4 Generative Models

Generative models produce novel data samples that closely resemble the original training data. Within agriculture, such models help mitigate data scarcity and facilitate advanced analytical approaches.

2.3.4.1 Generative Adversarial Networks (GANs)

GANs comprise two competing networks: a generator that produces synthetic samples and a discriminator that differentiates between real and synthetic data. Key agricultural applications include the following:

- *Data augmentation*: It involves generating synthetic images of diseased plants to balance training datasets.
- *Domain adaptation*: It refers to creating synthetic images that bridge the gap between controlled laboratory conditions and real field environments.
- *Super-resolution techniques*: These techniques enhance low-resolution satellite or drone imagery.

2.3.4.2 Variational Autoencoders (VAEs)

VAEs learn compressed representations of data and generate new samples by sampling from the learned latent space. In agricultural applications, VAEs are utilized for the following purposes:

- Learning phenotype representations
- Detecting anomalies, such as identifying unusual plant appearances
- Reducing dimensionality of high-dimensional sensor data

Recent research reviews emphasize the growing importance of GANs and VAEs for phenotype monitoring and for addressing data limitations in agricultural DL (Guntuka 2024).

2.3.5 Comparative Analysis of Architectures

Table 2.1 provides an overview of key architectures, their typical applications in agriculture, associated advantages, and representative performance metrics:

The selection of architecture is determined by the specific task, the nature of the available data, computational resources, and deployment constraints. Recent developments indicate that hybrid approaches, which combine multiple architectures such as CNNs, LSTMs, and Attention mechanisms, achieve superior performance by leveraging their complementary strengths (Rikendra and Chauhan 2025).

Table 2.1 Key architectures, their typical agricultural uses, advantages, and example performance

Architecture	Typical agricultural uses	Key advantages	Example performance
CNNs (ResNet, VGG, EfficientNet)	Disease detection, crop classification, fruit counting, and weed identification	Strong spatial feature extraction, excellent transfer learning, and well-established	ResNet-50: 98.98% accuracy on PlantVillage disease dataset (Islam et al. 2023)
RNNs/LSTM/GRU	Yield prediction, growth modeling, time-series analysis, sequential decision-making	Models' temporal dependencies; captures long-term patterns; handles variable-length sequences	LSTM in hybrid pipeline: 93.84% yield prediction accuracy (Rikendra and Chauhan 2025)
Transformers/Attention	Multimodal fusion, large-scale monitoring, complex scene understanding	Captures global context; scales to large datasets; effective multimodal integration	Attention-enhanced hybrid: 86.9% multicrop disease classification (Rikendra and Chauhan 2025)
GANs/VAEs	Data augmentation, domain adaptation, synthetic data generation, phenotype analysis	Addresses data scarcity; enables domain transfer; learns compact representations	Surveyed for phenotype monitoring and augmentation (Guntuka 2024)

2.4 Techniques and Methodologies

In addition to selecting suitable architectures, effective agricultural AI systems utilize diverse techniques and methodologies to improve performance, facilitate deployment, and address practical limitations.

2.4.1 Transfer Learning and Domain Adaptation

Transfer learning is now a foundational methodology in agricultural AI, as it addresses the critical challenge of limited labeled data.

2.4.1.1 Pretraining and Fine-tuning

The standard transfer learning workflow consists of the following stages:

- *Pretraining*: The model is initially trained on a large, general-purpose dataset, such as ImageNet, which contains 1.2 million images across 1000 categories.
- *Fine-tuning*: The pretrained model is adapted to agricultural data by further training on a smaller, task-specific dataset.

This approach utilizes low-level features, such as edges, textures, and colors, learned from general images that are also relevant to agricultural images. Simultaneously, higher layers are able to specialize in specific crops, diseases, or conditions (Islam et al. 2023).

2.4.1.2 Domain Adaptation Techniques

Agricultural data demonstrates substantial domain shifts across several factors:

- Laboratory and field conditions
- Distinct geographical regions
- Variable weather and lighting conditions
- Diverse camera types and resolutions

Domain adaptation techniques facilitate model generalization across these variations:

- *Adversarial domain adaptation*: It involves training models to extract features that are discriminative for the task while remaining invariant to domain differences.
- *Self-supervised learning*: It entails pretraining on unlabeled agricultural data prior to fine-tuning on labeled samples.
- *Multisource domain adaptation*: It takes advantage of the data from multiple farms or regions to enhance generalization.

2.4.2 Multimodal Data Fusion

Contemporary agricultural AI systems typically utilize multiple data sources. Multimodal fusion integrates complementary information types to enhance both accuracy and robustness.

2.4.2.1 Fusion Strategies

- *Early fusion*: It involves combining raw data from multiple modalities prior to processing, such as stacking RGB and near-infrared image bands.
- *Late fusion*: It refers to processing each modality independently and subsequently combining predictions or features at the final stage, for example, by averaging outputs from image-based and sensor-based models.
- *Hybrid fusion*: It combines features at intermediate stages, enabling the model to learn optimal integration strategies.

The hybrid CNN-LSTM-Attention pipeline for yield prediction demonstrates effective multimodal fusion. CNNs extract visual features from images, disease detection models generate severity scores, and environmental sensors supply weather and soil data. The LSTM-Attention module then integrates these heterogeneous inputs over time (Rikendra and Chauhan 2025).

2.4.2.2 Feature-Level Integration

Advanced fusion approaches assign weights to different modalities according to their assessed reliability and relevance.

- *Attention-based fusion*: It utilizes attention mechanisms to dynamically assign weights to modalities. For example, visual features may be emphasized during critical growth stages when satellite imagery provides the most informative data.
- *Gating mechanisms*: These mechanisms employ learned gates to determine which modalities contribute to decision-making processes under varying conditions.
- *Graph neural networks*: These networks represent relationships among different data sources as graphs, enabling structured fusion of multimodal information.

2.4.3 Data Augmentation and Synthetic Data Generation

Data augmentation artificially increases training set size and diversity, improving model robustness and reducing overfitting.

2.4.3.1 Traditional Augmentation

Common image augmentation techniques are as follows:

- Geometric transformations, such as rotation, flipping, cropping, and scaling, are frequently employed.
- Color space adjustments involve modifying brightness, contrast, and saturation.
- Noise injection consists of adding realistic noise patterns to images.
- Occlusion simulation is achieved by randomly masking regions of the image.

These augmentations help models become invariant to irrelevant variations while preserving task-relevant features.

2.4.3.2 Advanced Augmentation

Contemporary data augmentation strategies include the following approaches:

- *Mixup and CutMix*: They involve blending multiple training examples to generate synthetic samples.
- *AutoAugment*: It automatically learns optimal augmentation policies tailored to specific datasets.
- *GAN-based augmentation*: It involves the use of generative models to produce realistic synthetic images of diseased plants, thereby addressing class imbalance (Guntuka 2024).

2.4.4 Model Compression and Optimization for Deployment

Agricultural AI systems are frequently deployed on edge devices with limited computational resources, such as smartphones, drones, field robots, or embedded systems. Model compression techniques facilitate deployment while preserving model accuracy.

2.4.4.1 Pruning

Pruning eliminates redundant or less significant network connections, thereby reducing model size and computational demands. Structured pruning removes entire filters or layers, resulting in greater acceleration on standard hardware.

2.4.4.2 Quantization

Quantization decreases the numerical precision of model weights and activations from 32-bit floating-point to 8-bit integers or lower. This process substantially reduces memory requirements and accelerates inference, typically with minimal loss of accuracy.

2.4.4.3 Knowledge Distillation

Knowledge distillation trains a smaller "student" model to mimic a larger "teacher" model's behavior. The student learns from both ground-truth labels and the teacher's predictions, often achieving comparable accuracy with significantly fewer parameters.

2.4.4.4 Neural Architecture Search (NAS)

NAS enables the automatic discovery of efficient neural network architectures that are optimized for specific hardware constraints. EfficientNet, developed using NAS, demonstrates strong accuracy-efficiency trade-offs, making it particularly suitable for deployment in agricultural edge environments.

Real-time inference in agricultural systems, such as robotic applications requiring less than 100 milliseconds per sample, demands rigorous optimization. Effective implementations integrate multiple compression techniques to achieve sub-100-millisecond latency while preserving high accuracy (Rikendra and Chauhan 2025).

2.4.5 Ensemble Methods and Model Fusion

Ensemble methods integrate multiple models to enhance both prediction accuracy and robustness.

2.4.5.1 Traditional Ensembles

- *Bagging*: It involves training multiple models on distinct subsets of data and subsequently averaging their predictions.
- *Boosting*: It refers to the sequential training of models, where each model emphasizes examples that were misclassified by its predecessors.
- *Stacking*: It utilizes predictions from multiple base models as inputs to a meta-learner, which synthesizes the final output.

The yield prediction pipeline employs XGBoost, a gradient boosting method, to integrate DL features with environmental variables, thereby demonstrating effective ensemble integration (Rikendra and Chauhan 2025).

2.4.5.2 Deep Ensemble Techniques

- *Multicrop ensembles* involve training distinct models for each crop type and integrating their respective predictions.
- *Multiscale ensembles* process images at multiple resolutions and subsequently merge the resulting predictions.
- *Temporal ensembles* aggregate predictions from models that analyze the same field across various time points.

2.4.6 Explainability and Interpretability

Understanding why models make specific predictions is crucial for farmer trust and scientific insight.

2.4.6.1 Visualization Techniques

- Activation mapping involves visualizing the image regions that activate specific neurons, thereby revealing the features to which the model is most responsive.
- Gradient-weighted Class Activation Mapping (Grad-CAM) highlights the image regions that are most influential for the model's predictions.
- Saliency maps display the pixel-level importance that contributes to the model's decision-making process.

These techniques assist farmers in interpreting disease diagnoses, such as by identifying diseased leaf regions, and enable researchers to validate that models are learning scientifically meaningful patterns.

2.4.6.2 Rule Extraction and Feature Importance

In decision support systems, extracting interpretable rules or identifying the most influential features, such as specific growth stages that significantly affect yield, offers actionable insights that surpass those provided by black-box predictions.

2.5 Case Studies and Performance Metrics

This section reviews deployed systems and research prototypes, evaluating their performance, methodologies, and practical impact.

2.5.1 DeepCrop: Web-Based Disease Detection System

System overview: DeepCrop is a DL-based disease-detection platform designed for practical use by farmers, deployed as a web application (Islam et al. 2023).

Technical approach

- *Architecture*: ResNet-50 CNN with transfer learning from ImageNet
- *Dataset*: PlantVillage dataset containing labeled images of healthy and diseased leaves from multiple crop species
- *Preprocessing*: Image normalization, resizing to 224 × 224 pixels, data augmentation
- *Deployment*: Flask-based web application allowing image upload and real-time inference

Performance metrics

- *Classification accuracy*: 98.98% on PlantVillage test set
- *Inference time*: Real-time processing suitable for web deployment
- *Supported crops*: Multiple species, including tomato, potato, corn, and others
- *Disease categories*: Distinguishes multiple disease types per crop, plus healthy status

Impact and deployment: DeepCrop demonstrates that research-level accuracy can be achieved in accessible, user-friendly systems. The web interface lowers barriers to adoption, requiring only a smartphone camera and an internet connection. This democratization of AI technology enables smallholder farmers in developing regions to access sophisticated diagnostic capabilities previously available only through expert agronomists.

Limitations

- Performance on the curated PlantVillage dataset may not fully translate to unconstrained field conditions with varying lighting, backgrounds, and image quality
- Requires Internet connectivity, potentially limiting use in remote areas
- Does not integrate with broader farm management systems

2.5.2 Hybrid Multicrop Disease and Yield Prediction System

System overview: A comprehensive pipeline integrating disease detection with yield prediction across five major crops: maize, potato, rice, sugarcane, and wheat (Rikendra and Chauhan 2025).

Technical approach

Multistage architecture

- CNN-based disease detection and severity scoring
- LSTM processing of temporal sequences (multidate imagery and environmental data)
- Attention mechanism for adaptive feature weighting
- XGBoost ensemble for final yield regression

Multimodal data integration

- Visual features from satellite and drone imagery
- Disease severity scores from CNN classifiers
- Environmental variables: temperature, precipitation, solar radiation
- Soil characteristics: moisture, nutrient levels
- Management practices: planting dates, fertilization schedules

Performance metrics

- Disease classification accuracy: 86.9% overall across five crops
- Yield prediction accuracy: 93.84% average validation accuracy
- Inference latency: <100 ms per sample, enabling real-time deployment
- Temporal coverage: Processes entire growing season data

Methodological innovations

- Unified architecture handling multiple crop types reduces development and maintenance costs.
- Attention mechanism automatically identifies critical growth stages without manual specification.
- Integration of disease severity as an intermediate prediction improves yield forecasting,
- End-to-end training enables optimization of the entire pipeline.

Real-world applicability: The multicrop capability of the system makes it suitable for diverse agricultural operations and regional deployment. The <100 ms inference time enables integration with mobile applications and edge computing platforms for field use.

Challenges

- Lower accuracy (86.9%) for disease classification compared to single-crop specialized systems suggests trade-offs in multitask learning.
- Requires a comprehensive data collection infrastructure, including satellite access, ground sensors, and historical records.
- Validation accuracy may differ from operational performance in highly variable field conditions.

2.5.3 AI-IoT Precision Irrigation System

System overview: Integration of AI with IoT sensor networks for automated precision irrigation (Singh et al. 2025).

Technical approach

- *Sensor network*: Distributed soil moisture, temperature, and humidity sensors
- *DL models*: Process sensor streams to predict irrigation needs
- *Actuation system*: Automated valve control for precision water delivery
- *Integration*: Cloud-based processing with edge computing for low-latency control

Performance metrics

- *Water consumption reduction*: 20% compared to conventional irrigation
- *Yield increase*: 15% attributed to optimal water management
- *System reliability*: Continuous operation throughout the growing season
- *Response time*: Real-time irrigation adjustments based on soil conditions

Economic and environmental impact: The 20% water reduction translates directly into cost savings and environmental benefits, particularly critical in water-scarce regions. Simultaneously achieving yield increases demonstrates that precision irrigation optimizes rather than merely conserves resources.

Scalability considerations

- Initial infrastructure costs for sensor networks and automated irrigation systems
- Requires technical expertise for installation and maintenance
- Cloud connectivity needs may limit deployment in remote areas
- Potential for expansion to include weather forecasts and crop growth models for predictive irrigation scheduling

2.5.4 Field-Deployed Disease Detection in India

System overview: AI-powered disease detection system deployed in agricultural regions of India, targeting common crop diseases affecting smallholder farmers (Singh et al. 2025).

Technical approach

- Mobile application for image capture and upload
- Cloud-based DL inference
- Local language support and treatment recommendations

- Integration with agricultural extension services

Performance metrics

- *Disease detection accuracy*: 90% in field conditions
- *User adoption*: Deployed across multiple districts
- *Response time*: Near-instantaneous diagnosis via mobile app
- *Coverage*: Multiple crops and common regional diseases

Social impact: The system addresses critical needs in developing agricultural contexts where access to expert agronomists is limited. Mobile deployment leverages the widespread adoption of smartphones, while local-language support ensures accessibility. Integration with extension services creates a pathway from diagnosis to treatment implementation.

Challenges and lessons

- Field accuracy (90%) is lower than laboratory results, highlighting the importance of testing in real-world conditions.
- Connectivity requirements may limit use in areas with poor mobile network coverage.
- User education is needed to ensure proper image capture for accurate diagnosis.
- Continuous model updating is required to address new disease variants and regional variations.

2.5.5 Comparative Performance Analysis

Table 2.2 summarizes performance across multiple studies and systems:

Key observations

- *Laboratory vs. field performance*: Controlled datasets (PlantVillage: 98.98%) show higher accuracy than field deployments (India: 90%), emphasizing the importance of real-world validation.
- *Multitask trade-offs*: Unified multicrop systems (86.9%) sacrifice some accuracy compared to specialized single-crop models but offer greater practical flexibility.
- *Consistency across studies*: The ~92–94% accuracy range for yield prediction across multiple studies suggests this represents a reliable performance level for current methods.

Table 2.2 Performance across multiple studies and systems

System/study	Task	Crops	Accuracy/ performance	Deployment status
DeepCrop (Islam et al. 2023)	Disease detection	Multiple	98.98% accuracy	Web application deployed
Hybrid CNN-LSTM-Attention (Rikendra and Chauhan 2025)	Multicrop disease classification	5 crops	86.9% accuracy	Research prototype
Hybrid CNN-LSTM-Attention (Rikendra and Chauhan 2025)	Yield prediction	5 crops	93.84% validation accuracy	Research prototype
AI-IoT Irrigation (Singh et al. 2025)	Precision irrigation	General	20% water reduction, 15% yield increase	Field deployed
India Disease Detection (Singh et al. 2025)	Disease detection	Regional crops	90% field accuracy	Mobile app deployed
DL Methods Survey	Yield & bloom recognition	Various	~92.51% average accuracy	Meta-analysis

- *Integration benefits*: Systems combining multiple data modalities (images + sensors + environmental data) consistently outperform single-source approaches.
- *Deployment readiness*: The gap between research prototypes and deployed systems is narrowing, with multiple operational systems demonstrating real-world viability.

2.6 Challenges and Limitations

Despite impressive progress, DL and AI applications in agriculture face significant challenges that must be addressed for widespread adoption and continued advancement.

2.6.1 Data-related Challenges

2.6.1.1 Data Scarcity and Labeling

Agricultural datasets are typically much smaller than those available in other domains (Waqas et al. 2025) (e.g., ImageNet's 1.2 million images). Specific challenges include:

- *Limited labeled data*: Collecting and labeling agricultural data requires expert knowledge to identify diseases, growth stages, and other agronomic features.
- *Class imbalance*: Healthy plants vastly outnumber diseased examples in most datasets, leading to models biased toward majority classes.

- *Rare events*: Critical but infrequent phenomena (e.g., emerging diseases, extreme weather impacts) have minimal representation.
- *Temporal coverage*: Multiyear datasets capturing interannual variability are rare but necessary for robust yield prediction.

Mitigation strategies include transfer learning, data augmentation, synthetic data generation via GANs, and semisupervised or self-supervised learning approaches that leverage unlabeled data (Guntuka 2024).

2.6.1.2 Data Quality and Consistency

Agricultural data quality varies significantly

- Inconsistent image quality from different cameras and lighting conditions
- Sensor calibration drift over time
- Missing data from sensor failures or connectivity issues
- Inconsistent labeling standards across datasets

Standardization efforts and quality control protocols are needed to ensure reliable model training and evaluation.

2.6.1.3 Privacy and Data Sharing

Farmers may be reluctant to share data due to:

- Competitive concerns about revealing management practices or yields
- Privacy worries about location and operational details
- Unclear data ownership and usage rights
- Lack of trust in data security

Federated learning approaches, where models are trained on distributed data without centralizing it, offer potential solutions while preserving privacy (Singh et al. 2025).

2.6.2 Generalization and Domain Shift

2.6.2.1 Geographic and Climatic Variation

Models trained in one region often perform poorly in others due to

- Different crop varieties and cultivars
- Varying soil types and characteristics
- Distinct climate patterns and growing conditions
- Regional pest and disease pressures
- Different agricultural practices and management systems

This domain-shift problem means that models often must be retrained or adapted for each new deployment region, limiting scalability (Guntuka 2024).

2.6.2.2 Temporal Variation

Agricultural systems change over time

- New crop varieties are introduced.
- Diseases evolve, and new pathogens emerge.
- Climate change alters growing conditions.
- Management practices evolve.

Models require continuous updating to maintain accuracy, necessitating ongoing data collection and retraining infrastructure.

2.6.2.3 Controlled Versus Field Conditions

The performance gap between controlled laboratory datasets and real field conditions remains substantial. The PlantVillage dataset's uniform backgrounds and lighting enable 98.98% accuracy (Islam et al. 2023), while field deployment in India achieved 90% (Singh et al. 2025). Real-world challenges include:

- Variable lighting and weather conditions
- Complex backgrounds with soil, weeds, and other plants
- Occlusion and overlapping leaves
- Image quality variations from different devices

Addressing this gap requires training on diverse, field-collected data and developing robust models that handle real-world variability (Guntuka 2024).

2.6.3 Deployment and Infrastructure Challenges

2.6.3.1 Computational Constraints

Agricultural AI must often operate under resource constraints:

- *Edge devices*: Limited processing power, memory, and battery life in field-deployed sensors, drones, and robots (Botero-Valencia et al. 2025)
- *Real-time requirements*: Robotic systems need <100 ms inference for safe, effective operation (Rikendra and Chauhan 2025)
- *Connectivity limitations*: Intermittent or absent internet connectivity in remote agricultural areas precludes cloud-based processing

Model compression, optimization, and edge AI techniques are essential but may compromise accuracy.

2.6.3.2 Integration with Existing Systems

Successful deployment requires integration with

- Farm management software and databases
- Existing sensor networks and equipment
- Agricultural machinery and automation systems
- Supply chain and market information systems

Lack of standardized interfaces and protocols complicates integration (Singh et al. 2025).

2.6.3.3 Cost and Economic Viability

Adoption barriers include:

- High initial costs for sensors, cameras, drones, and computing infrastructure
- Ongoing costs for data storage, processing, and model maintenance
- Need for technical expertise to operate and maintain systems
- Uncertain return on investment, particularly for smallholder farmers

Economic analyses demonstrating clear value propositions are needed to drive adoption.

2.6.4 Interpretability and Trust

2.6.4.1 Black-box Nature of DL

DL models are often criticized as "black boxes" whose decision-making processes are opaque. In agriculture, this raises concerns:

- Farmers may distrust recommendations they don't understand.
- Inability to verify that models use scientifically sound reasoning.
- Difficulty diagnosing and correcting model errors.
- Limited scientific insight from model predictions.

Explainable AI (XAI) techniques, such as attention visualization and saliency mapping, help but don't fully resolve these concerns (Guntuka 2024).

2.6.4.2 Error Consequences

Agricultural decisions have significant consequences:

- Incorrect disease diagnoses may lead to inappropriate treatments, crop losses, or unnecessary pesticide use (Nawaz et al. 2025).
- Inaccurate yield predictions affect planning, marketing, and financial decisions.
- Irrigation errors may waste water or stress crops.

Understanding model uncertainty and providing confidence estimates alongside predictions is crucial for responsible deployment.

2.6.5 Regulatory and Ethical Considerations

2.6.5.1 Pesticide and Treatment Recommendations

AI systems providing treatment recommendations must navigate:

- Regulatory requirements for pesticide recommendations
- Liability concerns if recommendations cause harm

- Need for integration with local regulations and approved products
- Ethical considerations around promoting specific products

2.6.5.2 Data Ownership and Control

Questions of who owns and controls agricultural data remain contentious

- Farmers who generate the data
- Technology companies providing AI services
- Researchers using data for model development
- Governments and public institutions

Clear frameworks for data governance, ownership, and benefit-sharing are needed (Singh et al. 2025).

2.6.5.3 Equity and Access

There is a risk that AI technologies primarily benefit large-scale, well-resourced farms while excluding:

- Smallholder farmers in developing countries
- Farmers in remote or poorly connected regions
- Those lacking technical literacy or resources (Koli et al. 2025)

Ensuring equitable access and developing appropriate technologies for diverse agricultural contexts are ethical imperatives.

2.6.6 Scientific and Methodological Challenges

2.6.6.1 Benchmark Standardization

Agricultural AI research lacks

- Standardized benchmark datasets for fair comparison across studies
- Consistent evaluation protocols and metrics
- Agreed-upon train/test splits and cross validation procedures
- Reporting standards for model architectures and hyperparameters

This hampers reproducibility and makes it difficult to assess true progress (Guntuka 2024; Rai et al. 2025).

2.6.6.2 Causality Versus Correlation

DL excels at finding correlations but doesn't inherently understand causality. Agricultural applications require:

- Understanding causal relationships (e.g., how specific management practices affect yields)
- Ability to predict outcomes of interventions not seen in training data
- Robustness to confounding factors

Integrating causal inference methods with DL is an active research area.

2.6.6.3 Multiscale Integration

Agriculture involves processes at multiple scales:

- Plant physiology (cellular to organ level)
- Individual plant growth and development
- Field-level crop dynamics
- Regional climate and environmental patterns
- Global food systems and markets

Effectively integrating information across these scales remains a fundamental challenge.

2.7 Future Directions and Emerging Technologies

The field of AI in agriculture is rapidly evolving, with several promising directions poised to address current limitations and unlock new capabilities (Kumar et al. 2024).

2.7.1 Advanced Architectures and Methods

2.7.1.1 Foundation Models for Agriculture

Large-scale foundation models—pretrained on massive, diverse datasets and adaptable to many downstream tasks—are transforming AI. Agricultural foundation models could:

- Be pretrained on comprehensive agricultural imagery, sensor data, and scientific literature
- Enable few-shot learning for new crops, diseases, or regions with minimal labeled data (Devarajan et al. 2026)
- Provide unified representations across multiple agricultural tasks
- Incorporate agricultural domain knowledge and scientific principles

Development of such models requires coordinated data collection and computational resources but could democratize access to advanced AI capabilities.

2.7.1.2 Multimodal Transformers

Next-generation transformer architectures designed for multimodal data promise to:

- Seamlessly integrate images, text, sensor data, and structured information
- Capture complex spatio-temporal patterns across growing seasons and geographic regions
- Enable more sophisticated reasoning about agricultural systems
- Support natural language interfaces for farmer interaction

Research shows transformers' potential for agricultural spatio-temporal modeling and multimodal fusion (Sharma and Shivandu 2023).

2.7.1.3 Causal ML

Integrating causal inference with DL will enable:

- Understanding causal relationships between management practices and outcomes
- Predicting effects of novel interventions
- Robustness to distribution shift and changing conditions
- Scientifically interpretable models aligned with agronomic knowledge

2.7.2 Edge AI and Distributed Computing

2.7.2.1 On-device Intelligence

Advances in edge AI hardware and algorithms enable:

- Real-time inference on smartphones, drones, and field robots without cloud connectivity (Tian et al. 2020)
- Privacy-preserving local processing of sensitive farm data
- Reduced latency for time-critical applications
- Lower operational costs by minimizing cloud computing and data transmission

Specialized AI accelerators (e.g., Google Coral, NVIDIA Jetson) and efficient model architectures (e.g., MobileNets, EfficientNets) make sophisticated AI accessible on resource-constrained devices.

2.7.2.2 Federated Learning

Federated learning trains models across distributed devices while keeping data local:

- Farmers retain control and privacy of their data
- Models benefit from diverse data across many farms without centralization
- Reduced data transmission costs and bandwidth requirements
- Enables collaborative learning while respecting competitive concerns

This approach is particularly promising for agricultural AI, where data sharing barriers are significant (Singh et al. 2025).

2.7.3 Integration with Emerging Technologies

2.7.3.1 Digital Twins

Digital twins—virtual replicas of physical agricultural systems—combined with AI enable:

- Simulation of different management scenarios before implementation
- Prediction of outcomes under various weather or market conditions

- Optimization of complex multiobjective decisions (yield, sustainability, profitability) (Bayar et al. 2025)
- Continuous learning as digital twins are updated with real-world observations

Research identifies digital twins as a key emerging technology for precision agriculture (Singh et al. 2025).

2.7.3.2 5G and Advanced Connectivity

5G networks and satellite Internet (e.g., Starlink) promise:

- High-bandwidth, low-latency connectivity even in remote agricultural areas
- Real-time data streaming from sensors and cameras
- Cloud-edge hybrid architectures combining local processing with cloud resources
- Support for autonomous vehicle coordination and swarm robotics

2.7.3.3 Blockchain for Data Integrity

Blockchain technology offers:

- Immutable records of agricultural data for traceability and verification
- Transparent supply chains from farm to consumer
- Secure, decentralized data sharing among stakeholders
- Smart contracts for automated transactions and certifications

Integration of blockchain with AI and IoT creates comprehensive, trustworthy agricultural data ecosystems (Singh et al. 2025).

2.7.4 Robotics and Autonomous Systems

2.7.4.1 Agricultural Robots

AI-powered robots are being developed for:

- Selective harvesting of fruits and vegetables
- Targeted weeding and pest control (Aijaz et al. 2025)
- Autonomous planting and transplanting
- Continuous crop monitoring and data collection

DL provides the perception and decision-making capabilities necessary for robots to navigate complex, unstructured agricultural environments and interact safely with delicate plants.

2.7.4.2 Swarm Robotics

Coordinated teams of small, simple robots could

- Cover large areas more efficiently than individual large machines
- Provide redundancy and robustness to individual robot failures
- Enable fine-grained, plant-level interventions at scale
- Reduce soil compaction compared to heavy machinery

AI coordination algorithms enable effective swarm behavior and task allocation.

2.7.5 Sustainability and Climate Adaptation

2.7.5.1 Climate-resilient Agriculture

AI can support adaptation to climate change through

- Predicting climate impacts on crop suitability and yields
- Identifying climate-resilient crop varieties and management practices
- Optimizing resource use to minimize environmental footprint
- Early warning systems for extreme weather events

Integration of climate models with agricultural AI creates powerful tools for long-term planning and adaptation.

2.7.5.2 Precision Resource Management

Advanced AI systems will enable

- Plant-level precision in water, nutrient, and pesticide application
- Real-time optimization balancing yield, quality, and sustainability
- Circular agriculture systems minimizing waste and external inputs
- Carbon footprint tracking and optimization

These capabilities are essential for sustainable intensification—producing more food with less environmental impact.

2.7.6 Human-AI Collaboration

2.7.6.1 Decision Support Systems

Next-generation systems will

- Provide explainable recommendations with clear reasoning
- Allow farmers to incorporate local knowledge and preferences
- Support interactive exploration of trade-offs and scenarios
- Learn from farmer feedback to improve over time

Effective human-AI collaboration leverages both AI's data-processing capabilities and farmers' contextual expertise.

2.7.6.2 Farmer-centric Design

Future agricultural AI must:

- Be designed with and for farmers, not just for them
- Support diverse farming systems and scales
- Respect farmer autonomy and decision-making authority
- Provide value proportional to required investment and effort

Participatory design approaches involving farmers throughout development ensure practical relevance and adoption.

2.7.7 Data Infrastructure and Platforms

2.7.7.1 Open Agricultural Data

Development of open data platforms with:

- Standardized formats and metadata for interoperability
- Quality-controlled, diverse datasets for model training
- Benchmark datasets for fair comparison of methods
- Clear licensing and usage terms

Initiatives like the Open Agriculture Data Alliance aim to create such infrastructure.

2.7.7.2 AI-as-a-Service for Agriculture

Cloud platforms offering:

- Pretrained models for common agricultural tasks
- Easy customization for specific crops, regions, or conditions
- Scalable computing resources for training and inference
- Integration with farm management systems

These platforms lower barriers to AI adoption, particularly for small and medium operations.

2.7.8 Research Priorities

Key research directions identified in recent reviews include (Guntuka 2024; Singh et al. 2025):

- *Generalization and transfer learning*: Developing models that work across diverse conditions with minimal retraining
- *Multimodal integration*: Effectively combining heterogeneous data sources
- *Temporal modeling*: Better capturing dynamic agricultural processes
- *Uncertainty quantification*: Providing reliable confidence estimates
- *Explainability*: Making AI decisions transparent and trustworthy
- *Data efficiency*: Learning from limited labeled data
- *Real-world validation*: Rigorous testing in operational agricultural settings
- *Sustainability metrics*: Incorporating environmental and social impacts into optimization

Addressing these priorities will advance agricultural AI from promising research to transformative real-world impact.

2.8 Conclusions

2.8.1 Summary of Key Findings

This chapter has examined the landscape of DL and AI applications in agriculture, highlighting rapid advancements, notable achievements, and persistent challenges within the field.

Major application areas: AI has produced significant advancements in crop monitoring, disease detection, yield prediction, weed management, and soil analysis. Deployed systems in real-world environments achieve accuracies between 86% and 99%, depending on task complexity and environmental factors. For example, disease detection on curated datasets achieves 98.98% accuracy (Islam et al. 2023), while multicrop yield prediction attains 93.84% validation accuracy (Rikendra and Chauhan 2025).

Architectural diversity: The field utilizes a range of DL architectures tailored to specific agricultural tasks. CNNs are foundational for image-based applications, while RNNs and LSTMs are effective for modeling temporal dynamics. Attention mechanisms and transformers facilitate advanced multimodal data integration, and generative models help mitigate data scarcity. Hybrid architectures that combine multiple approaches are increasingly delivering improved performance (Rikendra and Chauhan 2025; Guntuka 2024).

Methodological maturity: Transfer learning, multimodal data fusion, advanced augmentation, and model compression are now established as standard practices. These methodologies allow agricultural AI to address data limitations, integrate heterogeneous information sources, and operate effectively on resource-constrained edge devices (Rikendra and Chauhan 2025; Islam et al. 2023).

Real-world impact: In addition to laboratory achievements, deployed systems provide measurable benefits. AI-driven precision irrigation reduces water consumption by 20% and increases yields by 15% (Singh et al. 2025). Web and mobile applications deliver advanced disease diagnostics to farmers globally, including smallholder farmers in developing regions (Islam et al. 2023).

Persistent challenges: Despite advancements, substantial obstacles persist. Data scarcity, domain shift, and the generalization gap between controlled environments and field conditions hinder large-scale deployment (Guntuka 2024). Infrastructure demands, economic constraints, and the need for explainable, reliable systems further limit adoption, especially among resource-limited farmers.

2.8.2 The Path Forward

The advancement of AI in agriculture depends on overcoming existing limitations and capitalizing on emerging opportunities.

Foundation models and transfer learning: The development of large-scale agricultural foundation models, pretrained on comprehensive datasets, will facilitate rapid adaptation to new crops, regions, and tasks with minimal labeled data. This democratization of AI capabilities may help equalize opportunities between well-resourced and resource-limited agricultural operations.

Edge intelligence and distributed systems: Advances in edge AI and federated learning will support sophisticated on-device processing, preserve data privacy, and reduce connectivity requirements. These capabilities are essential for serving remote agricultural regions and addressing concerns regarding data control.

Integration and interoperability: Seamless integration of AI with IoT sensors, robotics, digital twins, and farm management systems will enable comprehensive precision agriculture ecosystems. Standardization initiatives and open platforms will support this integration and mitigate the risk of system fragmentation.

Sustainability focus: Given the increasing pressure on agriculture to reduce environmental impacts while supporting a growing population, the role of AI in enabling sustainable intensification is critical. Key application areas include precision resource management, climate adaptation, and the optimization of multiobjective trade-offs such as yield, quality, sustainability, and profitability in various areas.

Human-centered design: Effective agricultural AI should be developed collaboratively with farmers rather than solely for them. Approaches such as participatory design, explainable AI, and systems that augment human expertise are likely to drive adoption and ensure that technology aligns with farmer needs and values.

2.8.3 Broader Implications

The integration of DL and AI into agriculture represents more than a technological advancement. It fundamentally transforms the relationship between society and food production as well as natural systems.

Food security: AI-enabled precision agriculture provides pathways to sustainably increase global food production. These approaches address food security challenges arising from population growth, climate change, and limited resources.

Environmental sustainability: AI optimizes resource use and enables targeted interventions, thereby supporting environmentally sustainable agricultural practices. Reductions in pesticide and fertilizer use, water conservation, and lower carbon emissions contribute to broader sustainability objectives.

Economic transformation: AI enables the development of new value chains and business models in agriculture, including AI-as-a-service platforms and data-driven advisory services. This transformation introduces both opportunities and risks related to market concentration, equity, and farmer autonomy.

Knowledge democratization: Mobile and web-based AI applications facilitate access to expert-level agricultural knowledge for farmers globally, which may reduce disparities in agricultural expertise. Nevertheless, achieving equitable access across economic and geographic boundaries continues to present significant challenges.

Scientific advancement: AI accelerates agricultural research by enabling high-throughput phenotyping, rapid variety screening, and identifying complex genotype-environment-management interactions. This progress supports faster crop improvement and enhances the responsiveness of agricultural science.

2.8.4 Final Reflections

DL and AI have progressed from experimental technologies to practical tools with demonstrated real-world impact in agriculture. High accuracies in disease detection, yield prediction, and related applications validate the potential of these methods. Implemented systems currently provide measurable benefits in water conservation, yield enhancement, and decision support for farmers.

However, agricultural AI remains in the early stages of development. The disparity between controlled laboratory performance and field deployment, challenges in generalizing across diverse agricultural contexts, and barriers to equitable access indicate that significant work remains. The field must balance technological optimism with rigorous real-world validation, assessment of economic viability, and careful consideration of social and ethical implications.

The most transformative potential of agricultural AI likely lies in augmenting, rather than replacing, human expertise. These technologies can provide farmers with unprecedented information, insights, and decision support while respecting their knowledge, autonomy, and values. Systems that effectively integrate AI's data-processing capabilities with farmers' contextual understanding and judgment are expected to be the most valuable and sustainable.

Continued progress will require interdisciplinary collaboration among computer scientists, agronomists, farmers, policymakers, and other stakeholders. Technical innovation should be informed by agricultural science, validated under real-world conditions, and designed to be economically viable, environmentally sustainable, and socially equitable. Addressing this multifaceted challenge necessitates not only algorithmic advancement but also institutional innovation, policy support, and a commitment to inclusive development.

The agricultural revolution driven by DL and AI is ongoing but remains incomplete. Its ultimate success will be determined not only by algorithmic performance metrics but also by its contribution to a food system that is productive, sustainable, resilient, and equitable, supporting global food security while preserving the planet for future generations.

References

Aijaz N, Lan H, Raza T et al (2025) Artificial intelligence in agriculture: advancing crop productivity and sustainability. J Agric Food Res 20:101762. https://doi.org/10.1016/j.jafr.2025.101762

Akkem Y, Biswas SK, Varanasi A (2023) Smart farming using artificial intelligence: a review. Eng Appl Artif Intell 120:105899. https://doi.org/10.1016/j.engappai.2023.105899

Bayar J, Ali N, Cao Z et al (2025) Artificial intelligence of things (AIoT) for precision agriculture: applications in smart irrigation, nutrient and disease management. Smart Agric Technol 12:101629. https://doi.org/10.1016/j.atech.2025.101629

Botero-Valencia J, García-Pineda V, Valencia-Arias A et al (2025) Machine learning in sustainable agriculture: systematic review and research perspectives. Agriculture 15(4). https://doi.org/10.3390/agriculture15040377

Devarajan D, Allafi R, Obayya M et al (2026) AI based real time disease diagnosis in plants using deep learning driven CNNs. Sci Rep 16(1):4587. https://doi.org/10.1038/s41598-025-34681-1

Guntuka S (2024) AI-driven precision agriculture: advancing crop yield prediction. Int J Multidiscip Res 6(5):29555

Han S, Wang H (2025) Application of deep learning technology in monitoring plant attribute changes. Sustainability 17(17). https://doi.org/10.3390/su17177602

Islam MM, Adil MAA, Talukder MA et al (2023) DeepCrop: deep learning-based crop disease prediction with web application. J Agric Food Res 14:100764. https://doi.org/10.1016/j.jafr.2023.100764

Koli S, Gehlot A, Singh R et al (2025) Sustainable edge AI for precision agriculture: a lightweight CNN model for Aloe vera leaf disease diagnosis. Expert Syst 42. https://doi.org/10.1111/exsy.70154

Kumar A, Gill AS, Sharma A et al (2024) Enhancing the explainability and interpretability of crop yield prediction models through precision agriculture. In: Proc Int Conf Agric-Centric Computation. Springer Nature Switzerland, Cham, pp 218–226

Nawaz U, Zaheer MZ, Khan FS et al (2025) AI in agriculture: a survey of deep learning techniques for crops, fisheries and livestock. arXiv preprint arXiv:2507.22101

Rai N, Boyd NS, Schumann AW (2025) Advancing site-specific disease and pest management in precision agriculture: from reasoning-driven foundation models to adaptive, feedback-based learning. arXiv preprint arXiv:2510.24650

Rikendra, Sharma M, Chauhan SS (2025) A deep learning-based framework for disease detection and yield enhancement in five major crops. Int J Appl Math 38:952–971. https://doi.org/10.12732/ijam.v38i4s.282

Sharma K, Shivandu SK (2023) Integrating artificial intelligence and Internet of Things (IoT) for enhanced crop monitoring and management in precision agriculture. Sens Int 5:100292. https://doi.org/10.1016/j.sintl.2024.100292

Singh AK, Barman S, Babu S et al (2025) Integration of AI and IoT for yield optimization in precision farming. J Exp Agric Int 47(3):233–241. https://doi.org/10.9734/jeai/2025/v47i33331

Tian H, Wang T, Liu Y et al (2020) Computer vision technology in agricultural automation—a review. Inf Process Agric 7(1):1–19. https://doi.org/10.1016/j.inpa.2019.09.006

Waqas M, Naseem A, Humphries UW et al (2025) Applications of machine learning and deep learning in agriculture: a comprehensive review. Green Technol Sustain 3(3):100199. https://doi.org/10.1016/j.grets.2025.100199

Agri-Tech Ecosystems: The Sustainable Agriculture by Smart Digital Innovations and Global Drivers

3

Srikanta Kumar Mohapatra, Jayashree Mohanty, Sumant Kumar Mohapatra, and Htet Ne Oo

Abstract

Agri-tech is experiencing significant growth through the integration of emerging technologies such as the Internet of Things (IoT), robotics, Artificial Intelligence (AI), blockchain, and cloud infrastructure. Decision-making in this sector increasingly relies on data related to climate change, food demand, resource management, and data analysis. Consequently, many developed countries are prioritizing support for smallholder farmers by leveraging these advanced digital technologies. However, key challenges persist, including data availability, cybersecurity concerns, infrastructure limitations, and user adoption. In this context, smart Agri-tech encompasses smart sensing networks, edge cloud infrastructures, predictive analytics frameworks, and data integration. These systems are designed to deploy and operate technologies efficiently, while integrating sustainable solutions to address future food production needs and enhance Agri-tech environments. The primary objective of the Agri-tech ecosystem is to establish a robust, technology-driven environment that meets user requirements and addresses ongoing challenges.

S. K. Mohapatra (✉)
Chitkara University Institute of Engineering and Technology, Chitkara University, Rajpura, Punjab, India

J. Mohanty
Department of Computer Science and Engineering, Chandigarh University, Mohali, Punjab, India

S. K. Mohapatra
Department of CSE (AI&ML), School of Engineering, Dayananda Sagar University, Bengaluru, Karnataka, India

H. N. Oo
Department of Computer Science and Engineering Chitkara University Institute of Engineering and Technology (CUIET), Chitkara University, Rajpura, Punjab, India

H. N. Oo et al. (eds.), *Transforming Agriculture with AI, IoT, and Blockchain*,
https://doi.org/10.1007/978-981-92-1491-4_3

Keywords

Agri-tech ecosystem · Smart contracts · UAVs · FEW framework

3.1 Introduction

The primary objective of Agri-tech is to enhance food production through the application of advanced digital technologies. This process encompasses the production, management, and distribution of food. The Agri-tech ecosystem integrates emerging technologies such as the Internet of Things (IoT), Artificial Intelligence (AI), robotics, and blockchain technologies (FAO 2019). These integrated digital technologies are designed to support sustainability management. The entire process forms the digital-agro-industrial-ecosystem, as illustrated in Fig. 3.1.

Generally, the digital-agro-industrial-ecosystem consists of

- Digital governance
- Infrastructure
- Advanced technologies

Digital governance provides regulatory support for land and resource management through smart contracts. It oversees operational processes from the initial entry of data-driven systems to their execution. These elements improve transparency by facilitating coordination among stakeholders and leveraging support from financial institutions, including those involved in registration, telecommunications,

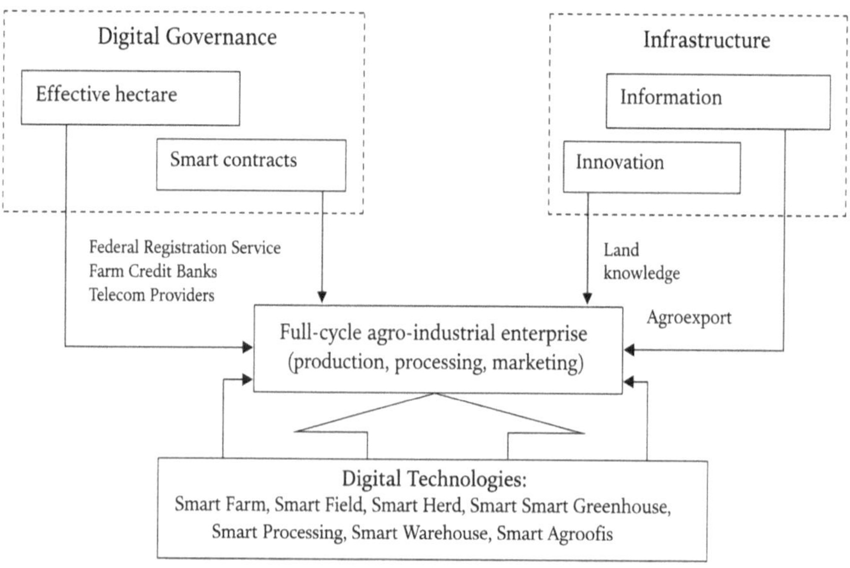

Fig. 3.1 Digital-agro-industrial-ecosystem

Table 3.1 Technologies and key benefits

Technology	Primary applications	Key benefits and impact
IoT and smart sensors	Soil moisture, nutrient, and pest monitoring	Real-time data-driven decisions; resource optimization
AI and Machine Learning (ML)	Predictive analytics, yield prediction, and disease diagnostics	Early detection of issues; improved accuracy in climate and pest forecasting
Robotics and automation	Autonomous tractors, robotic harvesters, and swarm drones	Addresses labor shortages; improves operational efficiency and precision
Blockchain	Supply chain tracking and secure smart contracts	Enhances transparency, trust, and food safety from "seed to shelf"
Remote sensing and UAVs	Crop health monitoring and early warning systems	Supports climate-smart practices and large-scale environmental monitoring
Agri-FinTech	Microloans, crop insurance, and digital payments	Empowers smallholders by improving access to finance and risk management

and other financial entities (Klerkx et al. 2019; Rose and Chilvers 2018). The primary objective is to monitor the availability of relevant information, land knowledge, and effective communication. Advanced technologies support both digital governance and the underlying infrastructure. Table 3.1 details the major advanced technologies and their key benefits.

The global population is increasing rapidly and is projected to reach nearly 10 billion by 2050. Food demand is growing exponentially, while the land available for cultivation is significantly decreasing. Additionally, water scarcity is becoming a critical issue for agriculture. Consequently, traditional farming methods now present a global challenge (United Nations (UN DESA) 2022) and are insufficient to meet current and future food demands. In response, stakeholders such as governments, industries, and research organizations are pursuing the digital transformation of the agricultural sector (Tilman et al. 2011). The primary challenges include ensuring food security, managing climate impacts, and optimizing supply chains. Effective management of these challenges and a strategic approach to digital transformation are essential to maximize agricultural productivity.

3.2 Literature Survey

The Agri-tech ecosystem represents a significant paradigm shift in contemporary farming, driven by the integration of advanced digital technologies, including the IoT, AI, cloud-edge computing, robotics, and blockchain. Early research highlights the critical role of Wireless Sensor Networks (WSNs) in enabling real-time monitoring of soil moisture, crops, and environmental conditions, which are fundamental to precision agriculture systems (Ruiz-Garcia et al. 2009). The evolution of smart farming practices emphasizes the necessity of big data analytics and digital platforms for promoting sustainable agriculture (Wolfert et al. 2017). AI and ML algorithms have been widely applied to enhance crop yield estimation, pest and disease detection, and irrigation optimization. Deep Learning (DL) algorithms further improve decision-making accuracy and reduce resource wastage (Kamilaris and Prenafeta-Boldú 2018). IoT-based system architectures support large-scale data acquisition and automation by integrating sensing, communication, and analytics components efficiently (Atzori et al. 2010). Recent investigations have explored the combined use of blockchain and AI to enable secure data exchange and reliable decision-making processes (Salah et al. 2019).

Digital twins and edge computing have been investigated to enhance adaptability and real-time responsiveness in smart agriculture environments (Verdouw et al. 2021). Edge computing is particularly advantageous in remote, resource-constrained farming areas (O'Grady et al. 2019). Security and privacy remain significant challenges in IoT-based agricultural systems due to increased exposure to cyber-attacks (Lin et al. 2017). Cloud-based smart agriculture platforms facilitate real-time data analysis and storage (Mekala and Viswanathan 2017), while blockchain applications improve traceability and transparency within agri-food supply chains (Tripoli and Schmidhuber 2018). The integration of smart city concepts and IoT has also influenced agricultural digitalization strategies (Kim et al. 2017). Recent research emphasizes sustainable IoT-based agriculture through low-cost, energy-efficient solutions (Dhanaraju et al. 2022; Ray 2017). Nevertheless, challenges such as interoperability, cybersecurity, and policy constraints persist (Gupta et al. 2020). Current studies advocate for holistic system development, supportive policy frameworks, and AgriTech-as-a-Service (AaaS) models to enhance accessibility for small-scale farmers and promote sustainability. A summary of some of the important studies referenced is presented in Table 3.2.

3.3 Operational Management and Framework

Effective operational management of the entire agri-ecosystem is essential to address the global challenges of food production and ecosystem sustainability. The primary objective is to implement data-driven technologies through intelligent and sustainable agricultural systems, as illustrated in Fig. 3.2.

It consists of four broad layers:

(a) Data acquisition layer
(b) Processing and intelligence layer
(c) Action and implementation layer
(d) Distribution and transparency layer

3.3.1 Data Acquisition Layer

In this layer, IoT sensors typically monitor key soil factors such as moisture and nutrient levels, as well as pest presence, to optimize agricultural processes. Remote sensors and unmanned aerial vehicles (UAVs) capture real-time data on crop health

Table 3.2 Literature survey

Basic study	Technology/focus area	Key contributions	Identified gaps/ limitations	Reference
Early studies	Wireless Sensor Networks (WSNs)	Real-time soil moisture, crop, and environmental monitoring for precision agriculture	Energy efficiency and scalability issues	Ruiz-Garcia et al. (2009)
Smart farming studies	Big Data & Digital Platforms	Data analytics support for sustainable agriculture	Interoperability and data integration challenges	Wolfert et al. (2017)
AI-based agriculture research	AI & DL	Crop yield prediction, pest/ disease detection, and irrigation optimization	High computational cost and data dependency	Kamilaris and Prenafeta-Boldú (2018)
IoT system design studies	IoT-based Smart Agriculture	Integrated sensing, communication, and analytics for automation	Security vulnerabilities and heterogeneous devices	Atzori et al. (2010)
Blockchain–AI integration studies	Blockchain + AI	Secure data exchange and trustworthy decision-making	Scalability and latency concerns	Salah et al. (2019)
Digital twin research	Digital Twins	Improved adaptability and real-time responsiveness	Complex modeling and high implementation cost	Verdouw et al. (2021)
Edge computing studies	Edge Computing	Low-latency processing for remote and resource-limited farms	Limited storage and computing capacity	O'Grady et al. (2019)
Security focused studies	IoT Security & Privacy	Identification of cyber threats in smart agriculture	Lack of comprehensive security frameworks	Lin et al. (2017)
Cloud agriculture platforms	Cloud Computing	Scalable storage and real-time data analysis	Network dependency and latency	Mekala and Viswanathan (2017)
Supply chain studies	Blockchain in Agri-Food Chains	Improved traceability and transparency	Energy consumption and regulatory challenges	Tripoli and Schmidhuber (2018)

(continued)

Table 3.2 (continued)

Basic study	Technology/focus area	Key contributions	Identified gaps/ limitations	Reference
Smart city integration research	Smart City–Agriculture Integration	Integration of agriculture with urban digital infrastructure	Policy and infrastructure limitations	Kim et al. (2017)
Sustainable IoT research	Low-cost IoT Solutions	Energy-efficient and affordable agriculture systems	Performance trade-offs	Dhanaraju et al. (2022)
Green agriculture studies	Sustainable Digital Farming	Eco-friendly and energy-aware farming solutions	Limited real-world deployment	Ray (2017)
Review studies	System-level Challenges	Identified interoperability, cybersecurity, and policy gaps	Lack of unified and holistic frameworks	Gupta et al. (2020)

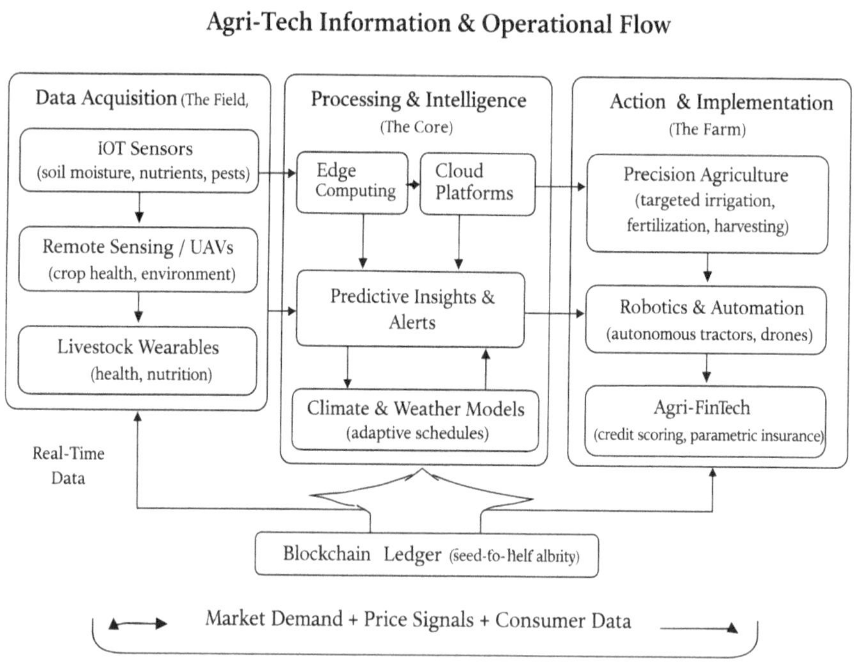

Fig. 3.2 Operational flow of Agri-tech ecosystem

and environmental conditions relevant to crop growth (Vashishth et al. 2024). Wearable devices for livestock monitor animal health and nutritional status. This system enables comprehensive data collection across both crop and livestock systems.

3.3.2 Processing and Intelligence Layer

Data collected through the data acquisition layer is transferred to the processing and intelligence layer, which serves as the core for processing and analyzing information. Typically, edge computing is employed for preliminary data processing to reduce latency and bandwidth requirements. Cloud platforms provide scalable storage and computational resources. AI and ML models are implemented to enhance overall system performance. Climate-related activities are integrated to enable effective control of farming operations and environmental conditions.

3.3.3 Action and Implementation Layer

This layer focuses on precision agriculture to enhance productivity, including irrigation, fertilization, and harvesting. It also aims to optimize resource management through automation. Automation is achieved by deploying autonomous tractors, drones, or robotics to increase accuracy and efficiency. Additionally, Agri-Fintech applications are utilized for climate data management, credit data management, and to support financial resilience (Mahbubi et al. 2024).

3.3.4 Distribution and Transparency Layer

This layer addresses market management by establishing trust through blockchain-based ledgers that secure high-value data throughout the supply chain, from seed to shelf. Farmers are integrated directly into the digital marketplace. A feedback loop enhances overall performance by informing decision-making across the entire agricultural technology ecosystem.

3.4 Agro-Ecosystem and Food–Water–Energy Framework

The framework is structured around three primary factors: food, water, and energy (FWE). The process emphasizes core ecological functions and the comprehensive flow of resources. At the center of the framework is the Agro-Ecosystem (Sánchez-Zarco and Ponce-Ortega 2023), which regulates biological and ecological processes to control agricultural production. This system operates cyclically, involving producers, consumers, decomposers, and nutrient pools. Producers depend on nutrient pools for essential elements, converting natural resources into biomass. Consumers utilize this biomass and return organic matter to the nutrient pools, thereby sustaining long-term soil fertility. The process is illustrated in Fig. 3.3.

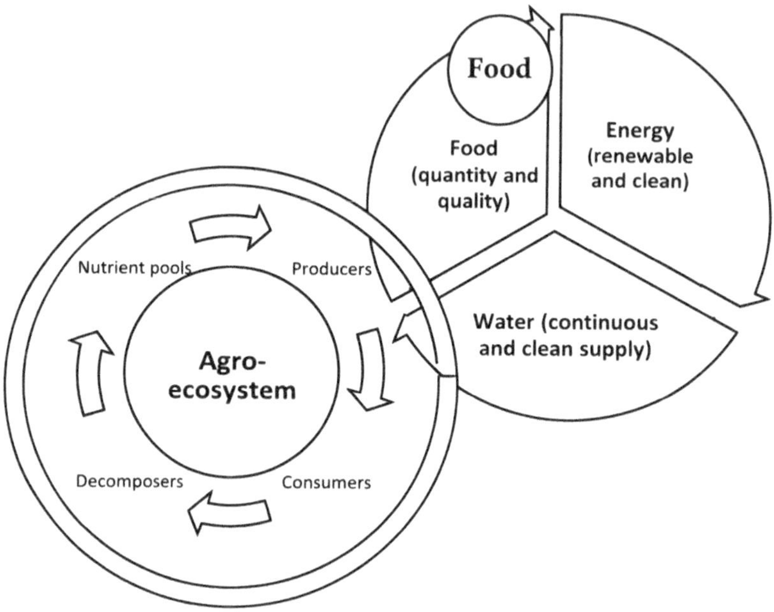

Fig. 3.3 Agro-ecosystem and food–water–energy framework

Within the Food–Energy–Water (FWE) subsystem, the components of food, energy, and water are interdependent. The food component is determined by both quantity and quality, which are directly proportional to the agroecosystem's efficiency. As process efficiency increases, both the quantity and quality of food production improve. Continuous water supply is required for irrigation to ensure effective distribution and processing within the system.

3.5 Challenges and Solution

The performance of the Agri-tech ecosystem relies on the integration of technologies such as AI, robotics, sensors, cloud infrastructure, and blockchain. However, effective implementation of these technologies presents significant challenges. Both real-world deployment and real-time execution require careful consideration. Figure 3.4 outlines the primary challenges and corresponding solutions for advancing these technologies.

3.5.1 Digital Infrastructure

Many regions, particularly rural and remote areas, continue to experience inadequate internet connectivity and unreliable electricity, which hinder the

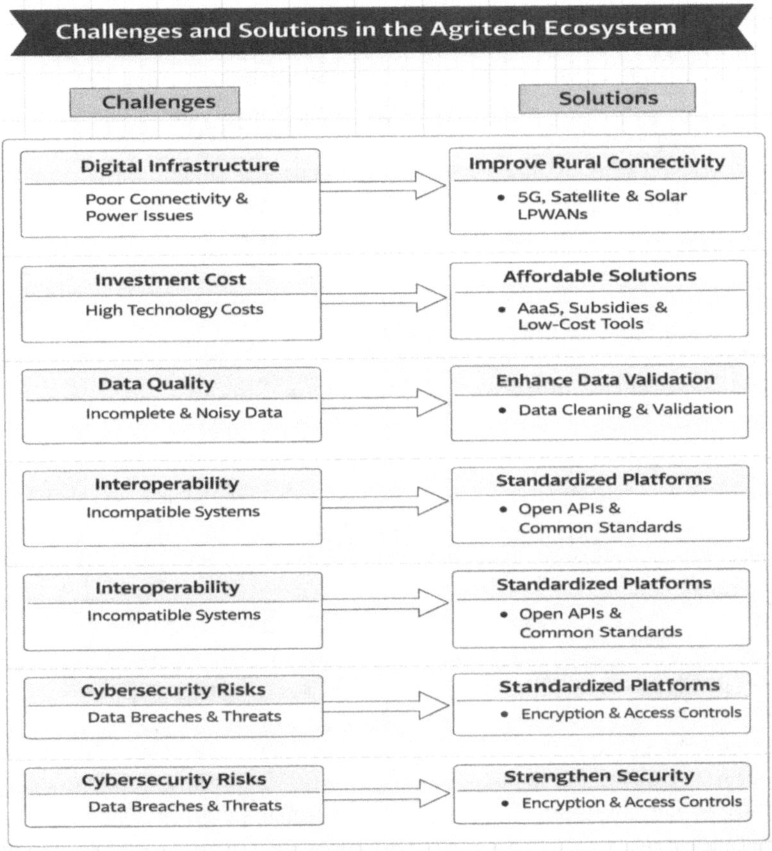

Fig. 3.4 Challenges and solutions

establishment of robust digital infrastructure. These limitations restrict real-time data transmission, cloud-based data management, and the adoption of smart farming technologies. Addressing these challenges requires targeted investment from governments and stakeholders in rural connectivity, with a focus on expanding 5G and satellite networks. Implementing Low-Power Wide-Area Networks (LPWANs) alongside solar technologies can facilitate the development of digital farming infrastructure in these underserved areas (Rafi et al. 2025).

3.5.2 Investment Cost

Digital infrastructure encompasses advanced technologies such as optimized sensors, drones, and robotics. Implementing these technologies requires significant investment to achieve optimal results, which is often unaffordable for small-scale or low-income farmers. Potential solutions include adopting low-cost sensors, using open-source software platforms, and providing financial support through nonprofit rental agreements. Additionally, pay-per-user models, AaaS, and government subsidies can help mitigate financial barriers.

3.5.3 Data Quality and Availability

High-quality agricultural data is essential for the effective implementation of advanced technologies. However, agricultural data is often incomplete, inconsistent, and noisy due to sensor errors, manual data entry, and variable environmental conditions. Such data quality issues negatively impact AI and decision support systems. Therefore, best practices include ensuring data accuracy and completeness through rigorous validation, preprocessing, and testing, thereby enhancing overall system reliability.

3.5.4 Interoperability

Agri-tech systems are typically heterogeneous, often utilizing proprietary software and integrated data technologies across incompatible platforms. This heterogeneity complicates the implementation of integrated technologies. A common solution involves adopting open Application Programming Interfaces (APIs), standardized data formats, and modular architectures to establish a secure environment.

3.5.5 Cybersecurity and Data Privacy Risks

The Agri-tech ecosystem relies extensively on digitization, which increases vulnerability to cyber-attacks and unauthorized access to market data. To mitigate these risks, data should be transmitted through encrypted channels, authentication protocols must be robust, access control mechanisms should be stringent, and data integrity must be maintained throughout all processes.

3.6 Conclusion

The Agri-tech ecosystem represents a significant transformation in contemporary agriculture by integrating advanced digital technologies, including the IoT, AI, edge and cloud computing, robotics, and blockchain. Recent literature indicates substantial advancements in precision agriculture, smart irrigation systems, pest management, and agri-food supply chain traceability. Despite these developments, challenges persist regarding interoperability, cybersecurity, energy sustainability, and technology accessibility, particularly for smallholder farmers in developing regions. Emerging paradigms, including edge computing, digital twin technology, and AgriTech as a Service, offer potential solutions to challenges related to latency, scalability, and inclusivity. Achieving sustainable and resilient digital agriculture requires a comprehensive system design that incorporates robust security architectures, shared platforms, and supportive policy frameworks. Future AgriTech innovations must balance technological advancement with affordability, adaptability, and sustainability to promote long-term food security and sustainable agricultural development.

References

Atzori L, Iera A, Morabito G (2010) The Internet of Things: a survey. Comput Netw 54(15):2787–2805. https://doi.org/10.1016/j.comnet.2010.05.010

Dhanaraju M et al (2022) Smart farming: IoT-based sustainable agriculture. Agriculture 12(10):1745. https://www.mdpi.com/2077-0472/12/10/1745

FAO (2019) Digital technologies in agriculture and rural areas – status report. FAO, Rome. https://www.fao.org/3/ca4887en/ca4887en.pdf

Gupta M, Abdelsalam M, Khorsandroo S et al (2020) Security and privacy in smart farming: challenges and opportunities. IEEE Access 8:34564–34584. https://doi.org/10.1109/ACCESS.2020.2975142

Kamilaris A, Prenafeta-Boldú FX (2018) Deep learning in agriculture: a survey. Comput Electron Agric 147:70–90. https://doi.org/10.1016/j.compag.2018.02.016

Kim T, Ramos C, Mohammed S (2017) Smart city and IoT. Future Gener Comput Syst 76:159–162. https://doi.org/10.1016/j.future.2017.03.034

Klerkx B, Jakku E, Labarthe P (2019) A review of social science on digital agriculture, smart farming and agrifood 4.0. NJAS Wageningen J Life Sci 90–91:100315. https://doi.org/10.1016/j.njas.2019.100315

Lin J, Yu W, Zhang N et al (2017) A survey on Internet of Things: architecture, enabling technologies, security and privacy, and applications. IEEE Internet Things J 4(5):1125–1142. https://doi.org/10.1109/JIOT.2017.2683200

Mahbubi A, Ruspendi I, Ibrohim MS (2024) Strengthening the value chain resilience of Indonesian agricultural Fintech in the aftermath of the Covid-19 pandemic. Int J Food Syst Dyn 15(1):44–54. https://doi.org/10.18461/ijfsd.v15i1.I4

Mekala MS, Viswanathan P (2017) A survey: smart agriculture IoT with cloud computing. In: Proceedings of IEEE ICCSP. https://ieeexplore.ieee.org/document/8282877

O'Grady M, Langton D, O'Hare G (2019) Edge computing: a tractable model for smart agriculture? Artif Intell Agric 3:42–51. https://doi.org/10.1016/j.aiia.2019.12.001

Rafi MSM, Behjati M, Rafsanjani AS (2025) Reliable and cost-efficient IoT connectivity for smart agriculture: a comparative study of LPWAN, 5G, and hybrid connectivity models. In: Proceedings of the international conference on Smart computing and informatics. Springer, Cham, pp 389–411

Ray PP (2017) Internet of Things for smart agriculture: technologies, practices and future direction. J Ambient Intell Smart Environ 9(4):395–420. https://doi.org/10.3233/AIS-170440

Rose J, Chilvers C (2018) Agriculture 4.0: broadening responsible innovation in an era of smart farming. Front Sustain Food Syst 2:00087. https://doi.org/10.3389/fsufs.2018.00087

Ruiz-Garcia L, Lunadei L, Barreiro P et al (2009) A review of wireless sensor technologies and applications in agriculture. Sensors 9(6):4728–4750. https://www.mdpi.com/1424-8220/9/6/4728

Salah K, Rehman MHU, Nizamuddin N et al (2019) Blockchain for AI: review and open research challenges. IEEE Access. https://ieeexplore.ieee.org/document/8611274

Sánchez-Zarco XG, Ponce-Ortega JM (2023) Water-energy-food-ecosystem nexus: an optimization approach incorporating life cycle, security and sustainability assessment. J Clean Prod 414:137534. https://doi.org/10.1016/j.jclepro.2023.137534

Tilman D, Balzer C, Hill J et al (2011) Global food demand and the sustainable intensification of agriculture. Proc Natl Acad Sci USA 108(50):20260–20264. https://www.pnas.org/doi/10.1073/pnas.1116437108

Tripoli S, Schmidhuber J (2018) Emerging opportunities for the application of blockchain in the agri-food industry. FAO, Rome. https://www.fao.org/3/CA2906EN/ca2906en.pdf

United Nations (UN DESA) (2022) World population prospects 2022. United Nations Report. https://www.un.org/development/desa/pd/content/world-population-prospects-2022

Vashishth TK, Sharma V, Sharma KK et al (2024) Integration of unmanned aerial vehicles (UAVs) and IoT for crop monitoring and spraying. In: Internet of Things applications and technology. Auerbach Publications, pp 95–117

Verdouw C, Wolfert J, Tekinerdogan B et al (2021) Digital twins in smart farming. Agric Syst 189:103046. https://doi.org/10.1016/j.agsy.2020.103046

Wolfert A, Ge L, Verdouw C et al (2017) Big data in smart farming – a review. Agric Syst 153:69–80. https://doi.org/10.1016/j.agsy.2017.01.023

AI and Deep Learning for Crop and Soil Health Monitoring

4

Rajwinder Kaur, Saravjeet Singh, and Rajender Kumar

Abstract

Artificial Intelligence (AI) and Deep Learning (DL) are transforming agricultural monitoring by enabling precise, data-driven assessments of crop and soil health. This chapter provides an overview of the evolution of agricultural monitoring systems, tracing the shift from traditional observation-based methods to contemporary digital agriculture supported by remote sensing, Internet of Things (IoT) sensors, and advanced computational analytics. Major agricultural data sources, including satellite imagery, Unmanned Aerial Vehicles (UAVs), ground-based sensors, laboratory soil analyses, weather datasets, and farmer-generated information, are examined within the framework of AI-enabled decision systems. Key DL architectures, such as Convolutional Neural Networks (CNNs), Recurrent Neural Networks (RNNs), Long Short-Term Memory (LSTM) networks, and transformer-based models, are analyzed for their applications in disease detection, nutrient stress diagnosis, irrigation optimization, crop phenotyping, and yield forecasting. This chapter also addresses challenges related to data availability, infrastructure, interpretability, and adoption, and outlines future directions that emphasize explainable AI, integrated digital platforms, and farmer-centric decision-support systems to promote sustainable and resilient agriculture.

Keywords

Artificial Intelligence · Deep Learning · Precision agriculture · Crop health monitoring · Soil health assessment · Internet of Things

R. Kaur (✉) · S. Singh · R. Kumar
Chitkara University Institute of Engineering and Technology, Chitkara University, Rajpura, Punjab, India

H. N. Oo et al. (eds.), *Transforming Agriculture with AI, IoT, and Blockchain*,
https://doi.org/10.1007/978-981-92-1491-4_4

4.1 Introduction

Agriculture has undergone a significant transformation due to the rapid advancement of Artificial Intelligence (AI) and Deep Learning (DL). Traditionally, agriculture was primarily governed by natural cycles and human intuition. However, it is now being redefined through data-driven intelligence, automated analytics, and computationally optimized decision-making. This transformation, often referred to as Agriculture 4.0 or smart farming, enables real-time monitoring of agricultural inputs, predictive analysis, and optimization, thereby revolutionizing the production, monitoring, and maintenance of food systems globally (Ge et al. 2025).

Agriculture in the twenty-first century faces complex and interconnected challenges. Rapid global population growth has increased food demand, placing significant pressure on already strained agro-ecosystems and intensifying concerns about global food security. Additionally, climate change has introduced greater variability in weather patterns, altered growing seasons and temperatures, and facilitated the spread of pests and diseases, all of which negatively affect agricultural productivity and stability. Concurrently, urbanization and industrialization have reduced the availability of arable land. Decades of intensive farming, excessive fertilizer use, and inefficient water management have resulted in severe soil erosion, salinization, nutrient depletion, reduced organic matter, and deterioration of soil structure (Padhiary 2025).

Water scarcity, once a regional issue, has become a global challenge as agriculture now competes with domestic and industrial sectors for limited freshwater resources. These converging pressures indicate that traditional agricultural management practices are no longer adequate to ensure future food and nutritional security. Consequently, the integration of AI-driven and data-centric technologies has become increasingly important for developing sustainable, resilient, and resource-efficient agricultural systems. Historically, agricultural health surveillance relied on manual field scouting, periodic laboratory-based soil tests, and decision-making informed by farmer experience. While these methods have been effective in traditional and small-scale farming, they are insufficient for contemporary large-scale agricultural systems. Crop stress is often detected only after significant deterioration, and physical sampling, though precise, is time-consuming, spatially limited, and costly. Farmer judgment, while practical, remains subjective and heavily dependent on individual experience (Reddy et al. 2025).

The rapid advancement and widespread adoption of digital agricultural systems represent a significant breakthrough in agricultural monitoring. Multispectral, hyperspectral, and thermal sensors mounted on satellite platforms and Unmanned Aerial Vehicles (UAVs) facilitate the large-scale, continuous monitoring of biophysical parameters such as canopy temperature, vegetation vigor, surface moisture, biomass density, and chlorophyll concentration without physical contact. Concurrently, sensor networks utilizing the Internet of Things (IoT) provide real-time monitoring of soil parameters, including pH, electrical conductivity, nutrient content, temperature, moisture, carbon levels, and evapotranspiration. Advanced imaging technologies now enable the early detection of physiological changes, such as pigment degradation, cellular dehydration, and initial fungal infections, well before they become visible to the human eye (Phan et al. 2024). These technological advancements have generated vast amounts of heterogeneous agricultural data, establishing a robust foundation for applying AI and DL in precision farming.

AI and DL approaches have become central to next-generation agricultural monitoring due to their ability to process nonlinear, multimodal, and complex data with high accuracy and efficiency. Unlike classical analytical methods, which require manual feature construction and specific assumptions, DL architectures can be trained to extract meaningful representations directly from raw data and identify hierarchical patterns that classical statistical models cannot detect. For example, Convolutional Neural Networks (CNNs) can identify subtle disease symptoms by analyzing vein deformation, texture, and color differences in crop images. Recurrent Neural Networks (RNNs) and Long Short-Term Memory (LSTM) networks effectively capture time-dependent, seasonal, and crop growth dynamics using time-series sensor and weather data. More recently, transformer-based architectures have demonstrated significant potential for integrating heterogeneous data sources, providing holistic field-level insights, and enabling real-time decision-making (Saleem et al. 2025).

Contemporary DL-powered systems support a wide range of agricultural functions, including disease classification, nutrient stress diagnosis, spatial assessment of soil fertility, yield forecasting, irrigation optimization, and pest outbreak prediction, with greater consistency and scalability than conventional methods. These capabilities enable site-specific input applications, leading to more efficient water use, optimized fertilizer distribution, and reduced pesticide usage, which yield both economic and ecological benefits. Overall, AI-driven systems facilitate timely farm-level responses, reduce uncertainty in farm planning, and enhance overall crop performance and sustainability (Buddi et al. 2025).

4.2 Technological Progress in the Field of Agricultural Monitoring

Technological advancements in field surveillance demonstrate a continued effort to understand the complex interactions among soil, vegetation, weather, and ecology. Initially rooted in observational and experience-based methods, agricultural surveillance has transitioned into a data-driven discipline supported by systematic measurement and automation. This transformation has occurred gradually through a series of significant developments, driven by challenges such as food security, environmental sustainability, labor shortages, and resource constraints. To contextualize this progression, the following sections will examine the evolution of agricultural surveillance from empirical judgment to real-time predictive analytics (Durach et al. 2025).

4.2.1 Traditional Agriculture: Ocular, Intuition, and Physical Observation

Throughout human history, agriculture has primarily relied on experiential knowledge and observation. Farmers assessed crop and soil health using intuitive indicators such as plant color, foliage wilting, root development, soil texture, and seasonal patterns informed by intergenerational experience and oral traditions. Elders and experienced farmers served as key decision-makers, offering guidance on irrigation schedules, sowing dates, seed selection, and pest management (Kerr 2020).

Observation at this stage was typified by:

- Agricultural practices relied on local knowledge that was highly adapted to specific areas and climate conditions, but lacked standardization.
- The predominant method of recordkeeping was analog, with weather variations, yields, and disease outbreaks primarily retained in memory rather than documented systematically.
- Diagnosis of crop stress often occurred late, as symptoms were typically observed only after significant damage had already taken place.
- Experimentation was primarily conducted through trial and error, which facilitated learning but was not supported by quantitative data or repeatable analysis (Pretty 2018).

Although these techniques demonstrated resilience and context sensitivity, they were insufficient to withstand the pressures of population growth and large-scale agricultural intensification.

4.2.2 Mechanization and Green Revolution: Surveillance by Intensification of Inputs

During the mid-twentieth century, agricultural production experienced a significant transformation through mechanization and scientific advancements. The adoption of tractors, tillers, and harvesters substantially reduced labor requirements and improved operational efficiency. The Green Revolution (1960s–1980s) further accelerated these changes by introducing high-yielding crop varieties, synthetic fertilizers, chemical pesticides, fungicides, and expanded irrigation infrastructure (Wik et al. 2012).

Although productivity increased substantially and widespread food shortages were averted, this period also revealed significant limitations. Extensive farming practices resulted in decreased soil organic matter, microbial degradation, waterlogging, salinization, and ecological imbalance due to excessive agrochemical use. Furthermore, farmers became increasingly dependent on external inputs, heightening economic vulnerability (Stupar et al. 2024). These challenges highlighted that sustainable productivity requires not only increased inputs but also the adoption of resource-efficient management strategies.

4.2.3 Remote Sensing, GPS, and GIS: Precision Agriculture Born

Precision agriculture emerged in the 1990s and early 2000s with the introduction of Geographic Information Systems (GIS), Global Positioning Systems (GPS), and satellite-based remote sensing. These technologies enabled spatial monitoring of agricultural fields and revealed intrafield variability that was previously undetectable through manual observation (Pan et al. 2022).

Key innovations during this period included the following:

- Assessment of crop vigor using vegetation indices such as the Normalized Difference Vegetation Index (NDVI).
- Mapping of soil fertility, salinity, and erosion risks through digital soil maps.
- Yield monitors are integrated into harvesting machines.
- Variable Rate Technology (VRT) enables the application of site-specific fertilizers and other agricultural inputs.
- UAVs are employed to collect high-resolution aerial imagery.

Impact of this phase

- Reduced input waste is achieved through the implementation of targeted measures.
- Enhanced understanding of spatial soil variability is obtained.
- Spectral data enables the detection of plant stress that is not visually apparent.
- Provides a foundation for digital mapping and data storage at the farm level.

Nonetheless, even advanced precision tools required an experienced analyst to extract relevant information from images and maps. Additionally, weather and satellite data were limited by cloud cover, insufficient temporal resolution, and inadequate spatial granularity for smaller farms. These limitations highlighted the need for tools capable of automatically interpreting large datasets.

4.2.4 Machine Learning (ML) and Decision Support: Toward Smart Analytics

A major advancement in agricultural monitoring emerged between 2005 and 2015, marked by the practical implementation of Machine Learning (ML) algorithms. During this period, field data such as weather records, soil properties, and crop imagery were increasingly analyzed using statistical learning methods to identify trends, classify crop conditions, and support decision-making processes (Liakos et al. 2018).

Classical ML techniques were applied to several common tasks during this period, as illustrated in Fig. 4.1:

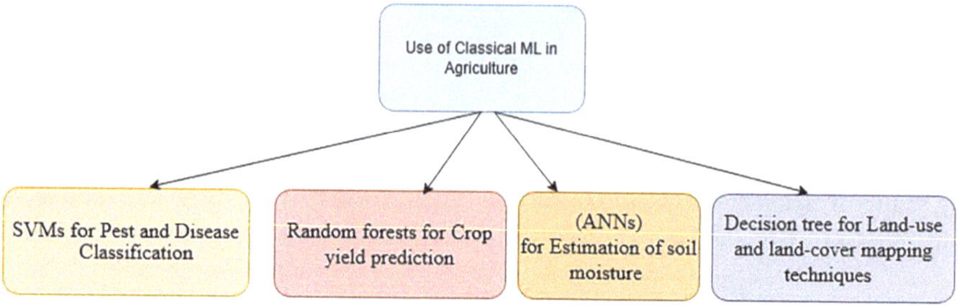

Fig. 4.1 Uses of classical ML in agriculture (Liakos et al. 2018)

- Support Vector Machines (SVMs) were employed for pest and disease classification.
- Random forests were utilized to develop predictive models for crop yield.
- Artificial Neural Networks (ANNs) were applied to estimate soil moisture and evapotranspiration.
- Decision tree algorithms were used for land-use and land-cover mapping.

This stage enhanced agricultural monitoring by enabling the following advancements:

- Detection of complex patterns that surpass human cognitive capabilities.
- Temporal analysis of interannual crop behavior and climatic variability.
- Implementation of early warning systems for disease outbreaks and yield anomalies.

Despite these advancements, classical ML methodologies exhibited several weaknesses. They relied heavily on manually engineered features, required significant assumptions about data distributions, and involved extensive data handling and normalization procedures. These limitations reduced scalability and adaptability as agricultural data became increasingly voluminous, diverse, and complex (Kamilaris and Prenafeta-Boldú 2018). These shortcomings prompted a transition to nonrepresentation learning methods, ultimately leading to the adoption of DL approaches.

4.2.5 DL, IoT, and Real-Time Surveillance: The Era of Smart Agriculture

Since 2015, AI and DL technologies, together with advancements in IoT technologies and advanced computing infrastructure, have fundamentally transformed agricultural monitoring. This transformation has been driven by rapid progress in computing capabilities, cloud and edge computing, affordable sensors, and increased access to big data.

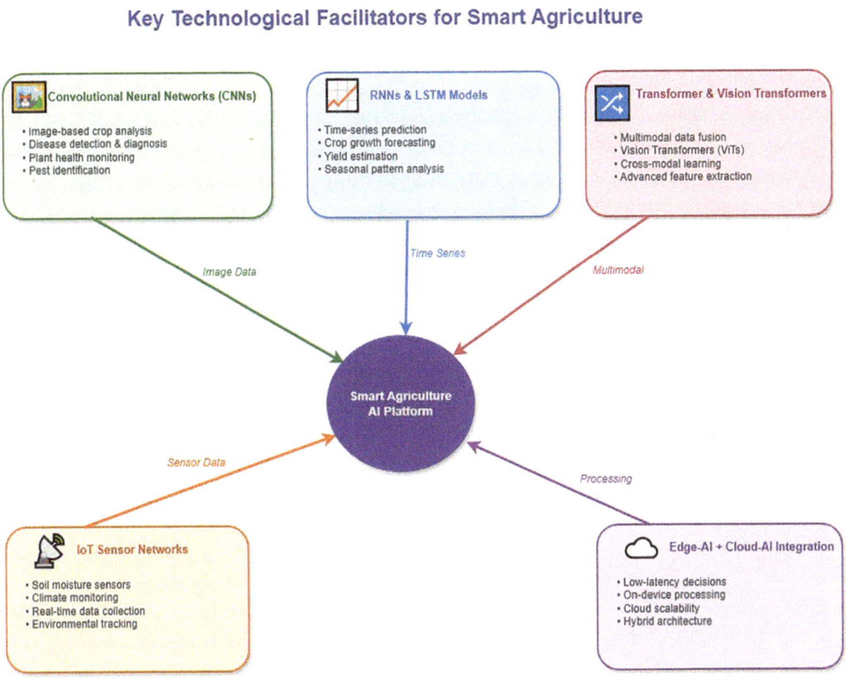

Fig. 4.2 Key technological facilitators for smart agriculture (Begum et al. 2025)

The key technological enablers of this stage are summarized below and illustrated in Fig. 4.2.

- Image-based crop and disease analysis utilizing CNNs.
- Time-series prediction of crop growth and yield using RNNs and LSTM models.
- Multimodal data fusion architectures, including transformer-based multimodal frameworks and Vision Transformers (ViTs).
- Soil and climate sensor networks based on IoT for real-time data acquisition.
- Combination of edge-AI and cloud-AI to make low-latency decisions.

DL has enabled several critical advancements, including the early detection of disease symptoms via microscopy, stress prediction using hyperspectral signatures, nondestructive estimation of soil nutrients, yield prediction from time-dependent data streams, and remote field surveying with UAVs and robots in agriculture (Begum et al. 2025). These developments have transformed agricultural monitoring into a real-time, predictive, and continuous process, replacing periodic observation and thereby enhancing responsiveness and resilience in farm management.

4.2.6 Future Directions: Farming Ecosystems: Cognitive and Autonomous

Recent advancements in agriculture are leading toward a future characterized by cognitive and autonomous systems. Monitoring and decision-making processes are becoming dynamic, incorporating adaptive intelligent capabilities rather than relying on static automation. These systems are expected to continuously learn from new data, respond dynamically to changing environmental and socio-economic conditions, and operate with minimal human intervention (Mohammed et al. 2026).

The new innovations expected during this stage are as follows:

- Simulation of crop growth under varying climatic and management conditions using virtual crop twins.
- Blockchain-based monitoring of agricultural supply chain transparency.
- Deployment of drone swarms for fertilization, pest monitoring, and field inspection.
- Implementation of autonomous machinery, including self-driving tractors, smart irrigation systems, and robotic sprayers.
- AI-driven socio-economic advisory platforms to support farmers' decision-making processes.
- Decision engines operated at a global scale, integrated weather data, and incorporated farm-level analytics.

Table 4.1 presents a comparative summary of the evolutionary stages of agriculture, ranging from the conventional stage to the autonomous stage.

Table 4.1 Comparative summary across evolutionary stages

Time	Measurement procedure	Merits	Limits
Conventional	Visual observation, experiential knowledge	Low cost, sustainable	Limited accuracy, nonscalable
Mechanization & Green Revolution	Laboratory testing input-driven management	High output	Environmental degradation
Remote sensing & GIS	Satellite/UAV imagery, GPS mapping	Spatial monitoring, precision	Requires expertise, costly
Smart agriculture based on ML	Data analytics and decision support	Predictive modeling	Needs feature engineering
AI & DL age	Pattern detection in autonomy, multimodal fusion	Account for high precision, early diagnosis	Data-hungry, infrastructure-intensive
Autonomous agriculture future	Cognitive systems, robotics, blockchain	Self-sustaining intelligence required	Elements of ethical barriers and adoption barriers must be overcome.

4.3 Data Sources Used for Crop and Soil Health Monitoring

High-quality and abundant farm data is fundamental to AI-based agricultural analytics. The agricultural sector relies heavily on data due to the biological complexity of plant physiology, the chemical heterogeneity of soil systems, and the highly dynamic interactions between climate and agriculture. Consequently, the digital transformation of agriculture depends on multisource, multiresolution, and multitemporal datasets, which collectively provide a comprehensive perspective on the entire crop lifecycle and soil ecosystem.

The following subsections discuss the primary data streams utilized for implementing AI and DL in crop and soil health monitoring.

4.3.1 Satellite Remote Sensing Data

Satellite remote sensing serves as the primary method for collecting long-term data on crop development. In similar scenarios, information can be acquired directly from satellites to monitor agricultural progress over extended periods.

Satellite sensing represents a transformative approach to collecting agricultural data at global, regional, and field scales. International satellites capture Earth observation imagery at various spectral, spatial, and temporal resolutions, generating extensive datasets that enable long-term monitoring of crops and soils.

In agricultural applications, key types of satellite-derived data include the following.

Multispectral Imagery Measurement of reflectance in specific spectral bands such as blue, green, red, near-infrared (NIR), and shortwave infrared (SWIR).

Vegetation indices

- Normalized Difference Vegetation Index (NDVI): Assesses plant greenness and biomass.
- Enhanced Vegetation Index (EVI): Detects canopy structure in regions with high biomass.
- Soil Adjusted Vegetation Index (SAVI): Evaluates vegetation health in areas with sparse crop cover.

Key applications include monitoring crop vigor, detecting drought stress and nutrient deficiencies, mapping postharvest residues, and identifying weed invasion. Table 4.2 presents examples of major satellite platforms and their agricultural uses.

Table 4.2 Major satellite platforms and their uses in agriculture

Agency/company	Satellite systems	Uses
NASA	Landsat-8/9, MODIS	Crop monitoring, biomass mapping, soil moisture
ESA (EU)	Sentinel-1, Sentinel-2	Disease stress, SAR moisture, vegetation indices
ISRO	Resourcesat, Cartosat	Yield estimation of crops in India.
Business-to-consumer	PlanetScope, WorldView, RapidEye	High-resolution farm-level insights

4.3.2 UAV (Drone) Aerial Photogrammetry and Imagery

UAVs have emerged as highly disruptive data-collection systems in precision agriculture. They address the limitations of satellite imagery by providing finer spatial resolution and overcoming the constraints associated with large-scale field scouting.

Advantages of UAV-based surveillance

- UAV imaging systems can achieve spatial resolutions of less than one centimeter.
- Flight schedules can be arranged daily or on demand, providing significant operational flexibility.
- UAVs can operate effectively in geographically fragmented regions and under cloudy conditions.
- They are particularly effective for variable rate applications and for localizing pest infestations and hotspots.

Analytical metrics enabled by UAVs

- Generation of plant height maps.
- Estimation of Leaf-Area Index (LAI).
- Assessment of canopy cover.
- Detecting cultivated products lodged and damaged by hail.
- Mapping weed dissemination to support precise herbicide application.

Photogrammetry with UAVs can also assist in creating Digital Elevation Models (DEMs), which are used to model water flow paths, drainage systems, and soil conservation layouts. Table 4.3 shows the sensor payloads used in agricultural drones.

4.3.3 Earth-Based IoT Sensors and Smart Agriculture Devices

The IoT enables real-time, high-resolution monitoring of soil conditions, crop microclimate, and farm infrastructure. Sensor networks can be remotely managed and automated through integration with cloud-based platforms.

Table 4.3 Sensor payloads used in agricultural drones

Type of sensor	What to measure	Uses
Three color cameras	Visible spectrum	Counting crop stand, defoliation, weed mapping
Multispectral cameras	Reflectance in 5–10 broad bands	Vegetation indices, stress maps
Hyperspectral cameras	100–400 narrow spectral bands	Nutrient imbalance, early detection of diseases
LiDAR	3D point clouds	Canopy height, biomass estimation, terrain modeling
Thermal cameras	Heat signatures	Irrigation schedule

Fundamental Types of IoT Sensors in Agriculture

Soil sensors

- Tensiometers (for measuring soil water potential or capacitance).
- Sensors for salinity and pH measurement.
- Sensors for NPK (nitrogen, phosphorus, potassium) nutrient content analysis.
- Sensors for estimating electrical conductivity and organic carbon content.

Crop physiological sensors

- Chlorophyll pigment fluorescence meters.
- Spectral reflectance spectrophotometers.
- Sap-flow and stem-diameter sensors for water transport analysis.

Environmental and microclimate sensors

- Humidity and air temperature sensors.
- Photosynthetically Active Radiation (PAR) and solar radiation sensors.
- CO_2 concentrations can be monitored at different wind speeds at a height of 10 m above ground level.

Intelligent irrigation and fertigation systems

- Automatic valve control systems.
- Drip-line monitoring systems.
- Automation of water pumps based on soil measurements.

IoT sensors form the foundation of digital twin agriculture, enabling each field to have a continuously updated virtual model for monitoring and simulation.

4.3.4 Soil Sampling, Laboratory Tests, and Field Survey

Laboratory analysis of soil remains essential because it provides reference-grade measurements necessary to verify AI model accuracy and calibrate it. Ground-truth datasets generated from these measurements are used to train DL algorithms to predict soil properties using remote sensing and sensor data.

Field survey data collection

- Manual observation of diseases and pest infestations.
- Assessment of crop performance using standardized scorecards.
- Evaluation of weed presence and lodging incidents.
- Documentation of local management practices, including sowing dates, seed varieties, and fertilizer application rates.

Socio-agronomic data are collected through surveys to facilitate the adaptation of AI systems to local agricultural conditions.

4.3.5 Weather and Climate Data Source

Every stage of the crop lifecycle, from germination to ripening, is significantly influenced by weather conditions. Therefore, access to accurate climate data is essential for predictive agricultural modeling.

Major sources of weather data

- Automatic Weather Stations (AWS).
- Government meteorological agencies, such as the India Meteorological Department (IMD) and the National Oceanic and Atmospheric Administration (NOAA).
- Satellite-based reanalysis datasets, including ERA5 and MERRA-2.
- IoT network microclimate samples.

AI in weather data

- Forecasting of crop yields.
- Prediction of pest and disease outbreaks, such as blight, rust, and armyworm.
- Modeling of Growing Degree Days (GDD).
- Development of advisory systems that account for climate-related risks.

Climate analytics assist farmers in getting out of the reactive response model and into proactive planning.

4.3.6 Farmer-Sourced and Participatory Data

Human observations continue to provide valuable data, especially in regions where digitalization remains limited.

Table 4.4 Important data repositories

Category	Examples
Image data of crop disease	PlantVillage, PlantDoc, IP102
Soil mapping datasets	SoilGrids, ISRIC, SoilWeb
Meteorology and climate	NASA POWER, ERA5, AgERA
Simulation and productivity of crops	AgMIP, GEOSHARE, USDA crop statistics

Categories of participatory agricultural data:

- Photographs documenting diseased leaves for diagnostic purposes.
- Comprehensive records of pest outbreaks and unanticipated pest-related damage.
- Field diaries and crop yield records.
- Indigenous knowledge regarding seasonal patterns and soil behavior.
- Advisory platforms delivered through voice and mobile extension services.

Participatory datasets help address regional biases in global datasets and support the development of more inclusive AI systems.

4.3.7 Agricultural Databases and Open-Access Repositories

Various organizations in the world and across other nations farm data in research and development. Table 4.4 shows the important data repositories available:

The algorithms are developed more quickly because these repositories provide standardized benchmarking data.

4.4 AI and DL Models in Agriculture

The application of AI and DL has transformed agricultural surveillance by enabling the analysis of large, heterogeneous datasets and providing accurate information for decision-making at both micro and macro levels. In contrast to traditional methods, which depend on manual examination and physical scanning, AI systems employ computational frameworks that take advantage of historical and real-time data to identify trends that may not be readily apparent to human observers. DL, in particular, is highly effective at capturing multispatial, multitemporal, and spectral interdependencies in images and sensor measurements. This capability makes it especially suitable for predictive systems related to plant diseases, nutrient deficiencies, water stress, crop classification, harvest forecasting, and soil characterization. The successful implementation of AI-based agricultural monitoring depends on integrating efficient computational models, large annotated datasets, and advanced sensors capable of detecting fine-scale variations in plant physiology and soil structure.

CNNs have emerged as the predominant DL approach for processing visual and spatially resolved data, particularly data acquired from satellites, UAVs, and field-based imaging systems. CNNs are effective due to their hierarchical capability to automatically extract features in diseased plants, such as texture, shape, color gradients, and lesion patterns, without the need for manual feature engineering. Detection of fungal infections, viral and bacterial outbreaks, leaf miners, and nutrient deficiencies using CNN-driven classification systems has been achieved by analyzing spectral signatures, thermal representations, and morphological changes in leaves. Furthermore, CNN-based architectures such as U-Net, Mask-RCNN, and DeepLab are frequently employed to accurately delineate infected areas, canopy edges, weed patches, and moisture-stressed regions. In soil health assessment, digitized soil surface and microscope images can be analyzed by CNNs to detect organic content, clay and sand composition, salinity levels, and the presence of harmful pathogens. As a result, CNNs constitute a foundational component of visual phenotyping platforms and digital crop surveillance systems.

RNNs and advanced variants, such as LSTM networks and Gated Recurrent Units (GRUs), are well-suited for analyzing the statistical properties of temporal patterns in sequential data. The agricultural domain is inherently dynamic; as crops are cultivated, their growth patterns, phenological stages, and associated soil characteristics evolve. RNN-based models can capture these temporal changes by learning from historical climate conditions, vegetation indices, periodic sensor measurements, and previous yield data. Utilizing weather-related variables, including temperature, rainfall, solar radiation, and evapotranspiration, LSTM and GRU models enable the prediction of yield, disease outbreaks, drought stress, and nutrient depletion up to several weeks in advance. Furthermore, hybrid architectures such as CNN-LSTM have gained prominence for integrating spatial and temporal learning, thereby enhancing time-series satellite image analysis to monitor crop development and detect anomalies across multiple seasons. These models are particularly valuable in regions with unpredictable climatic patterns, where sudden deviations can result in significant crop stress.

Transformer architectures have recently demonstrated exceptional performance in various agricultural monitoring tasks by enabling models to capture long-range dependencies and contextual relationships within multimodal datasets. Unlike RNNs, which process sequences sequentially, transformers utilize self-attention mechanisms to analyze entire sequences simultaneously, resulting in highly efficient and scalable representations. In applications such as remote sensing and soil monitoring, transformers integrate high-dimensional spectral data, multiyear weather records, geospatial layers, and sensor data to identify subtle correlations that previous models could not detect. This capability enhances accuracy in land-use classification, multicrop discrimination, pest outbreak prediction, and soil quality mapping. Vision Transformers (ViT) have already achieved state-of-the-art results in field type identification and disease detection using drone imagery, while Temporal Transformers enable precise forecasting with extended environmental data. This advancement marks the onset of foundation-model-driven agriculture, where a single large model can support multiple monitoring tasks concurrently.

In addition to large neural architectures, alternative representation learning methods such as Self-Supervised Learning (SSL) and contrastive learning are gaining prominence because they do not require extensive annotated datasets, which are often scarce in agriculture. SSL techniques extract meaningful patterns from unlabeled data by inferring missing information, generating pseudo-labels, or identifying similarities between transformed image pairs. These approaches are particularly valuable in developing regions where high-quality labeled data are limited due to resource constraints. Graph Neural Networks (GNNs) are increasingly utilized to model spatial relationships among fields, soil texture groups, irrigation zones, and crop rotations. Furthermore, explainable AI (XAI) systems, including Grad-CAM, SHAP, and LIME, are fostering trust in AI-driven agriculture by clarifying which features—such as leaf color, canopy temperature, vegetation index, or soil moisture—influence model predictions.

AI-powered decision support systems that integrate outputs from CNNs, RNNs, transformers, and related models provide comprehensive insights for precision agriculture. These systems analyze multivariate data to recommend optimal fertilizer dosages, irrigation schedules, pesticide application timings, and harvesting periods based on the spatial and temporal characteristics of each farm. Cloud-based applications enable real-time adaptation by automatically retraining models with newly available data, thereby supporting responsiveness to climate variability, pest dynamics, and evolving soil chemistry. The deployment of edge-AI devices and mobile applications allows for on-site predictions and immediate feedback, even in areas with limited connectivity. This technological advancement is gradually reducing barriers to the adoption of sophisticated AI in agriculture and promoting broader accessibility.

AI and DL methods represent a paradigm shift in computationally guided farming systems, resulting in reduced uncertainty, enhanced efficiency, and improved sustainability. Despite significant advancements in model accuracy and efficiency in recent years, future research should address domain adaptation, model interpretability, data fairness, energy-efficient architectures, and scalability within resource-limited farming environments. The continued digitalization of agricultural landscape development will require effective integration of AI technologies with field operations, which is essential for building resilient, climate-resistant, and productive food systems.

4.5 Applications of AI and DL in Crop Health Monitoring

DL and AI have become foundational to advancing precision agriculture by enabling continuous, automated, and highly accurate crop health monitoring. AI-based systems offer scalable solutions for real-time crop performance monitoring, in contrast to traditional methods that rely on field visits, manual scouting, and laboratory diagnostics. These technologies are transforming agricultural practices by providing actionable information that supports targeted interventions, reduces resource wastage, and minimizes crop losses and productivity declines. The rapid development of remote sensing platforms and high-resolution imaging systems, combined with AI's computational capabilities, now enables monitoring of crop conditions at both micro (plant-level) and macro (regional and national) scales without direct physical interaction.

Early disease detection and diagnosis represent critical applications of AI and DL in crop health monitoring. Numerous crop diseases, such as bacterial blight, rice blast, wheat rust, downy mildew, late blight, and viral mosaics, can cause irreversible physiological damage if not addressed promptly. DL-based disease detectors, trained on extensive datasets of diseased and healthy leaf images, can identify diseases using subtle indicators such as lesions, chlorosis, color shifts, wilting, and texture anomalies. CNNs and ViT networks can detect disease symptoms before they are visible to the naked eye by analyzing spectral and thermal variations in hyperspectral and multispectral images. Automated classification and segmentation tools enable both farmers and agricultural agencies to monitor diseases at the field and regional levels, facilitating rapid containment measures and reducing economic losses. Furthermore, mobile-based AI diagnostic applications have expanded access to plant health assessment by allowing farmers in remote areas to capture leaf images with smartphones and receive immediate diagnostic and treatment information.

AI plays a significant role in assessing plant stress and nutrient deficiencies, as well as in disease detection. Nutrient deficiencies often manifest as color distortions, stunted growth, or structural abnormalities in leaves and stems. These symptoms are frequently confused with pest damage or environmental stress when assessed manually. Deep neural networks can differentiate between deficiencies of nitrogen, phosphorus, potassium, magnesium, zinc, and iron by analyzing morphological and spectral patterns. Multispectral UAV imagery, thermal imagery, and vegetation indices such as NDVI, GNDVI, SAVI, and EVI provide detailed information on chlorophyll concentration, canopy temperature, leaf water content, and photosynthetic efficiency. AI systems utilize these data to generate precise nutrient stress maps, enabling location-specific fertilizer application. This approach optimizes input use, enhances crop vigor, and prevents excessive chemical application, thereby protecting soil and water quality.

Another essential application of AI is the detection of pests and weeds, which is vital for minimizing yield loss. Pests such as bollworms, aphids, fall armyworms, and whiteflies can cause significant damage to major crops within a short period. AI-based image recognition systems identify pest infestations by analyzing crop damage patterns and UAV-captured images of pest swarms. When direct observation of pests is challenging, DL models estimate the likelihood of infestation using spectral features and variations in canopy health. Similarly, AI-driven weed detection in crops such as wheat, rice, soybean, and cotton improves the precision of herbicide application. DL segmentation models accurately locate weed patches and distinguish them from crop plants, ensuring herbicides are applied only to unwanted vegetation. This targeted approach reduces chemical usage, supports soil biodiversity, lowers operational costs, and limits the development of herbicide resistance in weed species.

AI and DL play a significant role in plant growth monitoring and phenotyping, which involves continuous observation of plant morphology across growth stages.

Conventional phenotyping is often time-consuming and susceptible to human bias. In contrast, DL systems automate the measurement of plant height, leaf area index, canopy cover, flowering rate, branching patterns, fruit count, and biomass accumulation. These automated measurements enable researchers, seed developers, and plant breeders to compare genotype-by-environment interactions and assess phenotypic traits across various climatic conditions. DL-driven automated growth curve modeling provides insights into crop vigor and supports the prediction of maturity and harvest time. Computer vision systems are also implemented in intelligent greenhouse environments to monitor individual plants and regulate irrigation, nutrient levels, and microclimate conditions for optimal growth.

Water scarcity remains a critical global agricultural challenge, and AI technologies contribute to sustainable irrigation planning and the identification of water stress. DL algorithms analyze thermal images and evapotranspiration models to detect temperature variations indicative of water stress, such as water deficits, stomatal closure, and reduced canopy transpiration. Time-series AI models process environmental parameters, including soil moisture, rainfall patterns, humidity, and sunlight, to recommend optimal irrigation timing and frequency. These predictions are integrated into autonomous drip and sprinkler systems, enabling precise water delivery only where necessary. AI-based water management not only enhances crop resilience to drought but also promotes efficient resource utilization in water-deficient regions.

In recent years, yield and productivity forecasting has emerged as a critical application of AI in crop monitoring. DL systems trained on historical weather patterns, soil fertility metrics, vegetation indices, irrigation data, and disease incidence can predict anticipated yields early in the growing cycle with high precision. These forecasts inform government planning, market supply chain management, insurance policy development, and farm-level decision-making. Early yield prediction facilitates the timely provision of storage facilities, labor allocation, and fertilizer procurement. When integrated with economic models, AI-based yield prediction helps stabilize prices by balancing supply and demand.

AI and DL are also utilized in assessing harvest readiness and crop quality, enabling the identification of the optimal harvest time to maximize quality and nutritional value. Computer vision models analyze external characteristics such as color, firmness, and surface texture to estimate fruit maturity. AI-driven harvest prediction is implemented for crops including grapes, apples, tomatoes, pomegranates, sugarcane, and tea leaves, supporting the scheduling of manual, mechanical, or robotic harvesting. Postharvest grading systems equipped with DL algorithms detect defects and bruises, ensuring consistent quality for both export and domestic markets.

These programs highlight the strategic role of AI in fostering a sustainable and profitable agricultural environment. As climate change intensifies both biotic and abiotic pressures on crops, adopting AI-based crop health monitoring systems is essential to enhancing food security. The proliferation of affordable drones, imaging sensors, mobile platforms, and cloud computing is expected to further accelerate AI adoption among both small- and large-scale farmers. As agriculture transitions into a highly digitalized and data-driven sector, AI and deep learning will continue to serve as critical enablers of resilience, efficiency, and productivity.

4.6 Demanding Situations and Limitations

Despite the significant potential of AI and DL for crop and soil health monitoring, several challenges hinder their widespread adoption and effectiveness. A primary issue is data scarcity, as labeled agricultural datasets are often limited, fragmented, or inconsistent. Most DL models require large volumes of annotated images or sensor readings to achieve robust performance; however, obtaining such datasets is time-consuming, costly, and typically restricted to research institutions or well-funded agritech companies. Consequently, models trained on small or region-specific datasets may underperform when applied to broader contexts.

A lack of explainability further limits adoption, as many AI and deep learning models function as "black boxes," producing predictions that are difficult for farmers, agronomists, or policymakers to interpret. In the absence of transparent reasoning, stakeholders may hesitate to rely on AI recommendations for critical decisions such as irrigation, fertilization, or pesticide application. Additionally, hardware and connectivity limitations pose practical barriers, particularly in rural areas where internet access, reliable power supply, and modern sensor-deployment infrastructure are limited. These constraints impede real-time data acquisition, cloud-based analytics, and the implementation of AI-powered decision support systems in many developing regions.

Persistent barriers to adoption remain among farmers. Many smallholder farmers lack awareness of AI technologies, digital literacy, or confidence in integrating automated recommendations into traditional practices. Social, economic, and cultural factors can further slow adoption, highlighting the need for developers to design farmer-centric, intuitive, and cost-effective solutions that provide clear value and foster trust. Addressing these challenges requires coordinated efforts, including the development of large, diverse datasets, explainable AI frameworks, context-adaptive models, infrastructure improvements, and comprehensive farmer training programs to ensure that the potential of AI and DL translates into tangible benefits for agriculture.

4.7 Future Research Directions

To ensure that AI and DL are well-positioned for the future of agriculture, research and development must address current limitations while innovating to deliver scalable, sustainable, and farmer-centric solutions. One promising approach is the development of XAI frameworks, which provide transparent reasoning for model predictions. By revealing which features, such as leaf color, soil moisture, or spectral indices, drive specific recommendations, XAI fosters trust among farmers and agronomists and enables the confident adoption of AI-guided interventions for irrigation, fertilization, pest management, and yield optimization. The integration of AI with emerging technologies such as blockchain offers additional opportunities to enhance agricultural transparency and traceability. By linking AI-derived insights on crop health, soil conditions, and resource utilization with immutable blockchain records, stakeholders can verify quality, reduce fraud, and improve supply chain accountability. Additionally, climate-adaptive crop modeling, in which AI predicts the effects of extreme weather, shifting rainfall patterns, and temperature fluctuations on crop growth, can inform resilient agricultural practices and strengthen preparedness for climate change-related risks. Farmer-friendly, multilingual mobile advisory systems are essential for bridging the gap between advanced AI models and end users. These systems should provide actionable, locally contextualized recommendations in native languages, using intuitive interfaces and visualizations that support comprehension among farmers with varying literacy levels. By integrating AI, mobile computing, and user-centered design, such platforms can democratize access to advanced agricultural analytics, enabling farmers to make timely, informed decisions, enhance productivity, and implement sustainable practices. Collectively, these approaches provide a roadmap for transforming AI in agriculture into a transparent, scalable, and resilient technology that can support food security, resource efficiency, and environmental sustainability in the coming decades.

4.8 Conclusion

For AI and DL to achieve long-term success in agriculture, these technologies must be transparent, scalable, and applicable to real-world farming scenarios. Explainable AI (XAI) is essential for building trust, as it enables farmers and agronomists to understand the mechanisms behind model predictions, thereby supporting confident, informed decision-making. The integration of AI with complementary technologies, such as blockchain and climate-adaptive machinery, further enhances transparency, traceability, and resilience within agricultural systems. Additionally, multilingual mobile advisory systems that localize advanced analytics are crucial for translating complex data into actionable and contextually relevant information. Together, these advancements position AI-enabled agriculture as a reliable, inclusive, and sustainable system capable of promoting future food security, efficient resource management, and environmental sustainability.

References

Begum R, Arafat Y, Rahman M et al (2025) Detection of mango leaf disease using convolution neural network: a comparative analysis. Int J Stat Sci 25:143–157. https://doi.org/10.3329/ijss.v25i2.85777

Buddi T, Kandakatla R, Kotkunde NR et al (2025) Multi-disciplinary research and sustainable development. CRC Press, London. https://doi.org/10.1201/9781003675242

Durach CF, Simpson D, Wiengarten F et al (2025) Beyond the yield: enhancing agricultural sustainability through operations management. Wiley Periodicals, Boston

Ge C, Zhang G, Wang Y et al (2025) Research status and development trends of artificial intelligence in smart agriculture. Agriculture 15(21):2247. https://doi.org/10.3390/agriculture15212247

Kamilaris A, Prenafeta-Boldú FX (2018) Deep learning in agriculture: a survey. Comput Electron Agric 147:70–90. https://doi.org/10.1016/j.compag.2018.02.016

Kerr RB (2020) Agroecology as a means to transform the food system. Landbauforschung 70(2):77–82

Liakos KG, Busato P, Moshou D et al (2018) Machine learning in agriculture: a review. Sensors 18(8):2674. https://doi.org/10.3390/s18082674

Mohammed N, Tahir H, Othman S et al (2026) Smart irrigation systems: a comprehensive review of IoT, AI, and sustainable agriculture technologies. Kirkuk Univ J Agric Sci 16:266–281. https://doi.org/10.58928/ku25.16429

Padhiary M (2025) Integrating 3D printing, IoT, and AI for precision agriculture: automated sensing and smart farming applications. Academia Eng 2(4):8009

Pan Y, Zhang X, Liu H et al (2022) Remote sensing inversion of soil organic matter by using the subregion method at the field scale. Precis Agric 23:1813–1835. https://doi.org/10.1007/s11119-022-09914-2

Phan T, Vo Q, Nguyen H (2024) A novel method for identifying rice seed purity using hybrid machine learning algorithms. Heliyon 10(14):e33941. https://doi.org/10.1016/j.heliyon.2024.e33941

Pretty J (2018) Intensification for redesigned and sustainable agricultural systems. Science 362(6417):eaav0294. https://doi.org/10.1126/science.aav0294

Reddy GR, Kudtala U, Rao VS (2025) An IoT-driven machine learning framework for predictive crop health monitoring in precision agriculture. In: Proc 4th Int Conf Appl Artif Intell Comput (ICAAIC). IEEE, pp 1806–1812

Saleem A, Anwar S, Nawaz T et al (2025) Securing a sustainable future: the climate change threat to agriculture, food security, and sustainable development goals. J Umm Al-Qura Univ Appl Sci 11:595–611. https://doi.org/10.1007/s43994-024-00177-3

Stupar V, Živković Z, Stevanović A et al (2024) The effect of fertility control on soil conservation as a basic resource of sustainable agriculture. Not Bot Horti Agrobot Cluj-Napoca 52:13389. https://doi.org/10.15835/nbha52113389

Wik M, Pingali P, Brocai S (2012) Global agricultural performance: past trends and future prospects. World Bank

Predictive Analysis and Decision Support Systems in Agriculture

5

Nilotpal Das, Atin Kumar, and Sharad Sachan

Abstract

Agriculture plays a critical role in supporting the economies of developing nations. Nevertheless, it faces several challenges, with unpredictable natural factors among the most significant threats. To ensure the sustainability of agricultural production, the adoption of modern technologies for monitoring soil conditions and crop status has become a priority in recent years. Geospatial technologies, such as Geographic Information Systems (GIS) and remote sensing techniques, are particularly valuable in this context. These tools are essential for predicting crop maturity, managing crop stands, and optimizing agricultural yields. Additionally, access to agricultural weather information and meteorological data is crucial for effective crop management. The use of various sensors and modeling procedures further supports the evaluation of these factors. Remote sensing is also instrumental in detecting and assessing crop damage from biotic factors, enabling timely interventions and improved management practices. This chapter provides a concise overview of the applications and benefits of GIS and remote sensing technologies in agriculture, emphasizing their role in enhancing productivity and resilience.

Keywords

Agriculture · GIS · Remote sensing · Geospatial technology

N. Das (✉) · A. Kumar
School of Agriculture, Uttaranchal University, Dehradun, Uttarakhand, India

S. Sachan
Department of Agricultural Economics and Extension, School of Agriculture, Lovely Professional University, Phagwara, Punjab, India

H. N. Oo et al. (eds.), *Transforming Agriculture with AI, IoT, and Blockchain*,
https://doi.org/10.1007/978-981-92-1491-4_5

5.1 Introduction

The global population is projected to double by 2050, posing significant challenges for food production and distribution. Over the past century, agricultural processes have undergone substantial changes due to technological innovations, most notably during the Green Revolution. Spanning the 1960s to the 1980s, the Green Revolution introduced high-yielding crop varieties, synthetic fertilizers, pesticides, and improved irrigation methods, all of which contributed to increased food production (Pingali 2012). Despite these advancements, the demand for food and agricultural products is expected to rise by approximately 30% by 2025 and by over 70% by 2045. Currently, the agricultural sector is experiencing a fourth revolution, primarily driven by advancements in information and communication technologies (Boursianis et al. 2022).

Remote sensing is one of the key technologies driving this transformation. It involves monitoring objects or regions on Earth without direct physical contact. Remote sensing enables effective and accurate assessment of natural resources through satellite sensors and ground-based observations. This technology operates by analyzing features on Earth using the electromagnetic spectrum, including visible, infrared, and microwave wavelengths. Different land cover types, such as vegetation, bare soil, and water bodies, can be identified and characterized based on their unique patterns of reflection and absorption.

The integration of field data with remote sensing is widely utilized in agriculture. Applications include monitoring crop development, analyzing land use dynamics, detecting changes in field water cover, mapping water resources, assessing pest and disease prevalence, estimating crop yields and harvests, supporting precision farming, and improving weather predictions (Gebeyehu 2019). Combining crop simulation models with remote sensing data significantly enhances the accuracy of yield projections.

5.2 Remote Sensing and Geographic Information System (GIS)

Remote sensing refers to the acquisition of data about the Earth's surface and atmosphere using sensors mounted on aircraft or satellites. These sensors detect signals such as radar or sonar by responding to electromagnetic radiation across various wavelengths, including visible light, infrared, and microwaves. Remote sensing enables the analysis of natural resources and facilitates the mapping and monitoring of environmental changes. Industries such as agriculture, land use planning, environmental monitoring, and mineral mining rely heavily on this technology. In agriculture, remote sensing is extensively utilized for crop monitoring, particularly for assessing crop canopy. A primary advantage of remote sensing is its ability to provide consistent and unaltered data, reducing the need for physical crop testing and supporting precision agriculture. Additionally, satellites can collect data cost-effectively over large geographic areas (Sharma et al. 2023). In India, satellite-based remote sensing is commonly used to estimate crop production and acreage. The resulting data are crucial for estimating yield, monitoring crop phenology, and determining crop stress states (Gu et al. 2007). Recent advancements in remote sensing have been driven by the development of narrowband and hyperspectral sensors, as well as improved spatial resolution in satellite and airborne systems. Hyperspectral remote sensing, in particular, has enhanced crop characterization by providing greater spectral detail.

GIS support models that utilize data from maps and satellite imagery. In agriculture, GIS is employed to estimate crop yields, identify soil erosion, assess soil conditions, and remediate degraded soils. These applications enable farmers to better understand and manage agricultural resources, even when external factors are beyond their control. GIS enables users to create maps, integrate diverse datasets, model scenarios, address complex challenges, build evidence-based arguments, and develop practical solutions. The applications of GIS in agriculture are extensive and include resource management, corporate analytics, urban and rural planning, and defense. GIS is particularly valuable for natural resource management and inventory studies. These capabilities have contributed to the rapid adoption of GIS technology across multiple sectors.

5.3 Techniques Used in Predictive Analysis and Decision Support Systems in Agriculture

Predictive analysis and Decision Support Systems (DSS) in agriculture utilize a broad spectrum of advanced computational, statistical, and Artificial Intelligence (AI) methods to address challenges in crop management, resource optimization, disease detection, and yield forecasting. Machine Learning (ML) and Deep Learning (DL) algorithms are widely implemented for crop selection, yield prediction, disease and pest detection, and soil compatibility classification. ML techniques, including random forests, Support Vector Machines (SVMs), decision trees, k-nearest neighbors (k-NN), and ensemble methods, are commonly used for classification, regression, and time-series prediction, enabling accurate forecasts based on historical and current data (Mortazavizadeh et al. 2025; Aslan et al. 2024). DL models, such as Convolutional Neural Networks (CNNs) and Recurrent Neural Networks (RNNs), are particularly effective for image-based tasks, including plant disease detection, weed identification, yield estimation using satellite or UAV data, and processing hyperspectral and multimodal data to extract detailed information (Sakka et al. 2025). ARIMA and hybrid ML-statistical models are essential for time series analysis, supporting the prediction of crop yields, commodity prices, and irrigation requirements to facilitate proactive decision-making (Guindani et al. 2024). Geostatistical techniques, such as Kriging and Co-kriging, are integrated with ML to interpolate soil properties and optimize fertilization strategies in soil and nutrient management (Ennaji et al. 2023). The integration of Internet of Things (IoT) devices and remote sensing technologies enables the collection of real-time data, which, when processed by AI-based algorithms, supports automated and adaptive DSS for irrigation scheduling, water management, and environmental monitoring (Mortazavizadeh et al. 2025; Aslan et al. 2024; Espinel et al. 2024). Multicriteria Decision-Making (MCDM) models, including TOPSIS and VIKOR, are applied within DSS frameworks to compare and prioritize alternative actions based on agronomic, economic, and environmental criteria (Ali et al. 2023). Additionally, data preprocessing and augmentation methods, such as Generative Adversarial Networks (GANs) and Variational Autoencoders (VAEs), address challenges like class imbalance and limited labeled data, thereby enhancing model robustness (Miftahushudur et al. 2025).

5.4 Application of Remote Sensing and GIS in Different Facets of Agriculture

Remote sensing and GIS are revolutionizing modern agriculture by facilitating precise, data-driven decision-making. These technologies support real-time monitoring of crop health, soil properties, water availability, and environmental conditions across extensive areas, which enhances productivity and optimizes resource-use efficiency. Figure 5.1 illustrates the diverse applications of remote sensing and GIS in agricultural contexts.

5.4.1 Observing the Crop

Crop growth and development are affected by several factors, including soil moisture content, planting time, air temperature, day length, and soil conditions. To identify plant stress using remote sensing, it is important to note that plants experiencing adverse weather conditions often exhibit changes in their spectral reflectance and emission properties (Palanisamy et al. 2019). Droughts, for instance, can render land unusable and cause significant harm to both human and animal populations. In recent years, the use of the Vegetative Condition Index (VCI) and Normalized Difference Vegetation Index (NDVI) has become widely recognized for diagnosing agricultural drought in regions with diverse biological characteristics (Palanisamy et al. 2019).

5.4.2 Water Management and Estimation of Crop Nutrients

Nutrient and water stress are critical determinants of crop health and productivity, which can be addressed through precision farming using remote sensing and GIS. These technologies enable more accurate detection of soil nutrient deficiencies and targeted fertilizer application, resulting in financial savings and increased fertilizer efficiency. Mapping these indicators is essential in precision farming, as it facilitates the assessment of nutrient status across different field areas. In tractor-mounted systems, sensors are typically attached to the front of the spray boom to monitor

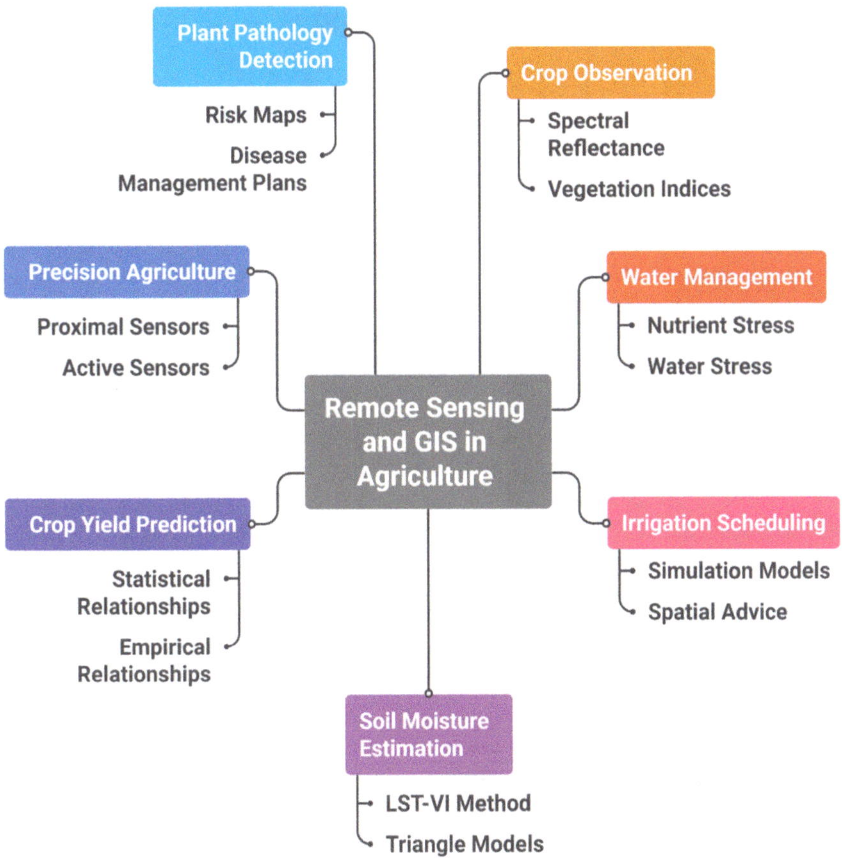

Fig. 5.1 Applications of remote sensing and GIS in modern agriculture

crop status in real time. These sensors use indices such as the NDVI to determine nitrogen requirements and communicate this information to nutrient spreaders, ensuring fertilizer is applied where needed. Sensor data are processed by algorithms to generate precise nitrogen-dose recommendations, supporting effective and adaptable in-season nutrition management. When combined with cost-effective solutions, vegetation indicators can optimize nitrogen application for various crops using different fertilizer rate algorithms (Ali et al. 2017).

Remote sensing is a valuable tool for measuring crop water stress. Stressed plants typically reflect more light than well-irrigated crops, enabling the identification of areas that require irrigation or other interventions. Microwave remote sensing has significantly improved the assessment of soil moisture availability. Through remote sensing, agricultural producers receive detailed information on soil water content and its impact on crop growth at various developmental stages, enabling more informed decisions about sustainable agricultural practices and efficient water use.

5.4.3 Calculation of Water Needs of Crops

The FAO Penman-Monteith model is widely used in GIS applications to estimate reference crop evapotranspiration, enabling spatial analysis of crop water requirements. This method facilitates the identification of regional crop water needs, thereby supporting improved planning for irrigation and water resource management.

Casa et al. (2009) established a correlation between monthly crop coefficient values and specific crop classes, and developed an evapotranspiration reference map to support the estimation of water requirements. Additionally, vegetation indices derived from satellite imagery were used to quantify crop water use across four growing seasons in northern Mexico, yielding detailed evapotranspiration maps for irrigation scheduling (Reyes-González et al. 2018).

Ramírez-Cuesta et al. (2018) integrated the dual-crop-coefficient approach with multisatellite imagery to develop a novel ArcGIS toolkit. This toolkit provides an effective method for determining crop water requirements and is particularly valuable for water balance assessments in agroecosystems. It supports informed irrigation decisions, thereby reducing water consumption and enhancing crop yield.

5.4.4 Irrigation Scheduling

Irrigation timing refers to the scheduling and quantity of water applied to crops. This process is critical for conserving water resources and ensuring that crops receive adequate moisture for optimal growth (Bwambale et al. 2022). GIS represent a valuable tool for improving irrigation management. To maximize water-use efficiency in agriculture, researchers have developed simulation models integrated with GIS. These models provide spatially-informed recommendations, thereby improving irrigation practices and promoting both water conservation and increased crop yields (Fortes et al. 2005).

5.4.5 Estimation of Soil Moisture

The LST-VI method estimates soil moisture by analyzing pixel distributions within plot-space generated from temperature and vegetation index data. When images encompass a wide range of soil moisture and vegetation density conditions, and outliers such as clouds or surface water are excluded, the resulting data points typically form a triangular or trapezoidal pattern. In this configuration, the apex corresponding to higher land surface temperatures represents drier conditions with low soil moisture (the dry side), while the opposite side indicates wetter conditions with higher soil moisture (the wet side) (Zhu et al. 2017).

Building on this concept, a new generation of triangle models was developed and validated to enable high-resolution soil moisture mapping (Carlson and Petropoulos 2019). These models have proven particularly beneficial in precision agriculture, where they support detailed assessments of soil moisture variability to facilitate predictive, targeted irrigation and improved crop management.

5.4.6 Crop Yield and Production Prediction

Remote sensing has been widely used to predict crop yields by establishing statistical and empirical relationships between vegetation indices and crop yields (Palanisamy et al. 2019). There are two primary approaches for inferring agricultural yields from remotely sensed data. These models facilitate the generation of crop yield maps for specific agricultural regions. To evaluate the relationship among maize crop yield, biomass accumulation, and spectral indicators measured at the V12 growth stage, a regression-based approach was employed (Maresma et al. 2016).

5.4.7 Precision Agriculture and the Role of Remote Sensing

Precision agriculture employs modern technologies and scientific principles to optimize agricultural production by accounting for spatial and temporal variability. The primary objectives of precision agriculture are to increase farm output, reduce operational costs, and minimize environmental impacts, such as chemical overuse and toxicity. Precision agriculture represents an information and management system that applies evidence-based approaches to improve agricultural practices. Proximal remote sensing systems, also referred to as ground-based sensors, can be handheld, field-installed, or mounted on agricultural machinery to monitor localized areas.

Passive sensors, commonly integrated into satellites, aircraft, and UAVs, rely on external light sources such as sunlight. In contrast, certain spacecraft, such as ERS-1 and ERS-2, are equipped with active sensors, such as the Active Microwave Instrument (AMI), that generate their own signals. Active proximity sensors are prevalent in ground-based systems and are integral to variable rate application technologies, such as Green Seeker and Crop Circle, which adjust fertilizer application based on crop requirements. Thermal infrared sensors measure energy emitted by crops, providing data on crop water stress, evapotranspiration, and irrigation needs (Mulla 2013). Similarly, microwave sensors assess ground emissions to determine irrigation water use and soil moisture levels (Khanal et al. 2017). A key advantage of microwave sensors is their ability to penetrate cloud cover, enabling reliable monitoring when visible and near-infrared sensors are ineffective.

5.4.8 The Use of RS and GIS in Plant Pathology

GIS can monitor disease outbreaks in agricultural regions by integrating spatial data on pests and diseases with crop distribution, weather patterns, and environmental variables. Through such analyses, GIS can generate risk maps that identify areas highly susceptible to specific plant diseases. This information is invaluable to

Fig. 5.2 Types of sensors in the detection of plant diseases

farmers and agricultural specialists, as it enables the development of targeted disease management strategies. For example, precise timing of fungicide application can reduce crop losses and minimize unnecessary chemical use. Notably, Prof. Pisharoth Rama Pisharoty was among the first researchers to utilize remote sensing to analyze coconut root-wilt disease in Kerala in the early 1970s (Dakshinamurti et al. 1971).

5.4.9 Sensors in the Detection of Plant Diseases

Remote sensing for detecting plant diseases in precision agriculture can be categorized into four sensor types: RGB (Red, Green, Blue), multispectral, hyperspectral, and thermal sensors (Fig. 5.2). These sensors capture high-resolution images that yield detailed information on various vegetation characteristics and stress indicators.

5.4.9.1 RGB Sensors

RGB sensors capture color images based on red, green, and blue light. These images are utilized for detecting disease symptoms, nutrient deficiencies, invasive plants, and weeds, and for identifying specific plant species in agricultural fields. Visualization in RGB images relies on light reflectance, optical properties, and human perception. RGB imaging is applied in weed identification, crop mapping, assessment of physiological changes on leaf surfaces, and enumeration of plant stands.

5.4.9.2 Multispectral Sensors

In addition to the visible red, green, and blue bands, multispectral sensors record data in the red edge and near-infrared regions. Band ratios and vegetation indices, such as the NDVI, are frequently applied in the analysis of multispectral images. These methods enable the identification of crop disease symptoms, pesticide damage, and weed species.

5.4.9.3 Hyperspectral Sensors

Hyperspectral sensors measure spectral reflectance at narrow bandwidths across the visible, near-infrared, and mid-infrared regions of the electromagnetic spectrum (typically around 510 nm). The spectral signature of a plant refers to its specific reflectance at the canopy or leaf level. By analyzing these spectral patterns, hyperspectral imaging can accurately identify weed species, crop types, and various disease symptoms.

5.4.9.4 Thermal Sensors

Thermal imaging operates on the principle that objects emit infrared energy proportional to their temperature; hotter objects emit more radiation than cooler ones. Thermal cameras detect electromagnetic radiation within the infrared spectrum (approximately 800–1400 nm) and represent temperature differences as false-color images. Each pixel corresponds to a specific temperature value. This technique is widely used to measure canopy or leaf temperatures in agricultural fields. Elevated temperatures may indicate plant disease symptoms, water stress, or pest presence, facilitating early detection and management.

5.4.10 The Use of Remote Sensing to Detect Plant Disease

Vegetation reflectance patterns across different spectral domains vary depending on remote sensing processes. Biophysical and biochemical properties of vegetation, including biomass, moisture content, pigment concentration, leaf area index (LAI), and spatial arrangement, significantly influence leaf reflectance and sunlight absorption (Sahoo et al. 2015). Remote sensing enables early detection of plant diseases by monitoring variations in plant color, canopy shape, and other indicators. Analysis of spectral data over time allows for effective tracking and monitoring of environmental changes across diverse landscapes, including agricultural fields, forests, and urban areas.

The ability to monitor and report changes in plant reflectance provides essential information for decision makers involved in resource management, disaster response, and environmental monitoring. In agriculture, remote sensing supports the early identification of emerging diseases and facilitates timely intervention measures.

5.4.10.1 Spectral Signature of Disease-Infected Plant

Interactions across different spectral ranges, including the visible (400–700 nm), near-infrared (700–1100 nm), and short-wave infrared (1100–2500 nm) regions, determine the reflection of sunlight by plant leaves. The absorption, scattering, and reflection of sunlight depend on factors such as leaf chemistry, surface structure, internal anatomy, water content, protein levels, and carbon composition (Mahlein et al. 2012). Spectral Vegetation Indices (SVIs) are critical for tracking, examining, and mapping vegetation changes over time and space. Ratios calculated among various spectral bands facilitate data interpretation and enhance the discrimination of plant diseases. Specific pigment indices are particularly useful for identifying stress caused by fungal pathogens, as changes in pigment concentrations are directly proportional to the physiological state of the leaf. For example, SVIs have been used to detect apple scab caused by *Venturia inaequalis* at distinct stages of disease progression (Delalieux et al. 2007). Hyperspectral imaging systems have also been applied in several host-pathogen systems, providing accurate detection and monitoring of diseases. Table 5.1 presents selected examples.

5.4.10.2 Use of RS and GIS in Insect-Pest Detection

Calibration of remotely sensed images often involves constructing large reference panels within the scene that possess known properties. These panels facilitate the adjustment of image data for accurate analysis. The NDVI is frequently employed as a primary index for analyzing multispectral data. For example, NDVI maps derived from multispectral imagery, combined with soil and crop condition data, have been used to investigate the activities of *Lygus lineolaris* in cotton fields (Willers et al. 2005). This approach improved field scouting by providing enhanced insight into pest distribution and hotspots. The reflectance profile of insect bodies,

Table 5.1 Table of plant disease and pathosystems analyzed using hyperspectral imaging

Host–pathogen system	Scale	Detection	Early detection	Quantification	Reference
Apple—*Venturia inaequalis*	Leaf	✅ Yes	❌ N.I.	✅ Yes	Mahlein et al. (2012), Delalieux et al. (2007)
Barley—*Blumeria graminis* f. sp. *hordei*	Tissue	✅ Yes	✅ Yes	❌ N.I.	Mahlein et al. (2012)
Cucumber—Cucumber mosaic virus, Cucumber green mottle mosaic virus, *Sphaerotheca fuliginea*	Leaf	✅ Yes	❌ N.I.	✅ Yes	Berdugo et al. (2014)
Oilseed rape—*Alternaria* spp.	Leaf	✅ Yes	❌ N.I.	✅ Yes	Baranowski et al. (2015)
Wheat—*Fusarium* spp.	Ear	✅ Yes	✅ Yes	✅ Yes	Mahlein et al. (2012)

✅ Yes, ❌ noninvestigated (N.I.) aspect

utilized in insect detection, can identify physiological, biochemical, or taxonomic differences among insects. In benchtop remote sensing studies of individual insects, variations in reflectance patterns are often attributed to metabolic changes induced by treatments, which alter the penetration depth of radiometric energy in specific spectral bands.

Numerous studies have demonstrated that insect herbivory and other plant stressors negatively affect photosynthesis, leading to increased leaf reflectance due to reduced pigment-mediated light absorption. GIS technologies have been employed by pest managers to map and monitor pest populations in agroforestry systems. For instance, GIS was utilized to track *Ceratitis capitata* populations, which led to a reduction in unnecessary pesticide applications when the pest did not pose a threat to citrus crops (Cohen et al. 2008). GIS has also been applied to examine the distributions of beetle families such as Elateridae, Silphidae, and Chrysomelidae (Toepfer et al. 2007). Additionally, locust management interventions based on GIS and remote sensing data have facilitated the prioritization of control regions with significant ecological value (Lazar et al. 2016). Ghobadifar et al. (2014) employed the NDVI, Standard Difference Index (SDI), and Ratio Vegetation Index (RVI) to establish outbreak thresholds, thereby enhancing pest outbreak monitoring.

5.5 Key Findings and Trends

AI, ML, DL, and big data analytics are significantly advancing agricultural productivity, sustainability, and resource efficiency. Predictive analytics and DSS facilitate accurate crop yield prediction, disease detection, nutrient management, and real-time monitoring, thereby supporting informed decision-making in both farming and policy contexts. CNNs and RNNs are the predominant ML and DL algorithms applied in crop classification, yield forecasting, pest and disease management, and smart irrigation, leading to increased yields and reduced environmental impact (Albahar 2023). DSSs, typically powered by these algorithms, process data from sensors, weather stations, and remote sensing to provide actionable recommendations on irrigation, fertilization, pest control, and harvest timing (Petraki et al. 2025; Gebresenbet et al. 2023). In dairy and livestock farming, predictive analytics and DSSs optimize feed regimes, disease detection, resource allocation, productivity, and animal welfare (Bang et al. 2023; Palma et al. 2025). The integration of IoT devices and wireless sensor networks enhances data collection and real-time analysis, enabling automated decision-making systems to improve field operations (Alaoui et al. 2024). However, challenges remain, including data integration, computational limitations, digital literacy gaps, and the lack of scalable, cost-effective solutions. Recent studies highlight the importance of multidisciplinary collaboration, system functionality, and ongoing training to enhance adoption and effectiveness (Petraki et al. 2025; Bang et al. 2023; Gebresenbet et al. 2023). Future research directions include developing robust multimodal foundation models, improving the integration of simulations with ML, and creating agroecological- and circular-economy-specific DSSs to ensure food security, environmental stability, and

Table 5.2 Applications and impacts of predictive analysis and DSS in agriculture

Application area	Key technologies/ methods	Main benefits/findings	Reference
Crop yield prediction	ML, DL, time series analysis	Accurate yield forecasts, improved planning	Albahar (2023)
Disease & pest detection	ML, DL, IoT, UAVs	Early detection, targeted interventions, reduced chemical use	Alaoui et al. (2024)
Nutrient & soil management	ML, AI, geostatistics	Optimized fertilization, efficient resource use	Ennaji et al. (2023)
Irrigation & water management	DSS, IoT, MPC	Water conservation, precise scheduling, higher yields	Bwambale et al. (2022), Petraki et al. (2025), Gebresenbet et al. (2023)
Livestock & dairy management	Predictive analytics, DSS, ML	Disease detection, feeding optimization, productivity gains	Bang et al. (2023), Palma et al. (2025)
Agroecology & sustainability	DSS, MCDA, decision analysis	Enhanced sustainability, circular economy, risk assessment	Petraki et al. (2025), Whitney et al. (2023), Lombardi and Todella (2023)
Commodity forecasting	ML, hybrid models, AI	Market risk mitigation, price prediction, supply chain support	Guindani et al. (2024)

economic sustainability in agriculture (Yin et al. 2025; Whitney et al. 2023; Lombardi and Todella 2023). The various applications and impacts of predictive analysis and DSS in agriculture are indicated in Table 5.2.

5.6 Future Directions in Predictive Analysis and DSS in Agriculture

The future of predictive analytics and DSS in agriculture is expected to expand significantly, driven by rapid advancements in AI, ML, DL, IoT, and big data analytics. A notable trend involves the development and deployment of foundation models (FM) and agricultural foundation models (AFM) that leverage multimodal data, including images, sensor outputs, weather data, and text, to provide detailed, context-aware decision support. These models are projected to enhance crop yield prediction, disease and pest detection, resource optimization, and supply chain management, thereby promoting the digitalization and intelligent transformation of agriculture (Yin et al. 2025). The integration of smart sensors and IoT devices will facilitate automated, real-time data collection and analysis, enabling adaptive, location-specific recommendations to improve resource efficiency and sustainability (Soussi et al. 2024; Mansoor et al. 2025). Future DSS are anticipated to incorporate more explainable AI (XAI) techniques, increasing model transparency and trust among farmers and stakeholders, which is likely to drive broader adoption (Dhal and Kar 2024). Emphasis on user-friendly interfaces and mobile platforms remains essential to ensure accessibility of advanced analytics for farmers with varying levels of digital literacy and technical expertise (Gebresenbet et al. 2023). Multidisciplinary collaboration among agronomists, data scientists, engineers, and end-users will be vital for developing robust, context-specific solutions that address the diverse needs of global agriculture (Bang et al. 2023; Gebresenbet et al. 2023). Research is also increasingly focused on integrating blockchain technology for secure data sharing, utilizing edge and cloud computing for scalable processing, and deploying drones and autonomous vehicles for automated field operations (Gebresenbet et al. 2023; Padhiary et al. 2024; Saini and Yadav 2025). Addressing challenges related to data privacy, interoperability, infrastructure limitations, and the initial high costs of technology adoption will be essential to maximize the impact of predictive analytics and DSS, particularly in resource-constrained and smallholder farming environments (Mansoor et al. 2025; Dhal and Kar 2024; Padhiary et al. 2024).

5.7 Conclusions

GIS have become integral to agricultural production worldwide, enabling farmers to optimize crop yields, reduce production costs, and efficiently manage land resources. To fully leverage remote sensing and GIS technologies, it is essential to develop structured information systems at the state or district level. These systems can integrate crop data obtained from satellite imagery and field surveys, providing stakeholders with actionable information to drive economic benefits and informed decision-making. GIS facilitates the comparison of historical and current data, thereby enhancing the ability to predict disease outbreaks and pest infestations. Additionally, GIS provides researchers, extension workers, and planners with advanced geospatial mapping and analytical tools to anticipate and manage agricultural threats. As technology continues to advance, the creation of innovative methodologies and tools is expected to further expand the applications of remote sensing and GIS. These advancements will enhance the monitoring, prediction, and management of agricultural systems, contributing to more sustainable and resilient agricultural practices.

References

Alaoui M, Amraoui K, Masmoudi L et al (2024) Unleashing the potential of IoT, artificial intelligence, and UAVs in contemporary agriculture: a comprehensive review. J Terrramech 115(3):100986. https://doi.org/10.1016/j.jterra.2024.100986

Albahar M (2023) A survey on deep learning and its impact on agriculture: challenges and opportunities. Agriculture 13:540. https://doi.org/10.3390/agriculture13030540

Ali MM, Al-Ani A, Eamus D et al (2017) Leaf nitrogen determination using non-destructive techniques: a review. J Plant Nutr 40(7):928–953. https://doi.org/10.1080/01904167.2016.1143954

Ali R, Hussain A, Nazir S et al (2023) Intelligent decision support systems—an analysis of machine learning and multicriteria decision-making methods. Appl Sci 13:12426. https://doi.org/10.3390/app132212426

Aslan M, Sabancı K, Aslan B (2024) Artificial intelligence techniques in crop yield estimation based on Sentinel-2 data: a comprehensive survey. Sustainability 16:8277. https://doi.org/10.3390/su16188277

Bang R, Guajardo M, Hansen B (2023) Recent advances in decision support for beef and dairy farming: modeling approaches and opportunities. Int Trans Oper Res 30:2807–2839. https://doi.org/10.1111/itor.13311

Baranowski P, Jedryczka M, Mazurek W et al (2015) Hyperspectral and thermal imaging of oilseed rape (*Brassica napus*) response to fungal species of the genus *Alternaria*. PLoS One 10(3):e0122913. https://doi.org/10.1371/journal.pone.0122913

Berdugo CA, Zito R, Paulus S et al (2014) Fusion of sensor data for the detection and differentiation of plant diseases in cucumber. Plant Pathol 63(6):1344–1356. https://doi.org/10.1111/ppa.12219

Boursianis AD, Papadopoulou MS, Diamantoulakis P et al (2022) Internet of things (IoT) and agricultural unmanned aerial vehicles (UAVs) in smart farming: a comprehensive review. Internet Things 18:100187. https://doi.org/10.1016/j.iot.2020.100187

Bwambale E, Abagale FK, Anornu GK (2022) Smart irrigation monitoring and control strategies for improving water use efficiency in precision agriculture: a review. Agric Water Manag 260:107324. https://doi.org/10.1016/j.agwat.2021.107324

Carlson TN, Petropoulos GP (2019) A new method for estimating evapotranspiration and surface soil moisture from optical and thermal infrared measurements: the simplified triangle. Int J Remote Sens 40(20):7716–7729. https://doi.org/10.1080/01431161.2019.1601288

Casa R, Rossi M, Sappa G et al (2009) Assessing crop water demand by remote sensing and GIS for the Pontina Plain, Central Italy. Water Resour Manag 23:1685–1712. https://doi.org/10.1007/s11269-008-9347-4

Cohen Y, Cohen A, Hetzroni A et al (2008) Spatial decision support system for medfly control in citrus. Comput Electron Agric 62(2):107–117. https://doi.org/10.1016/j.compag.2007.12.005

Dakshinamurti C, Krishnamurthy B, Summanwar AS et al (1971) Remote sensing for coconut wilt. In: Proceedings of international symposium on remote, Ann Arbor, pp 25–29

Delalieux S, Van Aardt JAN, Keulemans W et al (2007) Detection of biotic stress (*Venturia inaequalis*) in apple trees using hyperspectral data: non-parametric statistical approaches and physiological implications. Eur J Agron 27(1):130–143. https://doi.org/10.1016/j.eja.2007.02.005

Dhal S, Kar D (2024) Transforming agricultural productivity with AI-driven forecasting: innovations in food security and supply chain optimization. Forecasting 6:46. https://doi.org/10.3390/forecast6040046

Ennaji O, Vergütz L, Allali A (2023) Machine learning in nutrient management: a review. Artif Intell Agric 9:1–11. https://doi.org/10.1016/j.aiia.2023.06.001

Espinel R, Herrera-Franco G, García J et al (2024) Artificial intelligence in agricultural mapping: a review. Agriculture 14:1071. https://doi.org/10.3390/agriculture14071071

Fortes PS, Platonov AE, Pereira LS (2005) GISAREG—A GIS based irrigation scheduling simulation model to support improved water use. Agric Water Manag 77(1–3):159–179. https://doi.org/10.1016/j.agwat.2004.09.042

Gebeyehu MN (2019) Remote sensing and GIS application in agriculture and natural resource management. Int J Environ Sci Nat Resour 19(2):45–49. https://doi.org/10.19080/IJESNR.2019.19.556009

Gebresenbet G, Bosona T, Patterson D et al (2023) A concept for application of integrated digital technologies to enhance future smart agricultural systems. Smart Agric Technol 5:100255. https://doi.org/10.1016/j.atech.2023.100255

Ghobadifar F, Wayayok A, Shattri M et al (2014) Using SPOT-5 images in rice farming for detecting BPH (brown plant hopper). IOP Conf Ser Earth Environ Sci 20:012015. https://doi.org/10.1088/1755-1315/20/1/012015

Gu Y, Brown JF, Verdin JP et al (2007) A five-year analysis of MODIS NDVI and NDWI for grassland drought assessment over the central Great Plains of the United States. Geophys Res Lett 34(6):2006GL029127. https://doi.org/10.1029/2006GL029127

Guindani L, Oliveirai G, Ribeiro M et al (2024) Exploring current trends in agricultural commodities forecasting methods through text mining: developments in statistical and artificial intelligence methods. Heliyon 10:e40568. https://doi.org/10.1016/j.heliyon.2024.e40568

Khanal S, Fulton J, Shearer S (2017) An overview of current and potential applications of thermal remote sensing in precision agriculture. Comput Electron Agric 139:22–32. https://doi.org/10.1016/j.compag.2017.05.001

Lazar M, Piou C, Doumandji-Mitiche B et al (2016) Importance of solitarious desert locust population dynamics: lessons from historical survey data in Algeria. Entomol Exp Appl 161(3):168–180. https://doi.org/10.1111/eea.12505

Lombardi P, Todella E (2023) Multi-criteria decision analysis to evaluate sustainability and circularity in agricultural waste management. Sustainability 15:14878. https://doi.org/10.3390/su152014878

Mahlein AK, Oerke EC, Steiner U et al (2012) Recent advances in sensing plant diseases for precision crop protection. Eur J Plant Pathol 133(1):197–209. https://doi.org/10.1007/s10658-011-9878-z

Mansoor S, Iqbal S, Popescu S et al (2025) Integration of smart sensors and IoT in precision agriculture: trends, challenges and future prospectives. Front Plant Sci 16:1587869. https://doi.org/10.3389/fpls.2025.1587869

Maresma Á, Ariza M, Martínez E et al (2016) Analysis of vegetation indices to determine nitrogen application and yield prediction in maize (*Zea mays* L.) from a standard UAV service. Remote Sens 8(12):973. https://doi.org/10.3390/rs8120973

Miftahushudur T, Sahin H, Grieve B et al (2025) A survey of methods for addressing imbalance data problems in agriculture applications. Remote Sens 17:454. https://doi.org/10.3390/rs17030454

Mortazavizadeh F, Bolonio D, Mirzaei M et al (2025) Advances in machine learning for agricultural water management: a review of techniques and applications. J Hydroinf 27(3):474–492. https://doi.org/10.2166/hydro.2025.258

Mulla DJ (2013) Twenty five years of remote sensing in precision agriculture: key advances and remaining knowledge gaps. Biosyst Eng 114(4):358–371. https://doi.org/10.1016/j.biosystemseng.2012.08.009

Padhiary M, Saha D, Kumar R et al (2024) Enhancing precision agriculture: a comprehensive review of machine learning and AI vision applications in all-terrain vehicle for farm automation. Smart Agric Technol 8:100483. https://doi.org/10.1016/j.atech.2024.100483

Palanisamy S, Selvaraj R, Ramesh T et al (2019) Applications of remote sensing in agriculture: a review. Int J Curr Microbiol Appl Sci 8(1):2270–2283. https://doi.org/10.20546/ijcmas.2019.801.238

Palma O, Plà-Aragonés L, Mac Cawley A et al (2025) AI and data analytics in the dairy farms: a scoping review. Animals 15:1291. https://doi.org/10.3390/ani15091291

Petraki D, Gazoulis I, Kokkini M et al (2025) Digital tools and decision support systems in agroecology: benefits, challenges, and practical implementations. Agronomy 15:236. https://doi.org/10.3390/agronomy15010236

Pingali PL (2012) Green revolution: impacts, limits, and the path ahead. Proc Natl Acad Sci USA 109(31):12302–12308. https://doi.org/10.1073/pnas.0912953109

Ramírez-Cuesta JM, Mirás-Avalos JM, Rubio-Asensio JS et al (2018) A novel ArcGIS toolbox for estimating crop water demands by integrating the dual crop coefficient approach with multi-satellite imagery. Water 11(1):38. https://doi.org/10.3390/w11010038

Reyes-González A, Kjaersgaard J, Trooien T et al (2018) Estimation of crop evapotranspiration using satellite remote sensing-based vegetation index. Adv Meteorol 2018:4525021. https://doi.org/10.1155/2018/4525021

Sahoo RN, Ray SS, Manjunath KR (2015) Hyperspectral remote sensing of agriculture. Curr Sci 108:848–859

Saini A, Yadav AD (2025) A comprehensive review on technological breakthroughs in precision agriculture: IoT and emerging data analytics. Eur J Agron 163:127440. https://doi.org/10.1016/j.eja.2024.127440

Sakka M, Ivanovici M, Chaâri L et al (2025) A review of CNN applications in smart agriculture using multimodal data. Sensors 25:472. https://doi.org/10.3390/s25020472

Sharma T, Singh A, Kumar N et al (2023) Digital technologies and tools for the ensuing digital era. In: Applying drone technologies and robotics for agricultural sustainability. IGI Global, Hershey, pp 1–11. https://doi.org/10.4018/978-1-6684-6413-7.ch001

Soussi A, Zero E, Sacile R et al (2024) Smart sensors and smart data for precision agriculture: a review. Sensors 24:2647. https://doi.org/10.3390/s24082647

Toepfer S, Ellsbury MM, Eschen R et al (2007) Spatial clustering of *Diabrotica virgifera virgifera* and *Agriotes ustulatus* in small-scale maize fields without topographic relief drift. Entomol Exp Appl 124(1):61–75. https://doi.org/10.1111/j.1570-7458.2007.00549.x

Whitney C, Biber-Freudenberger L, Luedeling E (2023) Decision analytical methods for assessing the efficacy of agroecology interventions. CABI Agric Biosci 4:11. https://doi.org/10.1186/s43170-023-00151-9

Willers JL, Jenkins JN, Ladner WL et al (2005) Site-specific approaches to cotton insect control: sampling and remote sensing analysis techniques. Precis Agric 6:431–452. https://doi.org/10.1007/s11119-005-3680-x

Yin S, Xi Y, Zhang X et al (2025) Foundation models in agriculture: a comprehensive review. Agriculture 15:847. https://doi.org/10.3390/agriculture15080847

Zhu W, Jia S, Lv A (2017) A universal Ts–VI triangle method for the continuous retrieval of evaporative fraction from MODIS products. J Geophys Res Atmos 122(19):10206–10217. https://doi.org/10.1016/j.agwat.2024.109017

Intelligent Automation and Robotics for Smart Agriculture: Transforming Farming Through Technological Synergy

6

Anuradha Reddy, G. S. Pradeep Ghantasala, A. V. Lakshmi Prasuna, Shaik Sharmila, Aswathi Sheelan, and M. Arvindhan

Abstract

Smart agriculture is fundamentally transforming traditional farming systems by integrating advanced digital technologies into core agricultural operations. Computers, intelligent algorithms, and robotic systems are now widely utilized to manage key processes, including crop monitoring, growth regulation, and harvesting. This approach is being adopted globally to promote sustainable and profitable food production. Smart technologies have enabled agronomists to optimize resource use, address labor shortages, and enhance crop yields. Key innovations include drones for crop monitoring, robotic harvesters, artificial intelligence-powered irrigation systems, and autonomous tractors. These technologies reduce environmental impact, minimize human intervention, and facilitate more efficient farm management. The concept of "smart farms" relies on data analytics to improve supply chain management, detect diseases, and predict optimal harvest times. Case studies from leading agricultural regions demonstrate that effective implementation yields clear benefits, including improved sustainability, cost reduction, and increased productivity. Critical considerations

A. Reddy (✉) · S. Sharmila · A. Sheelan
Sridevi Women's Engineering College, Hyderabad, Telangana, India

G. S. Pradeep Ghantasala
Department of Computer Science and Engineering, Alliance School of Advanced Computing, Alliance University, Bengaluru, Karnataka, India

A. V. Lakshmi Prasuna
Department of Information Technology, Mahatma Gandhi Institute of Technology, Hyderabad, Telangana, India

M. Arvindhan
American School of Management, American University of Science, Hồ Chí Minh, Vietnam

H. N. Oo et al. (eds.), *Transforming Agriculture with AI, IoT, and Blockchain*,
https://doi.org/10.1007/978-981-92-1491-4_6

include the demand for skilled labor, data security, interoperability, and high initial investment costs.

Keywords

Smart agriculture · Agricultural robotics · Precision farming · Autonomous systems · IoT in agriculture · AI-driven irrigation · Drone monitoring · Sustainable farming

6.1 Introduction

Agriculture remains fundamental to human sustenance and economic stability. However, the twenty-first century presents unprecedented challenges, such as rapid population growth, climate variability, labor shortages, and depletion of natural resources. These pressures necessitate a transformation toward more efficient, resilient, and sustainable agricultural systems (Raj and Prahadeeswaran 2025; FAO 2022). Intelligent automation and robotics have become key drivers of this new agricultural paradigm by providing enhanced precision, operational efficiency, and environmental sustainability.

Recent advances in artificial intelligence (AI), machine learning (ML), the Internet of Things (IoT), and autonomous robotic systems have transformed the agricultural value chain by enabling data-driven decision-making and real-time monitoring (Yépez-Ponce et al. 2023; Hoque and Padhiary 2024; Nrip 2022). These technologies optimize essential operations, including soil assessment, crop health monitoring, irrigation, harvesting, and postharvest management. As a result, they reduce labor dependency and improve both productivity and consistency. Precision-based resource management (García et al. 2020), achieved through targeted application of water, fertilizers, and pesticides, increases resource-use efficiency and minimizes environmental impacts. The integration of sensor networks and cloud-based platforms provides farmers with actionable insights, which is especially transformative for smallholder systems in developing regions.

In this context, the chapter examines the convergence of robotics and intelligent automation in modern agriculture, focusing on their role in advancing smart farming systems (Patra 2023). It provides a critical analysis of the underlying technological frameworks, highlights real-world applications, and explores emerging research directions likely to shape the future development of sustainable and intelligent agriculture.

6.2 Background and State of the Art

The integration of intelligent automation and robotic systems into agriculture represents a recent yet rapidly advancing trend, offering pathways to enhanced sustainability, efficiency, and scalability in modern farming. These technologies are specifically engineered to address persistent challenges, including labor shortages, inefficient resource utilization, and the demand for precise management. By incorporating AI, the IoT, and advanced control systems, current agricultural robots can perform complex tasks with high precision, thereby reducing operational costs and mitigating environmental impacts.

Recent research demonstrates notable advancements in agricultural robotics. Han et al. (2021) developed a fruit-harvesting robot enabled by IoT and AI, which utilizes color, infrared, and ultrasonic sensors on an STM32 platform to detect fruit maturity in real time. This system employs streamlined algorithms for rapid decision-making, achieving efficient data acquisition with lower energy and cost demands, though further optimization is required for deployment at scale. Similarly, Orkweha (2022) presented a robust tracked robotic platform designed for challenging agricultural terrains. This system incorporates a Teensy 4.1 microcontroller for low-level control and a Jetson AGX Xavier for high-level processing, utilizing Model Predictive Control and the Robot Operating System (ROS) for navigation and SLAM-based localization, thus establishing a scalable foundation for autonomous agricultural operations.

Additional progress is evident in the work of Ghafar et al. (2022), who developed a cost-effective autonomous robot for crop monitoring, pesticide application, and fertilizer spraying. Featuring imaging systems and wireless control, the prototype achieves significant labor savings and improved resource-use efficiency, although its field coverage is marginally less than that of manual methods. Concurrently, Yang et al. (2023) introduced a comparable tracked robotic system that prioritizes integrated navigation and control for greater field adaptability and autonomy. In a complementary approach, Choudhary et al. (2023) demonstrated that enhancing seedling mat quality and mechanized transplanting efficiency through optimized soil mixtures with vermicompost and farmyard manure significantly advances climate-smart agriculture by increasing productivity and resource-use efficiency. Collectively, these studies highlight the transformative potential of robotics and intelligent systems in agriculture, while emphasizing the necessity for ongoing innovation, supportive policies, and equitable access to facilitate widespread adoption, especially among smallholder farmers (Yépez-Ponce et al. 2023; FAO 2022).

6.3 Conceptual Framework and System Architecture

Empirical validation is a foundational element in advancing intelligent automation and robotics in smart farming, as these technologies fundamentally transform agricultural systems rather than represent mutually exclusive paradigms. The integration of intelligent automation and robotics requires robust, evidence-based frameworks to assess performance, scalability, and sustainability outcomes across diverse agro-ecological contexts. Therefore, the establishment of governance protocols, standardized evaluation metrics, and data-driven validation methodologies is critical for ensuring reliable deployment and effective management of smart farming systems.

A comprehensive theoretical framework supporting these advancements is established by cyber–physical systems (CPS) integrated with digital twin technology (Montalvo et al. 2025). This approach enables the seamless connection of physical farm components with their virtual representations, facilitating real-time monitoring, simulation, and decision-making. Continuous data acquisition and synchronization through digital twins support predictive modeling, scenario analysis, and optimization of farm operations, thereby improving resource-use efficiency, system resilience, and operational sustainability. The integration of the IoT with advanced analytics (García et al. 2020) further reinforces this framework by enabling precise, real-time assessment of parameters such as soil moisture, microclimate conditions, and crop health, ultimately supporting informed and adaptive management strategies.

The growing integration of digital and physical domains has substantially improved real-time perception, autonomous response, and intelligent control within agricultural systems. Intelligent systems enhanced by ML and predictive analytics enable continuous monitoring and proactive intervention throughout the farming cycle, from pre-sowing planning to postharvest management. Digital twin environments further extend these capabilities by supporting virtual experimentation and forecasting without interrupting field operations, thereby reducing risks and increasing decision accuracy.

The foundation of smart farming architectures is the synergistic integration of key enabling technologies. The foundation of smart farming architectures is the synergistic integration of key enabling technologies, as illustrated in Fig. 6.1. AI supports pattern recognition, yield prediction, and autonomous decision-making (Hoque and Padhiary 2024; Nrip 2022). IoT networks enable distributed sensing of critical variables such as soil conditions, atmospheric parameters, and livestock

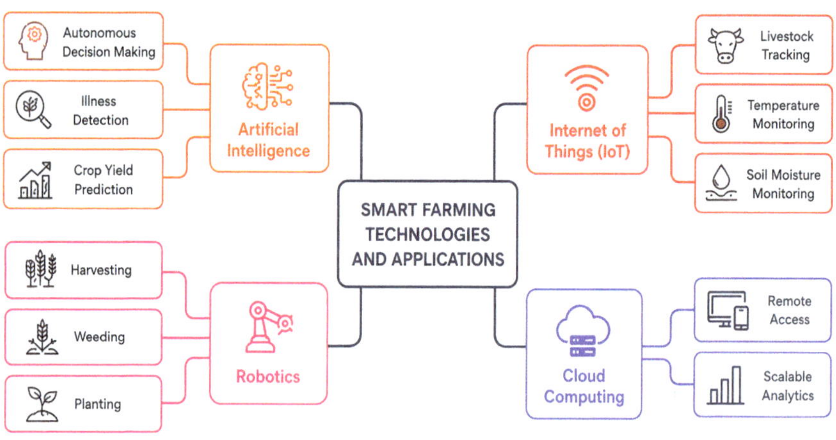

Fig. 6.1 Smart farming technologies and applications

health. Robotics provides precise execution of labor-intensive tasks, including planting, weeding, and harvesting. Cloud computing delivers scalable infrastructure for data storage, processing, and remote management. The close integration of these components within a unified platform ensures seamless data flow, interoperability, and synchronized operations, establishing the basis for next-generation sustainable agricultural systems.

6.4 Applications and Case Studies

This section analyzes the implementation of robotics and intelligent automation in various agricultural operations, emphasizing their contributions to operational efficiency, data-driven decision-making, and environmental sustainability. Real-world case studies demonstrate the transformative potential of smart farming technologies across different agroecological and socioeconomic settings.

6.4.1 Case Study 1: Deployment of Autonomous Field Robots in India

Autonomous agricultural robots (Agri-bots) (Yépez-Ponce et al. 2023; Nrip 2022) are increasingly deployed in major agricultural regions such as Maharashtra and Punjab to address labor shortages and enhance precision in farm operations. These systems perform repetitive and labor-intensive tasks, including seed sowing, weeding, and targeted pesticide application. Equipped with global positioning system (GPS), computer vision, and AI algorithms, these robots accurately navigate crop rows and adapt to heterogeneous field conditions.

A primary feature of these systems is real-time sensing and adaptive decision-making. Integrated sensors monitor soil parameters continuously, enabling dynamic adjustment of planting depth to optimize seed placement and germination. AI-enabled vision systems support precise identification of weeds and pest infestations, allowing site-specific pesticide application and significantly reducing chemical usage compared to conventional blanket spraying. The use of solar-powered energy systems further enhances operational sustainability by reducing reliance on fossil fuels and lowering long-term operational costs.

The implementation of robotic systems has yielded significant agronomic and economic advantages. Empirical data reveal up to a 60% reduction in manual labor requirements, as well as enhanced sowing accuracy and more consistent crop establishment. These results highlight the capacity of autonomous robotic technologies to increase productivity, improve resource efficiency, and support sustainable intensification in contemporary agriculture.

6.4.2 Case Study 2: Smart Greenhouse Automation in the Netherlands

Greenhouse systems in the Netherlands serve as leading examples of integrating robotics and intelligent automation in controlled-environment agriculture. These facilities function as fully automated units, utilizing extensive networks of IoT sensors to continuously monitor parameters including light intensity, temperature, relative humidity, and CO_2 concentration (García et al. 2020). Data collected from these sensors are analyzed by AI-driven control systems, which dynamically adjust the internal microclimate to sustain optimal crop growth conditions, thereby enhancing productivity and resource-use efficiency.

A central technological element in these systems is the use of digital twin models (Montalvo et al. 2025), which generate virtual representations of greenhouse environments and crop development processes. These models support simulation-based optimization, enabling operators to assess various cultivation strategies without interfering with ongoing operations. Concurrently, cloud-based analytics platforms support centralized data processing, remote monitoring, and large-scale decision-making. The addition of robotic manipulators further improves system efficiency, as these precisely controlled robotic arms can perform delicate tasks such as harvesting high-value crops while minimizing mechanical damage.

The implementation of smart greenhouse technologies has produced notable agronomic and environmental advantages. Empirical evidence demonstrates a reduction of approximately 25% in energy and water usage, as well as a significant increase of 30–40% in crop productivity. Furthermore, consistent environmental control leads to improved produce quality and longer shelf life. These results highlight the effectiveness of integrated automation frameworks in promoting resource efficiency, sustainability, and economic viability within modern intensive agricultural systems.

6.4.3 Case Study 3: Drone-Based Crop Surveillance in Brazil

Large-scale farms in Brazil have increasingly adopted drone-based crop surveillance systems that integrate advanced imaging technologies and AI to support precision agriculture. Unmanned aerial vehicles (UAVs) equipped with multispectral cameras are utilized to collect high-resolution spatial data across extensive agricultural areas. These datasets are then analyzed using deep learning (DL) methods, particularly convolutional neural networks (CNNs), to identify early-stage anomalies such as water stress, nutrient deficiencies, and disease incidence (Zhang and Li 2023). This process enables timely and targeted interventions.

The operational workflow consists of automated, scheduled drone flights that ensure consistent, frequent monitoring without manual intervention, as illustrated in Fig. 6.2. The collected imagery is processed through AI-driven analytical pipelines, in which CNN-based models detect deviations from standard crop signatures by analyzing spectral and spatial patterns. When anomalies are identified, the system

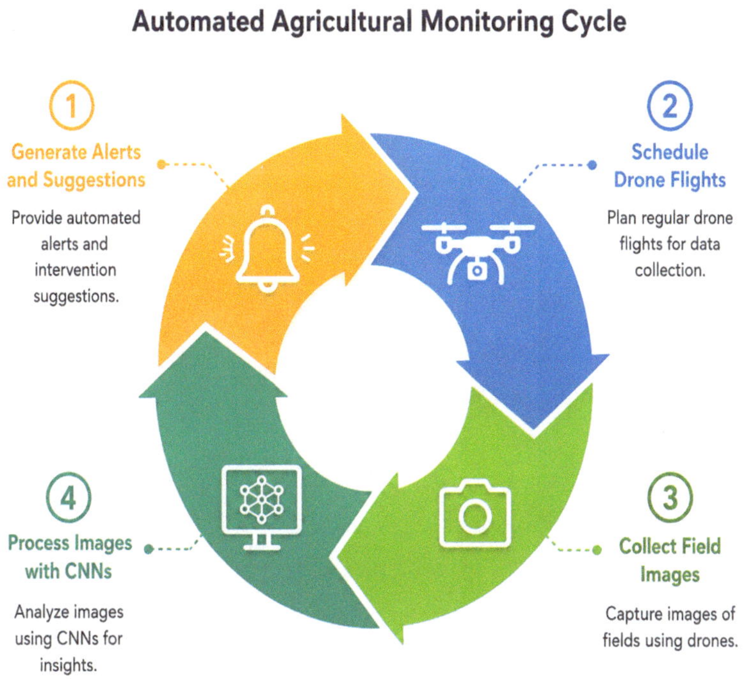

Fig. 6.2 Automated agricultural monitoring cycle

generates automated, real-time advisories that provide actionable insights and site-specific recommendations to farmers to take corrective action.

The adoption of these systems has yielded substantial agronomic and economic benefits. Improved precision in crop monitoring allows for accurate prediction of crop maturity, which optimizes harvest scheduling and reduces postharvest losses. Additionally, site-specific input application, guided by spatial intelligence, reduces the overuse of water, fertilizers, and pesticides, leading to lower input costs and enhanced environmental sustainability. These results highlight the essential role of AI-enabled aerial surveillance in promoting data-driven and resource-efficient agricultural practices.

6.4.4 Case Study 4: Automated Livestock Monitoring and Management

Contemporary livestock production systems are increasingly utilizing integrated automation frameworks that incorporate IoT-enabled wearable sensors, robotic feeding mechanisms, and AI-driven analytics to facilitate precision livestock farming (FAO 2022). Livestock are fitted with RFID tags and biometric sensors that continuously monitor physiological and behavioural parameters, including body

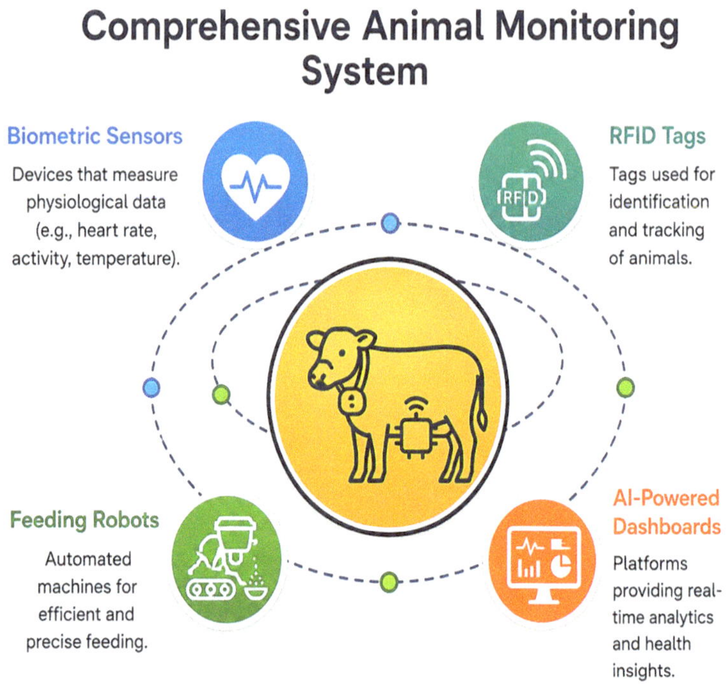

Fig. 6.3 Comprehensive animal monitoring system

temperature, activity levels, and movement. These data streams are transmitted in real time to centralized platforms, where ML algorithms assess deviations from established baselines to identify early indicators of stress, disease onset, or nutritional imbalance. This process enables proactive and preventive health management.

The system architecture consists of three primary functional components, as illustrated in Fig. 6.3: (i) identification and continuous monitoring using RFID and biosensors to ensure individual-level traceability; (ii) automated nutrition management through robotic feeding systems that optimize feed allocation according to specific animal requirements; and (iii) AI-enabled health management interfaces that provide real-time alerts and decision support for veterinarians and farm managers. This integrated framework improves operational efficiency and supports compliance with sanitary and traceability standards.

The adoption of intelligent livestock systems has produced notable benefits, including enhanced traceability and regulatory compliance, reduced veterinary expenses through preventive healthcare, and improved overall herd productivity and health. Collectively, this case study and previous examples illustrate that the effectiveness of smart farming technologies depends on context-specific implementation. Adapting technological solutions to local production systems, resource availability, and management objectives is essential for achieving operational relevance, scalability, and long-term sustainability in agriculture.

6.5 Impacts

The integration of robotics and intelligent automation has fundamentally transformed agricultural systems by increasing productivity, enhancing resource-use efficiency, and advancing sustainability. Empirical studies across various farming systems demonstrate that these technologies yield measurable improvements across multiple performance metrics. Autonomous machinery and robotic systems have resulted in a 20–35% increase in crop productivity through precise planting, harvesting, and crop management (Raj and Prahadeeswaran 2025; FAO 2022). In controlled environments such as smart greenhouses, automated climate regulation has further increased productivity, with yield improvements of up to 40% per unit area.

Resource optimization constitutes a significant area of impact. AI-driven irrigation systems enable precise water application based on real-time soil and crop data, resulting in water savings of 30–50%, particularly in regions facing water scarcity (García et al. 2020). Similarly, site-specific application of fertilizers and pesticides reduces input use and environmental impact by approximately 25%. Labor efficiency has improved markedly, with robotic systems decreasing reliance on manual labor by up to 60%, thereby mitigating labor shortages and reducing operational costs. In livestock production, automated monitoring technologies support early disease detection, reducing veterinary interventions by nearly 20% and enhancing overall herd health (FAO 2022). Economic analyses indicate that most smart farming technologies achieve a return on investment within 2–4 years for medium to large-scale operations, while modular solutions allow smallholders to increase profitability by 15–20% through higher yields and lower input costs.

Despite these considerable benefits, several critical factors affect the widespread adoption and long-term sustainability of these technologies. Accessibility and scalability remain significant challenges, especially for smallholder farmers, due to high initial investment requirements and limited technical expertise. Policy interventions, such as subsidies and the development of cost-effective, modular solutions, are necessary to address these barriers. Data-driven agriculture, facilitated by the IoT and AI, raises concerns regarding data privacy, ownership, and interoperability, which require the establishment of secure data infrastructures and standardized protocols. Although smart technologies promote environmental sustainability through efficient resource use, comprehensive life-cycle assessments must also account for the environmental impacts of producing and disposing of electronic components.

The human dimension is central to the effective implementation of intelligent systems. These technologies are intended to augment human decision-making, making capacity building, digital literacy, and skill development among farmers essential. Regional adaptation is also crucial, as technological solutions must be tailored to specific agroclimatic, socioeconomic, and cultural contexts to ensure their effectiveness and adoption. Overall, a balanced, integrated approach to smart agriculture is essential to align technological innovation with socioeconomic inclusivity and environmental sustainability. Figure 6.4 depicts the multidimensional impacts of smart farming technologies on agricultural productivity and sustainability.

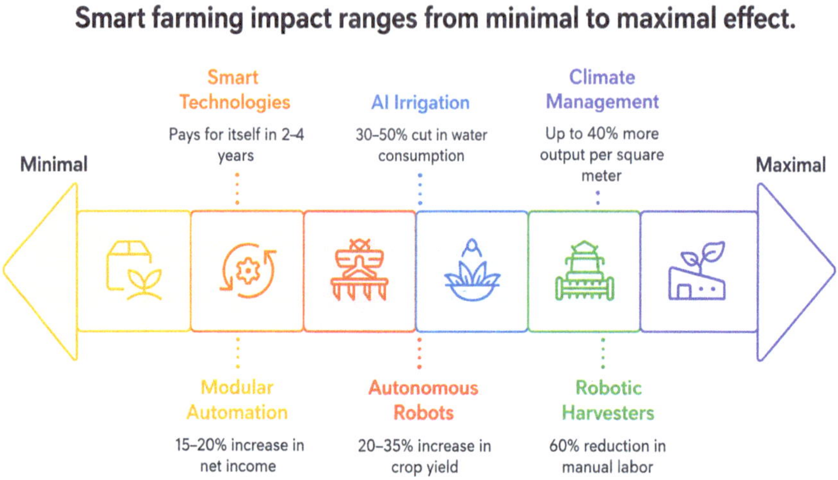

Fig. 6.4 Schematic representation of the multidimensional impacts of smart farming technologies on agricultural systems

6.6 Conclusion

Smart automation and robotics, supported by advanced digital technologies, have fundamentally transformed modern agricultural systems by shifting operations from reactive and fragmented approaches to predictive, data-driven, and adaptive management frameworks. The integration of AI, the IoT, and autonomous machinery enables real-time monitoring and anticipatory decision-making, which enhances productivity, operational efficiency, and system resilience (Yépez-Ponce et al. 2023; FAO 2022). Technologies such as drones, precision irrigation systems, and automated greenhouse platforms have led to significant increases in yield, reduced input costs, and improved adaptability to global challenges, including climate variability and rising food demand. Additionally, predictive analytics and integrated digital platforms facilitate the assimilation of multisource data, including weather patterns, soil conditions, and pest dynamics, supporting both short-term optimization and long-term sustainability in agricultural practices.

Despite these advancements, the widespread adoption of smart farming technologies is limited by structural and socioeconomic challenges, particularly in developing regions. High initial investment costs, limited digital literacy, data privacy and ownership concerns, and inadequate infrastructure present significant barriers for smallholder farmers. Overcoming these obstacles requires coordinated efforts among governments, private sector stakeholders, research institutions, and farming communities. Strategic interventions, including policy support, capacity building, the development of affordable, modular technologies, and the creation of secure data ecosystems, are essential for inclusive adoption. Therefore, agricultural digitization should be regarded not only as technology transfer but as a transformative pathway for building resilient, equitable, and sustainable food systems.

6.7 Future Perspectives

Recent developments in smart farming highlight the necessity for technologies that are sophisticated, accessible, scalable, and sustainable. Future research should prioritize the creation of cost-effective, modular, and interoperable solutions that allow incremental adoption by farmers, especially smallholders. Open-source platforms and community-driven innovation models are key to democratizing access to smart farming technologies, facilitating adaptation to local contexts, and reducing reliance on expensive proprietary systems. These strategies will improve adaptability and ensure long-term viability across varied agroecological and socioeconomic environments.

The integration of advanced digital technologies, such as blockchain, is anticipated to enhance transparency, traceability, and trust within agricultural value chains (Torres and Oliveira 2024). In combination with IoT-enabled sensing and AI-driven analytics, blockchain frameworks can ensure data integrity and product authentication, which is essential for organic certification, export compliance, and consumer confidence. Concurrently, future AI systems should prioritize climate resilience by utilizing multisource data, including weather forecasts, satellite imagery, and climate models, to deliver early warnings for droughts, pest outbreaks, and soil degradation (Hoque and Padhiary 2024; Zhang and Li 2023). These predictive capabilities will facilitate a transition from reactive to proactive farm management, thereby strengthening adaptive capacity in the face of changing climatic conditions.

The future of agriculture depends on the convergence of technological innovation, ecological sustainability, and human expertise. Smart farming systems should be regarded not only as technological solutions but as integral elements of a broader transformation toward resilient and inclusive food systems. Achieving equitable access, ethical implementation, and environmental stewardship will be crucial in guiding this transition. Collaboration among researchers, policymakers, industry stakeholders, and farming communities is essential to establish agriculture that is intelligent, sustainable, and responsive to both human and environmental needs.

References

Choudhary V, Machavaram R, Soni P (2023) Optimizing mat quality and transplanter performance using soil mix with vermicompost and farmyard manure in paddy tray nursery: a sustainable smart farming approach in India. Farming Syst 1(3):100046. https://doi.org/10.1016/j.farsys.2023.100046

FAO (2022) The state of food and agriculture: digital automation technologies and robotics in agriculture. Food and Agriculture Organization of the United Nations. https://openknowledge.fao.org/server/api/core/bitstreams/1c329966-521a-4277-83d7-07283273b64b/content

García L, Parra L, Jimenez JM et al (2020) IoT-based smart irrigation systems: an overview on the recent trends on sensors and IoT systems for irrigation in precision agriculture. Sensors 20(4):1042. https://doi.org/10.3390/s20041042

Ghafar ASA, Hajjaj SSH, Gsangaya KR et al (2022) Design and development of a robot for spraying fertilizers and pesticides for agriculture. Mater Today Proc 81:242–248. https://doi.org/10.1016/j.matpr.2021.03.174

Han J, Liu L, Zeng H (2021) Design and implementation of intelligent agricultural picking mobile robot based on color sensor. J Phys Conf Ser 1757:012157. https://doi.org/10.1088/1742-6596/1757/1/012157

Hoque A, Padhiary M (2024) Automation and AI in precision agriculture: innovations for enhanced crop management and sustainability. Asian J Res Comput Sci 17(10):95–109. https://doi.org/10.9734/ajrcos/2024/v17i10512

Montalvo A, Camacho O, Chavez D (2025) Cyber-physical systems for smart farming: a systematic review. Sustainability 17(14):6393. https://doi.org/10.3390/su17146393

Nrip N (2022) Applications of artificial intelligence and robotics for Indian agriculture development. Int J Adv Eng Manag 4:100–104. https://doi.org/10.35629/5252-0401100104

Orkweha P (2022) Trajectory control design for a custom-built agricultural robot considering slippage effects. Doctoral dissertation, Southern Illinois University at Edwardsville

Patra B (2023) Smart agriculture: technologies and applications. ISBM University. https://www.-int-jecse.net/media/article_pdfs/CSE_Smart_Agriculture_Technologies_and_Applications.pdf

Raj M, Prahadeeswaran M (2025) Revolutionizing agriculture: a review of smart farming technologies for a sustainable future. Discov Appl Sci 7:937. https://doi.org/10.1007/s42452-025-07561-6

Torres L, Oliveira M (2024) Blockchain applications in agricultural supply chains: enhancing traceability and trust. Agric Syst 225:103115. https://doi.org/10.1016/j.agsy.2024.103115

Yang Q, Du X, Wang Z et al (2023) A review of core agricultural robot technologies for crop productions. Comput Electron Agric 206:107701. https://doi.org/10.1016/j.compag.2023.107701

Yépez-Ponce DF, Salcedo JV, Rosero-Montalvo PD et al (2023) Mobile robotics in smart farming: current trends and applications. Front Artif Intell 6:1213330. https://doi.org/10.3389/frai.2023.1213330

Zhang Y, Li H (2023) Deep learning for crop disease detection using UAV imagery. Comput Electron Agric 205:107118. https://doi.org/10.1016/j.compag.2023.107118

IoT for Precision Farming and Resource Optimization

7

Prachi Arora and Sandeep Mahato

Abstract

The integration of intelligent technologies within the agricultural sector is essential for minimizing resource use, reducing environmental impact, and enhancing productivity, sustainability, and efficiency. Traditionally, agricultural practices relied on manual labor, which posed significant challenges. To address these issues, agriculture has evolved through stages from 1.0 to 4.0, with precision farming representing a key component of Agriculture 3.0. The adoption of the Internet of Things (IoT) has addressed challenges such as pest control, weed detection, fertilization, harvesting, and sorting. This chapter explores the emergence of precision farming, examining the challenges and benefits of IoT integration in agriculture and how these advancements help farmers operate more profitably. Additionally, this chapter identifies opportunities for researchers, academic institutions, and undergraduate students engaged in projects or research on technology applications in precision farming, with a focus on sustainability and productivity-enhancing practices. Furthermore, IoT integration contributes to achieving the objectives of the EU Green Deal by reducing emissions that drive climate change and supports the Sustainable Development Goal of "Zero Hunger" by meeting global food demand.

Keywords

Smart farming · Precision farming · Industry 3.0 · Internet of things · Sustainability

P. Arora (✉)
Indian Institute of Information Technology, Una, Himachal Pradesh, India

S. Mahato
Chaibasa Engineering College, Kelende, Jharkhand, India

H. N. Oo et al. (eds.), *Transforming Agriculture with AI, IoT, and Blockchain*,
https://doi.org/10.1007/978-981-92-1491-4_7

7.1 Introduction

Agriculture remains a fundamental component of human society, with populations in developed countries highly dependent on its outputs. Over time, significant changes in the agricultural sector have led to higher yields. Technological advancements in agriculture, forestry, and horticulture now enable the monitoring of plant development, disease, and insect populations. The introduction of modern technologies such as artificial intelligence (AI), machine learning (ML), and the Internet of Things (IoT) has led to a wide range of applications that benefit farmers by improving decision-making and transforming the working environment. These technologies address challenges such as soil condition assessment, harvest time prediction, optimal crop selection, crop monitoring, weather forecasting, pest detection, and postharvest management. Additionally, the integration of modern technologies in agriculture facilitates real-time monitoring of parameters, including water content, soil quality, humidity, and temperature. IoT solutions are implemented across various industries, such as healthcare, retail, transportation, security, smart homes, smart cities, and agriculture. This chapter focuses specifically on the application of these technologies in agriculture (Miller et al. 2025).

Smart farming enhances precision in agricultural operations by enabling more effective management and informed decision-making based on collected data. The integration of IoT with multiple sensors enables efficient crop and vegetable management across large areas in shorter time frames (Boursianis et al. 2022). This integration also facilitates optimal timing and application of fertilizers and pesticides, reducing environmental contamination and health risks associated with overuse. The implementation of IoT technologies results in cost savings and minimizes resource waste (Tzounis et al. 2017). Additionally, IoT and sensor technologies enable precise disease detection in crops, making them compatible with a variety of existing agricultural devices and further improving productivity. IoT is utilized throughout the agricultural production chain, including precision farming, livestock management, and greenhouse operations, to monitor key parameters. Equipment such as unmanned aerial vehicles (UAVs) accelerates the irrigation process by collecting essential agricultural data, supporting the development of smart irrigation systems. These tools also detect pest infestations by capturing high-resolution images, which are stored in cloud databases for further analysis. Integrating IoT with intelligent decision-making systems reduces the need for human intervention, thereby automating agricultural processes and enhancing data processing accuracy and efficiency to support decision-making (Sravanthi and Moparthi 2024). The bar chart illustrating the projected growth of the IoT market is shown in Fig. 7.1. This chapter examines the use of IoT and sensors to monitor key plant growth parameters, including temperature, humidity, moisture content, and nitrogen levels.

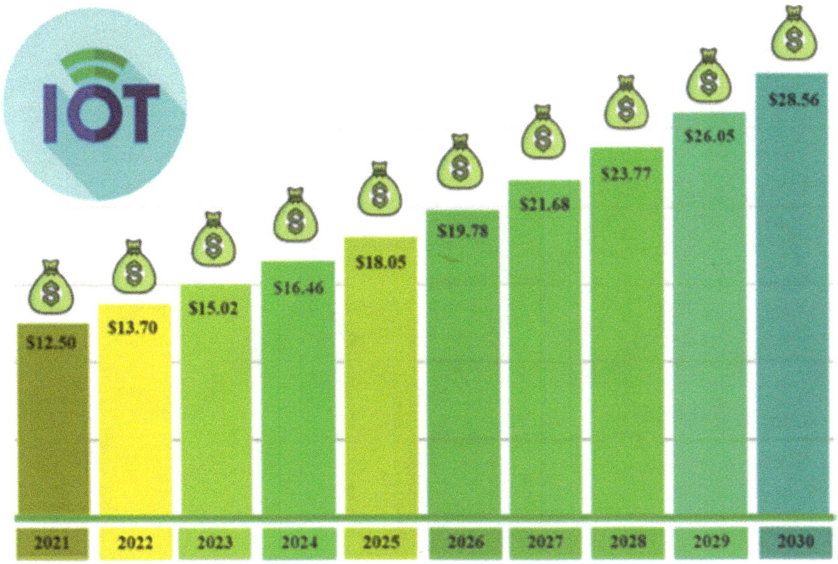

Fig. 7.1 Use of IoT in agriculture from 2021 to 2030 (Duguma and Bai 2025)

7.2 Types of Sensors

Sensors are machines, devices, or modules that detect events or changes in their environment and communicate this information to other devices, typically a computer processor or controller. The various types of sensors used in agriculture are shown in Fig. 7.2. In agriculture, sensors play a critical role in collecting data to enhance sustainability and efficiency. These smart devices transfer the collected data to designated storage locations. In the agricultural sector, data gathered by sensors includes plant disease detection, animal tracking, automated farming, grain storage monitoring, pest detection, smart greenhouse systems, soil moisture levels, and temperature. This information enables farmers to allocate and optimize resources, facilitating well-informed decision-making. Additionally, sensor-collected data in smart agriculture provides valuable insights for precision farming and sustainable agricultural practices, leading to increased yields and profits (Steeneken et al. 2023).

Various types of IoT sensors, such as optical, electrochemical, photosynthetically active radiation (PAR), and location and position sensors, are used in precision agriculture alongside communication technologies, including Wi-Fi, ZigBee, Bluetooth, 2G, 3G, 4G, and LoRaWAN (Alahmad et al. 2023). This section reviews some of the most commonly used sensors in the agricultural sector. Table 7.1 presents several sensors and their respective applications.

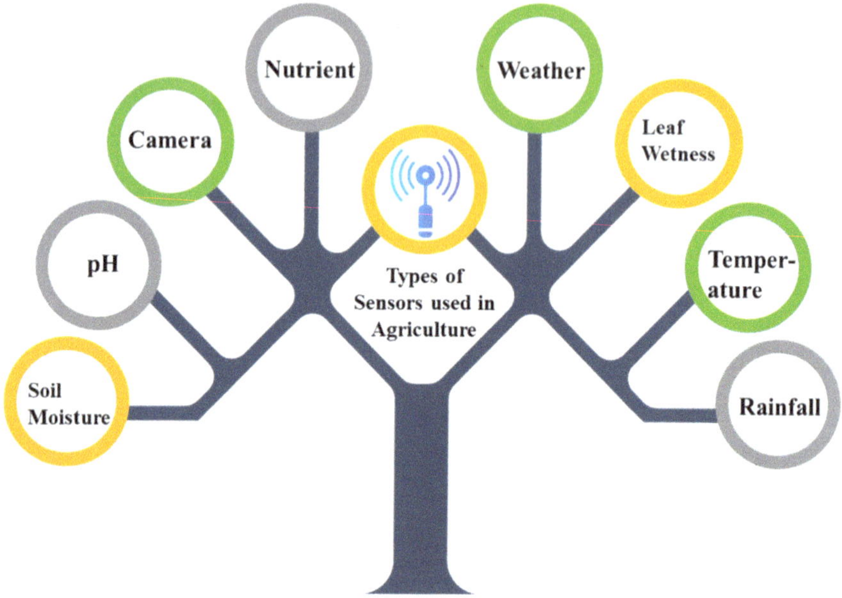

Fig. 7.2 Different types of sensors used in agriculture

Table 7.1 Different types of sensors along with their applications

Different types of sensors	Type of sensors	Applications
Arduino	IoT platform	Data acquisition, custom sensor integration
GPS module	GPS	Location tracking, precision farming
BMP180	Pressure	Weather monitoring, weather prediction and analysis
DHT11	Temperature/ humidity	Crop monitoring, environmental control, monitoring temperature and humidity
Raspberry Pi	IoT platform	Data processing & communication, data aggregation for multiple sensors
Capacitive soil moisture sensor	Soil	Measuring soil moisture, irrigation management, soil health
LM35	Temperature	Crop health monitoring, measuring temperature
SH1145	Light	Plant growth monitoring, measuring UV and IR light
MQ-135	Air quality	Air quality monitoring, pest detection, environmental monitoring
LoRaWAN module	Communication device	Data transmission, long range data transmission

7.2.1 Camera Sensor

Images captured by camera sensors play a vital role in agricultural applications such as security enhancement, crop health assessment, and the determination of optimal harvest time. Typically, these images are in RGB format and are subjected to analysis and feature extraction using specialized algorithms. The extracted features are compared with those obtained during the harvesting stage. This process enables timely and efficient harvesting by providing objective assessments of crop health. Additionally, integrating image processing techniques with intelligent tracking systems has been shown to improve monitoring of crop health, growth trends, and leaf color (Shimonomura 2019).

7.2.2 Nutrient Sensor

Soil nutrient sensors are used to detect key nutrients such as potassium, phosphorus, and nitrogen. These sensors operate based on various principles, with each type designed to measure specific nutrients. Examples include optical sensors, electrochemical sensors, spectroscopic methods, and ion-selective electrodes. Soil nutrient sensors enable real-time mapping and monitoring of nutrient concentrations, thereby supporting precise and effective fertilizer application. The advancement of nutrient sensor technology represents a significant innovation in agriculture, enhancing sustainable farming practices by optimizing plant growth, increasing crop yields, improving resource management, and reducing environmental impacts, fertilizer usage, and preserving water quality and soil health (Burton et al. 2020).

Visible near-infrared (Vis–NIR) sensors are currently utilized in both research laboratories and field environments to measure soil pH and nutrient concentrations. Vis–NIR technology is specifically applied to determine nitrogen concentrations in soil minerals. Additional instruments, such as attenuated total reflectance (ATR), Raman spectroscopy, and ion-selective transistors, are also employed for assessing soil nutrients and pH. These sensors collectively provide critical information for soil health assessment and fertilizer management, thereby supporting precision agriculture (PA) practices (Burton et al. 2020).

7.2.3 Weather Sensor

This tool enables farmers to remain informed about current weather conditions and potential future impacts. The sensor accurately detects various external parameters, such as temperature, solar radiation, wind speed, rainfall, wind direction, precipitation, and humidity. Although farmers cannot control weather conditions, they can reduce crop losses by implementing additional farm management strategies. Consequently, weather monitoring is a critical consideration in planning agricultural operations, managing pest and disease risks, and forecasting weather patterns to mitigate adverse outcomes based on precise and reliable weather information (Selvam and Saud 2023).

7.2.4 Leaf Wetness Sensor

This device provides accurate and cost-effective measurement of leaf wetness. Numerous fungal and bacterial infections occur only when the leaf surface is damp. To protect plant canopies and predict disease and pest outbreaks, sensors are used to measure surface moisture (Patle et al. 2021).

7.2.5 Temperature Sensor

Temperature sensors are widely used in precision farming to characterize soil and air conditions. These sensors convert analog voltage outputs into digital values using an analog-to-digital converter. The resulting digital readings of soil temperature are subsequently utilized to regulate temperature and irrigation systems. Soil temperature significantly influences plant growth by affecting nutrient uptake, root development, respiration, transpiration, water potential, microbial activity, soil translocation, photosynthesis, and seed germination. Common types of temperature sensors include resistance temperature detectors (RTDs), thermocouples, and thermistors (Kuzubasoglu and Bahadir 2020).

Effective crop management requires careful monitoring of the temperature surrounding plants. Both elevated and reduced temperatures can either promote or hinder plant development and health. High temperatures induce heat stress, while low temperatures may damage crops. Temperature data can be integrated with weather sensors for comprehensive environmental monitoring. An IoT-enabled farming system delivers real-time information on local conditions, enabling farmers to respond proactively to adverse weather. This information enables adjustments to greenhouse environments, pest control, and informed irrigation decisions. By combining temperature monitoring with IoT infrastructure, agriculturists gain a deeper understanding of the impact of weather on farming practices. The use of smart devices and advanced technologies enables farmers to identify weather patterns and implement measures to maintain optimal conditions for crop production and growth (Mahan et al. 2010).

Temperature monitoring also enables farmers to optimize irrigation schedules, ensuring that crops receive adequate water at appropriate times. By closely observing temperature fluctuations and implementing preventive measures, pest infestations can be minimized. Additionally, it is essential to position temperature sensors correctly within greenhouses, as plants thrive within specific temperature ranges. Consistent and accurate temperature monitoring allows growers to maintain optimal plant growth throughout the year, independent of external weather conditions.

7.2.6 Rainfall Sensor

This sensor measures the amount of rainfall at specific times and dates, enabling farmers to plan proactively and minimize potential crop damage. A rain gauge is commonly used to determine both the hourly rate and total amount of precipitation.

Rainfall sensors are often integrated into advanced flood and water control systems on farms, facilitating more efficient irrigation scheduling and water management. This integration conserves water and prevents over-irrigation (Nielsen et al. 2024). Consequently, farmers can more accurately allocate water to crops and maintain crop health.

7.2.7 pH Sensor

Soil pH is a key determinant of soil health and plant growth, reflecting the balance between acidity and alkalinity. Soil pH is typically measured by inserting two electrodes into the soil and calculating the electrical potential difference between them. Monitoring pH enables the early identification of field areas with pH imbalances. In precision agriculture, pH monitoring is applied in several ways, including the use of pH sensors to maintain soil pH within the optimal range for specific crops. The optimal pH varies by crop; some species require acidic conditions, while others thrive in alkaline soils. Excessively acidic or alkaline soils can result in stunted leaves with dark patches and increased plant susceptibility to disease, rendering them unsuitable for cultivation. The literature indicates that the ideal pH range for most crops is between 5.5 and 7.5, with values below 5.5 being incompatible with plant growth. A common application of pH monitoring is in hydroponic farming, where maintaining the water's pH is essential for optimal plant growth. By regularly measuring and adjusting soil or water pH, farmers can maintain conditions conducive to healthy crop growth (Dutta et al. 2024).

7.2.8 Soil Moisture Sensor

Soil moisture sensors measure the amount of water present in the soil and are essential for evaluating plant water use, thereby improving irrigation efficiency. Common types of moisture sensors include gypsum blocks, capacitance sensors, tensiometers, electromagnetic sensors, densiometric sensors, remote sensing devices, and gravimetric sensors. Among these, remote sensing is particularly useful for measuring surface soil moisture across large areas. Moisture sensors can be classified as either contact or noncontact devices. Contact sensors measure soil moisture directly, while noncontact sensors use techniques such as microwave or electromagnetic waves. These technologies assist farmers in managing water resources and optimizing irrigation schedules (Leone 2022).

Moisture sensors deliver real-time data on soil moisture levels, enabling farmers to optimize irrigation practices and prevent both under-watering and overwatering. Insufficient or excessive watering can reduce agricultural yields and increase unnecessary water consumption. By providing precise and timely information, these sensors support improved water management systems and help identify specific areas of the field that require more or less water. As a result, moisture sensors promote healthier crop growth and reduce water waste by informing irrigators about conditions within the crop root zone.

7.2.9 Optical Sensor

Optical sensors operate by emitting light onto the soil and measuring the reflected light. This approach enables the assessment of various soil characteristics, including organic matter content, moisture levels, color, mineral presence and composition, and clay content. These sensors utilize different regions of the electromagnetic spectrum to evaluate the soil's reflectance properties. By detecting variations in the reflected light, they can reveal changes in soil density and other critical parameters. In addition to soil analysis, fluorescence-based optical sensors are employed for basic plant assessments, particularly in monitoring fruit maturation. When integrated with microwave scattering technology, these optical instruments are also effective for mapping the dense canopy structures of groves, such as those found in olive and similar crops (Navulur and Prasad 2017; Freidenreich et al. 2019; Wijesinghe et al. 2017; Rajak et al. 2023).

7.2.10 Ultrasonic Ranging Sensor

Ultrasonic sensors are devices that utilize sound waves at frequencies higher than the human ear can detect (ultrasound) to measure distance. These sensors emit a sound pulse and precisely measure the time required for the echo to return after reflecting off an object. By applying the known speed of sound, the distance to the target can be accurately calculated. This straightforward, reliable, and non-contact technique is well-suited for detecting object presence and continuously measuring levels and distances across a range of industrial and consumer applications (Abbas et al. 2020; Ayaz et al. 2019).

7.2.11 Optoelectronic Sensor

Optoelectronic sensors function as devices that mediate the interaction between light and electricity. These sensors operate by converting light energy into electrical signals or, alternatively, transforming electrical signals into light. The category encompasses components such as light-emitting diodes (LEDs), which convert electrical energy into light, and photodetectors, including photodiodes and phototransistors, which detect light and generate electrical current. Due to their high sensitivity to variations in light intensity, color, or presence, optoelectronic sensors are essential in a wide range of applications, including remote controls, optical communication, bar code scanners, and advanced imaging systems (Ayaz et al. 2019).

7.3 Evolution of Agriculture: From Traditional to Smart Farming

Traditional farming presents several drawbacks that pose significant challenges for farmers globally. Historically, these tasks were performed manually, requiring substantial time and effort and resulting in inconsistencies in crop management and output. A major disadvantage is that essential activities such as planting, irrigation, and pest control continue to be conducted manually using outdated techniques (Gupta et al. 2022; Ikram et al. 2022; Maheswary et al. 2024; Bouni et al. 2024). This section examines the limitations of traditional farming and explores its evolution into modern agricultural practices.

1. Traditional farming practices often suffer from a lack of precision and inefficiency, which contribute to reduced crop yields. For instance, uniform application of fertilizers and pesticides across extensive fields can result in either overuse or underuse of these inputs. Such practices may harm the environment, elevate production costs, and further diminish yields. Additionally, reliance on experiential knowledge alone exposes farmers to unpredictable factors, including volatile market conditions, variable weather patterns, and pest outbreaks.
2. Traditional farming systems frequently prioritize short-term gains over long-term sustainability. Practices such as monocropping, excessive tilling, and reliance on chemical inputs contribute to environmental degradation. These approaches may also jeopardize food security and human health while diminishing the overall resilience of agricultural systems.
3. Conventional methods of disease and pest control often depend on chemical applications, which can harm and contaminate the environment. Increased reliance on chemicals also raises production costs for farmers seeking to protect crops from disease.
4. Unpredictable and fluctuating weather and climatic conditions present additional challenges for farmers. Adverse events such as floods, droughts, and extreme temperatures can cause crop damage, reduce productivity, and create financial instability.
5. Traditional farming methods typically depend on general plans rather than precise, data-driven strategies. An aging workforce is contributing to a growing labor shortage in the agricultural sector, while various factors are accelerating urbanization. These challenges highlight the need for smart and precision farming practices aided by advanced technology.

7.3.1 Evolution of Smart Farming from 1.0 to 4.0

Agriculture 1.0 It refers to historical agricultural practices characterized by the use of basic tools, such as sickles and shovels, for food production intended for human and animal consumption. During this period, manual labor predominated, leading to relatively low productivity.

Agriculture 2.0 In the nineteenth century, the introduction of the steam engine transformed agriculture. Farmers began utilizing machinery and chemicals to enhance productivity and efficiency. However, these advancements also resulted in negative consequences, including depletion of natural resources, chemical pollution, environmental degradation, and increased energy consumption.

Agriculture 3.0 It emerged in the twentieth century alongside significant advancements in electronics and computational technologies. These technological developments enabled greater efficiency in agricultural operations through the adoption of robotics, automation, and other advanced tools. This era addressed challenges from the nineteenth century, such as excessive chemical use, inadequate pest management, and inefficient use of resources, by introducing precision farming, precision irrigation, and targeted fertilizer applications. As a result, Agriculture 3.0 established more efficient and sustainable farming systems.

Agriculture 4.0 It refers to the current era in which modern technologies such as cloud computing, big data, AI, ML, remote sensing, artificial neural networks (ANNs), and the IoT are employed to increase yields and reduce water and energy consumption. These advancements minimize negative environmental impacts by developing low-cost sensors and network platforms. This period is characterized by significant changes and improvements in agricultural activities. Big data technologies facilitate effective decision-making by providing comprehensive summaries of current agricultural conditions. Real-time programming, built on AI concepts and IoT integration, helps farmers make informed decisions. Collectively, these technologies are driving a technological revolution that transforms conventional farming practices into automated systems (Liu et al. 2020; Huang et al. 2020; Arora et al. 2026).

In this context, smart farming takes advantage of the information and communication technologies (ICT), including global positioning systems (GPSs), sensors, robotics, drones, actuators, data analytics, precision equipment, IoT, AI, ML, and edge computing, to optimize labor use and improve both the quantity and quality of agricultural products. These technologies deliver tailored solutions to meet specific farmer requirements, thereby enhancing productivity and enabling more timely and accurate decision-making. For instance, sensors continuously monitor crops with high accuracy to detect adverse conditions. Intelligent tools are employed throughout the agricultural process, from planting and harvesting to storage and transportation. Additionally, rapidly collecting sensors provide crop- and site-specific information that is immediately accessible online for further analysis (Friha et al. 2021).

Smart agriculture is therefore considered an evolution of precision agriculture, characterized by the adoption of modern techniques to collect diverse data on farm

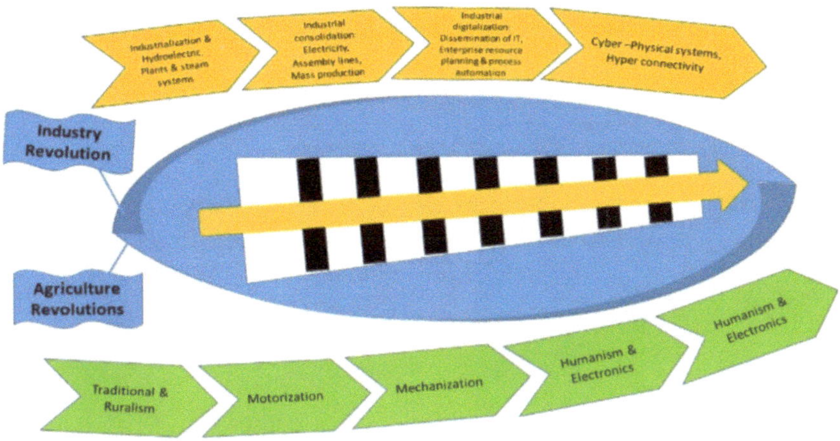

Fig. 7.3 Roadmap for the transition between the agricultural and industrial revolutions

activities. These data are remotely managed and supported by real-time farm maintenance solutions. A roadmap for the transition from the agricultural to the industrial revolution is shown in Fig. 7.3.

7.4 Role of Technology in Precision Farming

Precision farming is distinguished by its integration of multiple advanced technologies, rather than reliance on a single tool or technique, across nearly all aspects of crop management. Many innovations enabling precision farming originated outside the agricultural sector, reflecting a historical pattern in which inventions such as electricity, the internal combustion engine, and weather satellites were later adapted to transform farming practices. The contemporary precision farming toolbox includes sophisticated technologies such as remote sensing, geographic information systems (GISs), GPS, big data, machine vision, the IoT, and nanotechnology, each of which has also found significant applications in other industries (Karunathilake et al. 2023). The various technologies used in precision farming are shown in Fig. 7.4.

7.4.1 Remote Sensing

Remote sensing refers to the collection of information about a subject without direct physical contact. In agriculture, this process uses specialized equipment that detects various forms of light and energy, including visible light, infrared light, electromagnetic radiation, and near-infrared light, to measure key environmental and crop-related factors. This technology provides valuable data on parameters such as soil and air temperature, humidity, crop height, diameter, width, and local wind

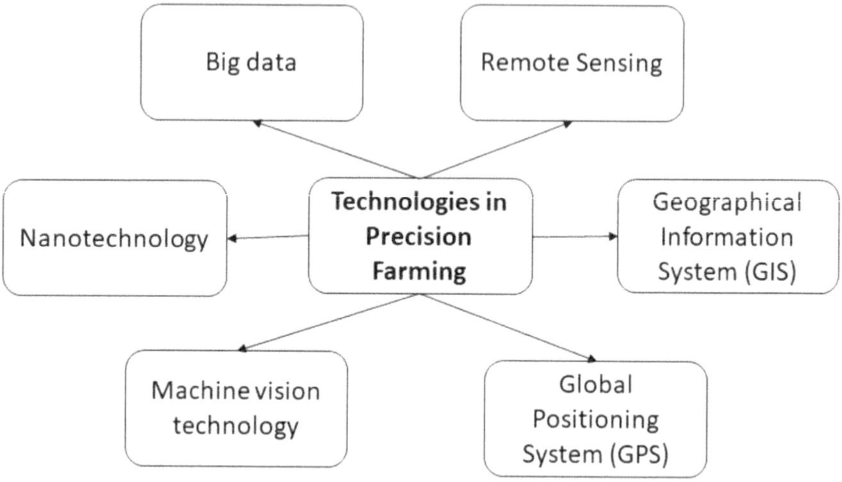

Fig. 7.4 Technologies in precision farming

conditions. The required instruments are typically mounted on aerial platforms, including UAVs, drones, high-altitude balloons, blimps, and global positioning system (GPS) satellites in orbit (Singh et al. 2021).

7.4.2 Geographical Information System (GIS)

A GIS is a computer-based database and software suite that visualizes data in a map-based format. This system allows users to input, store, retrieve, and analyze information associated with specific geographic locations, referred to as spatially referenced data. By processing and displaying spatial data as maps, GIS serves as a critical tool for understanding and managing variability between farms or fields. As it provides the essential framework for extracting insights from collected data and translating them into actionable management zones, GIS is often described as the core of precision farming (Raj et al. 2021).

7.4.3 Global Positioning System (GPS)

The GPS, a network of orbiting satellites, is fundamental to modern precision agriculture. GPS enables the precise location of farm machinery within a field to within a meter. This technology provides reliable navigation and positioning capabilities regardless of location, time, or weather conditions. In agricultural applications, GPS enables automated tractor guidance; once a route is recorded during plowing, the tractor can be programmed to follow the same optimized path for subsequent operations such as cultivating, fertilizing, harvesting, and pest control. This repeatable accuracy reduces operational costs by minimizing overlap and increasing efficiency.

Additionally, when integrated with mapping software, GPS is essential for systematic tasks such as soil sampling, allowing for accurate identification and recording of specific test locations. GPS also supports mapping yield variations by dividing farmland into zones and precisely plotting performance for each area (Raj et al. 2021).

7.4.4 Big Data

Precision farming systems rely extensively on data and information. In a manner analogous to large industries such as social media, which utilize substantial volumes of unstructured data to predict customer behavior, Big Data analytics in agriculture extracts insights from data-intensive farming operations to support critical decision-making. Analytical tools manage and interpret large datasets through data mining, statistical analysis, AI, and predictive analytics. In agricultural contexts, Big Data science is frequently integrated with technologies such as ML, cloud computing, image processing, and GIS. The comprehensive analysis of these data sources throughout the agricultural production process enables product traceability and substantially enhances product quality, including safety and taste (Karunathilake et al. 2023).

7.4.5 Machine Vision Technology

Successful precision agriculture depends on the availability of precise and accurate data. Increasingly, image analysis is replacing manual, time-consuming, and costly data collection methods. This shift is driven by Machine Vision (MV), also referred to as agro-vision or the "eyes" of agricultural robots. MV systems enable machines to interpret real-world environments by processing pixel-level images, thereby generating accurate, site-specific information through nondestructive means. This technology is already being applied to tasks such as crop species identification, weed and disease detection, and the assessment of crop stress or seed quality. Although many MV applications remain in the prototype stage, the integration of advanced deep learning (DL) techniques—a rapidly expanding subset of ML—is facilitating the development of intelligent robots capable of complex analysis of multispectral imagery and real-time adjustments for variable-rate applications in the field (Karunathilake et al. 2023).

7.4.6 Nanotechnology

Nanotechnology is a multidisciplinary field that is increasingly investigated for its potential to enhance agricultural yields. The previous green revolution led to the excessive use of chemical fertilizers and pesticides, degrading soil biodiversity and contributing to pest and pathogen resistance. Nanotechnology addresses these challenges by enabling the precise delivery of materials to plants through nanoparticles or nanochips and by facilitating the development of advanced biosensors essential for precision farming. Rather than applying large quantities of conventional fertilizers, precision farming employs nanomaterials to optimize nutrient application. In particular, nanomaterial-coated fertilizers demonstrate significant benefits, as their higher surface tension enables gradual and sustained nutrient release, thereby improving nutrient uptake by plant roots and promoting sustainability (Raj et al. 2021).

7.5 Open Issues and Challenges

Numerous unresolved problems and challenges arise from the integration of IoT in the agricultural sector. This section outlines key challenges identified in the existing literature (Fig. 7.5) (Farooq et al. 2020; Dhal et al. 2024; Rana et al. 2023; Seetha et al. 2023; Kumar et al. 2024).

7.5.1 Lack of Technological Knowledge

A primary obstacle for farmers in rural areas is insufficient technological knowledge. This challenge is widespread, as most farmers in underdeveloped nations lack formal education. Consequently, implementing IoT solutions poses significant challenges, particularly due to the substantial investment required to train farmers before establishing IoT infrastructure (Elijah et al. 2017; Sinha and Dhanalakshmi 2021).

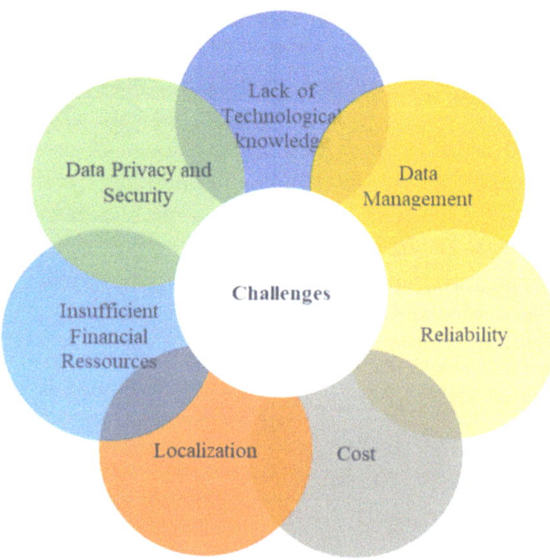

Fig. 7.5 Challenges of IoT in the agricultural sector

7.5.2 Data Management

The organization and manipulation of sensor-collected data present significant challenges for farmers. Although weather stations generate substantial data, many farmers lack the knowledge to effectively utilize or convert this information into accessible formats. The complexity of these systems often leads to computational inaccuracies and issues related to data acceptability and usability.

7.5.3 Reliability

IoT devices utilized in agricultural practices are frequently exposed to harsh environmental conditions in open settings, which may result in communication failures and sensor malfunctions. Consequently, ensuring the physical protection of deployed IoT devices and sensors against adverse weather conditions is essential (Asikainen et al. 2013).

7.5.4 Insufficient Financial Resources

In instances where farmers do not achieve the predicted yield due to unforeseen environmental disasters such as drought, floods, or pest infestations, financial institutions may provide funds or loans to help farmers resume agricultural activities.

7.5.5 Cost

The implementation and deployment of IoT technologies present several financial challenges, encompassing both initial setup costs (such as IoT sensors, gateways, and base station infrastructure) and ongoing operational expenses (including information sharing, service subscriptions, and centralized data collection services). Although current technologies offer rapid and accurate performance with minimal labor requirements, their adoption remains limited. This limitation is particularly evident in regions where poverty persists, and the agricultural sector relies heavily on manual labor as a primary source of income. Farmers seeking to transition from traditional agricultural practices to advanced technological solutions often face significant financial barriers, as the costs of these devices and technologies are frequently prohibitive (Farooq et al. 2020, 2022; Elijah et al. 2018).

7.5.6 Infrastructure for Telecommunication

Most agricultural activities occur in rural areas, where arable land is preferable to contaminated soil. However, inadequate telecommunications infrastructure often leads to unreliable data transmission when using smart devices such as smartphones and tablets. Smart farming depends on a stable internet connection to transmit real-time data. Furthermore, various control system operations, including the management of seed quantities, fertilizers, pesticides, and water, require a consistent and high-quality internet connection. Although the recent proliferation of mobile phones has increased access to mobile internet in rural regions, limitations in speed and signal quality persist.

7.5.7 Localization

The deployment of IoT technologies requires consideration of multiple factors. Devices and sensors must be capable of supporting nationwide operations without necessitating additional installations or complex configurations. Selecting optimal deployment locations is essential to maintain continuous device communication and effective information sharing (Biral et al. 2015).

7.5.8 Interoperability and Data Integration

The four principal components of interoperability are semantic, technological, syntactic, and organizational. Semantic interoperability concerns the understanding and interpretation of information shared among users. Technological interoperability enables efficient interaction among IoT devices, encompassing both hardware and software, as well as communication protocols and infrastructure. Syntactic interoperability addresses data structures and types, facilitating electronic data sharing. Organizational interoperability involves frameworks and rules that ensure effective data transfer across diverse technologies and locations (Perera et al. 2014). Significant challenges arise regarding data compatibility and standardization when integrating information from heterogeneous sources, such as sensors, weather stations, machinery, and satellites. Data incompatibility often results from differences in data formats and communication protocols, impeding data integration and exchange. Consequently, robust data management and integration tools are essential for connecting multiple systems and consolidating data into a unified platform for efficient analysis and informed decision-making (Ahmed and Shakoor 2025; Patel and Patel 2016).

7.5.9 Scalability and Adaptability

Precision agriculture technologies are predominantly developed for large-scale commercial farms and may not be easily adapted or directly implemented in smallholder or subsistence farming contexts. For small organizations, the potential benefits of adopting modern technologies are often offset by their complexity and the associated costs. Scalability challenges also arise in data management and processing. Farmers and service providers frequently encounter difficulties in expanding their infrastructure and expertise to efficiently manage the substantial volumes of agricultural data (Al-Fuqaha et al. 2015; Misra 2016).

7.5.10 Data Privacy and Security

Agricultural data generated by these technologies, such as yield projections, field maps, and sensor readings, raises significant questions regarding data ownership and control. Farmers often hesitate to share their data with third parties or external platforms due to concerns about privacy breaches and potential competitive disadvantages. Additionally, ambiguity surrounding data rights, usage, and monetization frequently arises from unclear agreements between farmers and technology providers. This lack of clarity undermines confidence and discourages cooperation in data sharing within the agriculture industry (Asplund and Nadjm-Tehrani 2016; Chen et al. 2017; Varga et al. 2017; Duan et al. 2014; Newell et al. 2015).

7.5.11 Cybersecurity Risks

Significant risks are associated with precision agriculture systems, including unauthorized access, susceptibility to cyberattacks, and data breaches. Precision farming relies on interconnected devices and cloud platforms, which are vulnerable to cybercriminals seeking to steal confidential data or disrupt operations. Agricultural data may be exposed to theft and tampering if network security, encryption, and data protection measures are inadequate. Data breaches can have severe consequences for farmers, undermining their competitiveness and damaging their reputation. While precision farming technologies offer substantial benefits for agricultural operations, they also introduce notable challenges. Addressing technological, financial, and security concerns is essential to fully realize the advantages of precision farming and to ensure its equitable and large-scale implementation.

7.6 Benefits of Smart Farming

The integration of smart farming in agriculture offers several advantages (Fig. 7.6). These include the following: (i) reduced costs associated with pesticides, insecticides, and water consumption; (ii) increased productivity through the adoption of advanced technologies such as robotics and drones, which minimize manual labor; (iii) improved crop quality and yield through resource optimization and effective crop management; (iv) enhanced decision-making capabilities, enabling farmers to make informed choices regarding crops and livestock by utilizing ML algorithms; and (v) greater sustainability, achieved by mitigating negative environmental impacts such as greenhouse gas emissions and water pollution (Idoje et al. 2021; Alwis et al. 2022; Elbasi et al. 2024; Badoni et al. 2023; Riskhan et al. 2024; Digra et al. 2025).

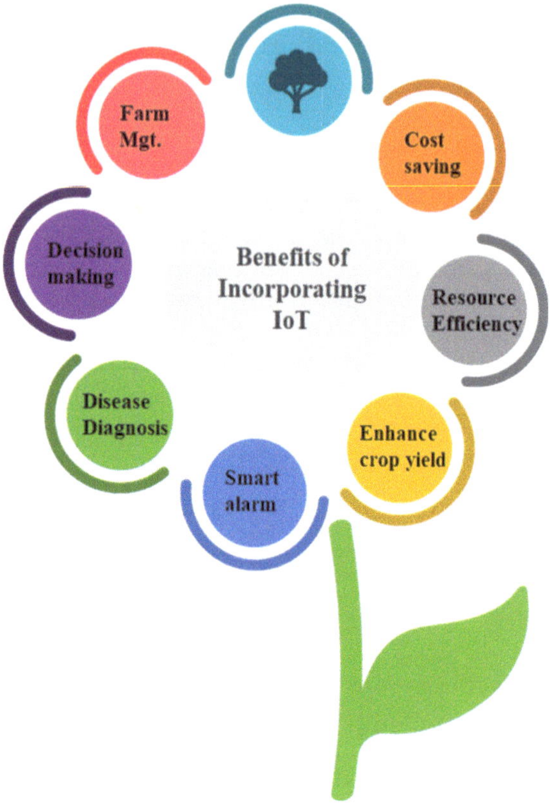

Fig. 7.6 Benefits of integrating IoT in the agricultural sector

7.7 Conclusion

This chapter has traced the evolution of agriculture from manual labor (1.0) to the data-driven optimization of precision farming. This transformation is enabled by a sophisticated ecosystem in which diverse sensors, including pH and nutrient detectors as well as weather and soil moisture monitors, serve as critical data sources for the farm. The revolutionary impact emerges from the synthesis of these granular data points through advanced technologies. Tools such as remote sensing, Big Data analytics, and machine vision allow farmers to move beyond guesswork and implement highly specific, real-time strategies. These advancements yield measurable benefits, including significant reductions in waste (water, fertilizer, and pesticides), improved crop quality, enhanced resource efficiency, and positive environmental outcomes. The integration of the IoT in precision farming represents more than a technological enhancement; it constitutes a foundational investment in global resilience. By maximizing resource efficiency and promoting sustainable productivity, smart farming contributes to achieving global objectives, notably the EU Green Deal's climate targets and the United Nations' Sustainable Development Goal of Zero Hunger. While the technology is available, the subsequent priority is to foster a collaborative environment that ensures this precision revolution is accessible and equitable for farms of all sizes.

References

Abbas I, Liu J, Faheem M et al (2020) Different sensor based intelligent spraying systems in agriculture. Sens Actuators A Phys 316:112265. https://doi.org/10.1016/j.sna.2020.112265

Ahmed N, Shakoor N (2025) Advancing agriculture through IoT, big data, and AI: a review of smart technologies enabling sustainability. Smart Agr Technol 10:100848. https://doi.org/10.1016/j.atech.2025.100848

Alahmad T, Neményi M, Nyéki A (2023) Applying IoT sensors and big data to improve precision crop production: a review. Agronomy 13(10):2603. https://doi.org/10.3390/agronomy13102603

Al-Fuqaha A, Guizani M, Mohammadi M et al (2015) Internet of things: a survey on enabling technologies, protocols, and applications. IEEE Commun Surv Tutor 17(4):2347–2376. https://doi.org/10.1109/COMST.2015.2444095

Alwis SD, Hou Z, Zhang Y et al (2022) A survey on smart farming data, applications and techniques. Comput Ind 138:103624. https://doi.org/10.1016/j.compind.2022.103624

Arora P, Khalandar Basha D, Kumar VSRSP (2026) Evolution of AI in smart agriculture. In: Poonia RC, Thakur D, Thakur I (eds) Intelligent systems for sustainable infrastructure: AI solutions shaping a green future. Engineering cyber-physical systems and critical infrastructures, vol 17. Springer, Cham. https://doi.org/10.1007/978-3-031-95240-1_2

Asikainen M, Haataja K, Toivanen P (2013) Wireless indoor tracking of livestock for behavioural analysis. In: Proceedings of the 9th international wireless communications and mobile computing conference (IWCMC), Sardinia. IEEE, pp 1833–1838

Asplund M, Nadjm-Tehrani S (2016) Attitudes and perceptions of IoT security in critical societal services. IEEE Access 4:2130–2138. https://doi.org/10.1109/ACCESS.2016.2560919

Ayaz M, Ammad-Uddin M, Sharif Z et al (2019) Internet-of-things (IoT)-based smart agriculture: toward making the fields talk. IEEE Access 7:129551–129583. https://doi.org/10.1109/ACCESS.2019.2932609

Badoni P, Kaur S, Sharma B et al (2023) Enhancing water efficiency and crop yield in agriculture sector using IoT. In: Proceedings of the international conference on advances in computation, communication and information technology (ICAICCIT). IEEE, pp 1039–1044

Biral A, Centenaro M, Zanella A et al (2015) The challenges of M2M massive access in wireless cellular networks. Digit Commun Netw 1(1):1–19. https://doi.org/10.1016/j.dcan.2015.02.001

Bouni M, Hssina B, Douzi K et al (2024) Integrated IoT approaches for crop recommendation and yield prediction using machine learning. IoT 5(4):634–649. https://doi.org/10.3390/iot5040028

Boursianis AD, Papadopoulou MS, Diamantoulakis P et al (2022) Internet of things (IoT) and agricultural unmanned aerial vehicles (UAVs) in smart farming: a comprehensive review. Internet Things 18:100187. https://doi.org/10.1016/j.iot.2020.100187

Burton L, Jayachandran K, Bhansali S (2020) The "real-time" revolution for in situ soil nutrient sensing. J Electrochem Soc 167(3):037569. https://doi.org/10.1149/1945-7111/ab6f5d

Chen L, Thombre S, Järvinen K et al (2017) Robustness, security and privacy in location-based services for future IoT: a survey. IEEE Access 5:8956–8977. https://doi.org/10.1109/ACCESS.2017.2695525

Dhal S, Wyatt BM, Mahanta S et al (2024) Internet of things (IoT) in digital agriculture: an overview. Agron J 116(3):1144–1163. https://doi.org/10.1002/agj2.21385

Digra M, Rajput P, Sharma P (2025) Advances in smart farming for precision agriculture: green-IoT and machine learning as a solution. In: Next generation computing and information systems. CRC Press, London, pp 49–56

Duan J, Gao D, Yang D et al (2014) An energy-aware trust derivation scheme with game theoretic approach in wireless sensor networks for IoT applications. IEEE Internet Things J 1(1):58–69. https://doi.org/10.1109/JIOT.2014.2314132

Duguma AL, Bai X (2025) How the internet of things technology improves agricultural efficiency. Artif Intell Rev 58:63. https://doi.org/10.1007/s10462-024-11046-0

Dutta L, Bharali S, Barman P et al (2024) An IoT-enabled smart pH monitoring and dispensing system for precision agriculture application. Agric Res 13:309–318. https://doi.org/10.1007/s40003-024-00702-0

Elbasi E, Mostafa N, Zaki C et al (2024) Optimizing agricultural data analysis techniques through AI-powered decision-making processes. Appl Sci 14(17):8018. https://doi.org/10.3390/app14178018

Elijah O, Orikumhi I, Rahman TA et al (2017) Enabling smart agriculture in Nigeria: application of IoT and data analytics. In: Proceedings of the IEEE 3rd international conference on electro-technology for national development (NIGERCON), Owerri. IEEE, pp 762–766

Elijah O, Rahman TA, Orikumhi I et al (2018) An overview of internet of things (IoT) and data analytics in agriculture: benefits and challenges. IEEE Internet Things J 5(5):3758–3773. https://doi.org/10.1109/JIOT.2018.2844296

Farooq MS, Riaz S, Abid A et al (2020) Role of IoT technology in agriculture: a systematic literature review. Electronics 9(2):319. https://doi.org/10.3390/electronics9020319

Farooq MS, Riaz S, Helou MA et al (2022) Internet of things in greenhouse agriculture: a survey on enabling technologies, applications, and protocols. IEEE Access 10:53374–53397. https://doi.org/10.1109/ACCESS.2022.3166634

Freidenreich A, Barraza G, Jayachandran K et al (2019) Precision agriculture application for sustainable nitrogen management of *Justicia brandegeana* using optical sensor technology. Agriculture 9(5):98. https://doi.org/10.3390/agriculture9050098

Friha O, Ferrag MA, Shu L et al (2021) Internet of things for the future of smart agriculture: a comprehensive survey of emerging technologies. IEEE/CAA J Autom Sin 8(4):718–752. https://doi.org/10.1109/JAS.2021.1003925

Gupta S, Geetha A, Sankaran KS et al (2022) Machine learning- and feature selection-enabled framework for accurate crop yield prediction. J Food Qual 2022:6293985. https://doi.org/10.1155/2022/6293985

Huang K, Shu L, Li KL et al (2020) Photovoltaic agricultural internet of things towards realizing the next generation of smart farming. IEEE Access 8:76300–76312. https://doi.org/10.1109/ACCESS.2020.2988663

Idoje G, Dagiuklas T, Iqbal M (2021) Survey for smart farming technologies: challenges and issues. Comput Electr Eng 92:107104. https://doi.org/10.1016/j.compeleceng.2021.107104

Ikram A, Aslam W, Hama Aziz RH et al (2022) Crop yield maximization using an IoT-based smart decision. J Sens 2022:2022923. https://doi.org/10.1155/2022/2022923

Karunathilake EMBM, Le AT, Heo S et al (2023) The path to smart farming: innovations and opportunities in precision agriculture. Agriculture 13:1593. https://doi.org/10.3390/agriculture13081593

Kumar V, Sharma KV, Kedam N et al (2024) A comprehensive review on smart and sustainable agriculture using IoT technologies. Smart Agr Technol 8:100487. https://doi.org/10.1016/j.atech.2024.100487

Kuzubasoglu BA, Bahadir SK (2020) Flexible temperature sensors: a review. Sens Actuators A Phys 315:112282. https://doi.org/10.1016/j.sna.2020.112282

Leone M (2022) Advances in fiber optic sensors for soil moisture monitoring: a review. Results Opt 7:100213. https://doi.org/10.1016/j.rio.2022.100213

Liu Y, Ma XY, Shu L et al (2020) From industry 4.0 to agriculture 4.0: current status, enabling technologies, and research challenges. IEEE Trans Ind Inform 17:4322–4334. https://doi.org/10.1109/TII.2020.3003910

Mahan JR, Conaty W, Neilsen J et al (2010) Field performance in agricultural settings of a wireless temperature monitoring system based on a low-cost infrared sensor. Comput Electron Agric 71(2):176–181. https://doi.org/10.1016/j.compag.2010.01.005

Maheswary A, Nagendram S, Kiran KU et al (2024) Intelligent crop recommender system for yield prediction using machine learning strategy. J Inst Eng India Ser B 105:979–987. https://doi.org/10.1007/s40031-024-01029-8

Miller T, Mikiciuk G, Durlik I et al (2025) The IoT and AI in agriculture: the time is now—a systematic review of smart sensing technologies. Sensors 25(12):3583. https://doi.org/10.3390/s25123583

Misra P (2016) Build a scalable platform for high-performance IoT applications. Technical report, TCS Exper Certainty, Mumbai

Navulur S, Prasad MG (2017) Agricultural management through wireless sensors and internet of things. Int J Electr Comput Eng 7(6):3492–3499. https://doi.org/10.11591/ijece.v7i6.pp3492-3499

Newell A, Yao H, Ryker A et al (2015) Node-capture resilient key establishment in sensor networks: design space and new protocols. ACM Comput Surv 47:1–34. https://doi.org/10.1145/2636344

Nielsen J, Van de Beek C, Thorndahl S et al (2024) Merging weather radar data and opportunistic rainfall sensor data to enhance rainfall estimates. Atmos Res 300:107228. https://doi.org/10.1016/j.atmosres.2024.107228

Patel KK, Patel SM (2016) Internet of things (IoT): definition, characteristics, architecture, enabling technologies, application and future challenges. Int J Eng Sci Comput 6:6122–6131. https://doi.org/10.4010/2016.1482

Patle KS, Saini R, Kumar A et al (2021) IoT enabled leaf wetness sensor on flexible substrates for in-situ plant disease management. IEEE Sensors J 21(17):19481–19491. https://doi.org/10.1109/JSEN.2021.3089722

Perera C, Liu CH, Jayawardena S et al (2014) A survey on internet of things from industrial market perspective. IEEE Access 2:1660–1679. https://doi.org/10.1109/ACCESS.2015.2389854

Raj EF, Appadurai M, Athiappan K (2021) Precision farming in modern agriculture. In: Advances in smart farming technologies. Springer, Singapore, pp 61–87. https://doi.org/10.1007/978-981-16-6124-2_4

Rajak P, Ganguly A, Adhikary S et al (2023) Internet of things and smart sensors in agriculture: scopes and challenges. J Agric Food Res 14:100776. https://doi.org/10.1016/j.jafr.2023.100776

Rana V, Sharma S, Thakur KK et al (2023) Internet of things in livestock farming: implementation and challenges. https://doi.org/10.21203/rs.3.rs-2559126/v1

Riskhan B, Hanass N, Sheikh MAU et al (2024) IoT irrigation system using Arduino. https://doi.org/10.20944/preprints202405.1642.v1

Seetha J, Nruthya K, Rao NKK et al (2023) Enhancement of agriculture productivity and automation of agriculture using IoT. In: Proceedings of the international conference on inventive computation technologies (ICICT). IEEE, pp 1411–1417

Selvam A, Saud S (2023) The impact of IoT and sensor integration on real-time weather monitoring systems: a systematic review. https://doi.org/10.21203/rs.3.rs-3579172/v1

Shimonomura K (2019) Tactile image sensors employing camera: a review. Sensors 19(18):3933. https://doi.org/10.3390/s19183933

Singh RK, Berkvens R, Weyn M (2021) AgriFusion: an architecture for IoT and emerging technologies based on a precision agriculture survey. IEEE Access 9:136253–136283. https://doi.org/10.1109/ACCESS.2021.3116814

Sinha BB, Dhanalakshmi R (2021) Recent advancements and challenges of internet of things in smart agriculture: a survey. Futur Gener Comput Syst 126:169–184. https://doi.org/10.1016/j.future.2021.08.006

Sravanthi G, Moparthi NR (2024) An efficient IoT based crop disease prediction and crop recommendation for precision agriculture. Clust Comput 27:5755–5782. https://doi.org/10.1007/s10586-023-04246-w

Steeneken PG, Kaiser E, Verbiest GJ et al (2023) Sensors in agriculture: towards an internet of plants. Nat Rev Methods Primers 3(1):60. https://doi.org/10.1038/s43586-023-00250-x

Tzounis A, Katsoulas N, Bartzanas T et al (2017) Internet of things in agriculture, recent advances and future challenges. Biosyst Eng 164:31–48. https://doi.org/10.1016/j.biosystemseng.2017.09.007

Varga P, Plosz S, Soos G et al (2017) Security threats and issues in automation IoT. In: Proceedings of the IEEE 13th international workshop on factory communication systems (WFCS), Trondheim, Norway. IEEE, pp 1–6

Wijesinghe RE, Lee SY, Kim P et al (2017) Optical sensing method to analyze germination rate of *Capsicum annum* seeds treated with growth-promoting chemical compounds using optical coherence tomography. J Biomed Opt 22(9):091502. https://doi.org/10.1117/1.JBO.22.9.091502

Sensor Networks and Real-Time Farm Data Acquisition

8

Ruchika Bindal, Mandeep Kaur, and Righa Tandon

Abstract

The integration of sensor networks and real-time data acquisition systems serves as a transformative foundation for modern precision agriculture by enabling continuous monitoring of soil, crop, environmental, livestock, and machinery parameters. This chapter analyzes the technological framework and operational principles reinforcing real-time farm data collection, with a focus on Internet of Things (IoT)-enabled sensing platforms, multi-layered communication networks, and data-driven decision-support systems. The functional roles of various sensors, including those for soil moisture, pH, temperature, nutrients, humidity, rainfall, radiation, and crop health, are examined in capturing high-resolution spatial and temporal data essential for adaptive farm management. The architecture of real-time agricultural monitoring systems is described using a layered model that integrates sensor nodes, edge devices, communication gateways, cloud-based storage, and artificial intelligence and machine learning analytics. Key operational parameters, such as latency, reliability, and accuracy, are assessed to evaluate system performance and data integrity. This chapter also details the processes of data sampling, synchronization, transmission, preprocessing, calibration, and storage within distributed edge, fog, and cloud computing environments. Case studies on precision irrigation, greenhouse monitoring, and large-scale farm data acquisition demonstrate the practical benefits of sensor-enabled agriculture in enhancing resource-use efficiency, crop productivity, and operational automation. Finally, critical challenges such as network connectivity constraints, sensor durability, data security, and cost scalability are addressed, underscoring the need for robust, affordable, and secure technological solutions to achieve sustainable smart farming.

R. Bindal (✉) · M. Kaur · R. Tandon
Chitkara University Institute of Engineering and Technology, Chitkara University, Rajpura, Punjab, India

H. N. Oo et al. (eds.), *Transforming Agriculture with AI, IoT, and Blockchain*,
https://doi.org/10.1007/978-981-92-1491-4_8

Keywords

IoT in agriculture · Real-time farm data acquisition · Precision agriculture · Sensor networks in smart farming · Edge–Fog–Cloud computing for agriculture

8.1 Introduction

Technological advancements have driven a rapid transition from manual agricultural practices to technology-based approaches. This shift has improved productivity and facilitated more sustainable resource management. Advanced sensors now enable the collection of detailed field data in real time. The recorded information is subsequently analyzed to assess various crop and soil parameters. Based on real-time data and advanced analytics, farmers can make informed decisions regarding agricultural inputs such as water and nutrients. Additionally, these analytics offer evidence-based recommendations for crop selection.

8.1.1 Need for Real-Time Information for Data-Driven Smart Farming

Recent years have witnessed significant advancements in agricultural technology. The integration of data analytics and real-time control has facilitated a shift from labor-intensive traditional farming to smart farming systems. The real-time collection of data and the deployment of sensor networks now form the foundation of smart farming (Shahab et al. 2025). These technologies enable continuous monitoring and reporting, allowing farmers to make timely, effective, and resource-efficient decisions. Escalating global challenges in food production, environmental sustainability, and climate change have highlighted the necessity for adaptive and resilient farming systems that optimize resource utilization (Prasanna et al. 2024).

8.1.2 Benefits of Sensor-Based Farming

The automation and enhanced visibility provided by sensor networks in agricultural environments facilitate the integration of precision and monitoring. These networks enable automated precision irrigation systems to optimize water usage and improve irrigation efficiency by accurately determining the timing and location of water application. Real-time monitoring of soil nutrients further contributes to improved soil health and minimizes adverse effects during irrigation and fertilization processes (Shahab et al. 2025). Predictive modeling and timely interventions are expected to significantly reduce agricultural losses and enhance operational efficiency. The implementation of sensor networks is anticipated to promote greater autonomy and sustainability in agriculture (Colaço et al. 2021).

8.1.3 Fundamentals of Real-Time Data Acquisition

Real-time data collection forms the foundation of the modern smart agriculture paradigm. This approach employs networked sensing tools to automatically and continuously gather data on crops, soil, and environmental conditions. Unlike traditional systems, which rely on manual and infrequent on-site data collection, real-time acquisition systems capture data as soon as relevant events occur. This immediacy enables efficient responses to rapidly changing field conditions and enhances both the accuracy and efficiency of agricultural operations. As a result, farmers and decision-support systems can respond promptly to changes in field conditions (Chourlias et al. 2025).

8.1.4 Real-Time Data Collection Concepts

A network of distributed sensor nodes enables real-time data collection through the continuous monitoring of key agricultural and environmental variables. Each sensor records specific variables at programmed intervals and transmits the data using cellular protocols, Wi-Fi, LoRaWAN, or Zigbee. The process includes data acquisition, transfer to a primary gateway, and subsequent processing, analysis, and visualization on cloud or edge computing platforms (Dahiphale et al. 2025). The primary objective of real-time data gathering is to minimize the latency between data collection and access. Timely availability of information allows farmers or automated systems to implement preventive measures. For example, irrigation systems can be activated, or automated alerts can notify field operators of potential pest activity. Continuous data collection and monitoring facilitate enhanced precision agriculture, thereby improving farmers' decision-making processes and outcomes (Cheema and Pires 2025).

8.1.5 Key Performance Parameters

The effectiveness of any monitoring system depends on the accurate measurement of performance parameters, which is achieved through efficient real-time data acquisition.

Latency It is inferred to be the time difference between the availability of data and the point at which it becomes usable. In agricultural contexts, latency is typically low due to the need for rapid responses, such as adjusting irrigation levels to prevent crop stress and addressing severe environmental conditions. In these scenarios, data from high-latency systems is often irrelevant because the information is outdated.

Reliability Consistency and trustworthiness in data communication and overall system performance are critical in remote agricultural systems. Sensors in these environments are exposed to adverse weather conditions, power loss, and interference from wildlife. A reliable system ensures that data remains operational and that the system functions without interruptions or data gaps, a concept known as operational continuity. In such systems, sensors are expected to operate continuously for extended periods, providing uninterrupted data streams. Enhancing system reliability, operational continuity, and data accessibility can be achieved through redundancy, robust communication protocols, and fault-tolerant designs.

Accuracy Accuracy describes the degree to which sensor readings correspond to the true value of the measured variable. In applications such as disease prediction and diagnosis, fertilizer optimization, and water resource management, high accuracy is essential. Data quality below a certain threshold can result in incorrect decisions, including excessive or insufficient irrigation, nutrient deficiencies, and delays in disease detection. Maintaining accuracy over time requires high-quality sensing elements, regular calibration, and adaptive algorithms. In agriculture, the combined latency, reliability, and accuracy of real-time data capture determine the value and applicability of sensor systems. Achieving these standards in sensor monitoring enables the agricultural industry to enhance efficiency, sustainability, and productivity for farmers.

8.2 Sensors for Real-Time Farm Monitoring

Instantaneous farming surveillance relies on a comprehensive array of sensors that monitor soil, environmental, crop, animal, and equipment parameters. These devices form the foundation of modern digital agriculture, enabling both farmers and automated systems to respond to real-time observations. By converting physical phenomena into quantifiable data, sensors provide a comprehensive overview of operational variables and available resources. This section discusses the primary types of sensors used in smart farming and examines their contributions to precision agriculture, sustainable farming practices, and operational efficiency on farms (Dahiphale et al. 2025; Marjiya et al. 2025). Figure 8.1 illustrates various sensors employed in real-time farming.

8.2.1 Soil Sensors

Soil sensors are essential for understanding subsurface conditions that influence plant growth and crop yields. These devices enable the optimization of irrigation, fertilizer application, and soil management practices (Kashyap and Kumar 2021a).

8.2.1.1 Soil Moisture Sensors

Soil moisture sensors measure the water content within the soil profile. Examples of such technology include capacitance sensors, tensiometers, and time-domain reflectometry (TDR).

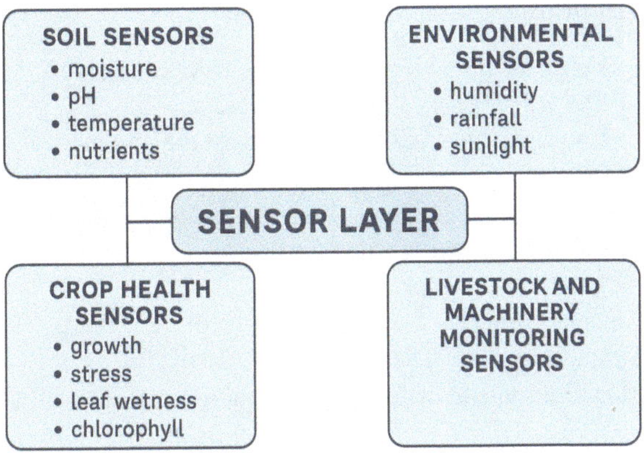

Fig. 8.1 Sensors for real-time farming

Applications

- Scheduling irrigation with precision
- Avoid under- and over-irrigation
- Maintain moisture levels according to crop requirements

8.2.1.2 Soil pH Sensors

These sensors measure soil acidity and alkalinity. Soil pH has a considerable impact on the availability of plant nutrients, soil biota, and overall soil fertility (Niranjan 2018).

Applications

- Directing lime or sulfur applications
- Effective pH control for desired nutrient absorption
- Long-term soil health monitoring

8.2.1.3 Soil Temperature Sensors

Soil temperature has an impact on the rate of seed germination, root respiration, and soil biota, as well as the rate of release of nutrients contained in the soil (Kashyap and Kumar 2021a).

Applications

- Determining the best time to sow
- Understanding thermal conditions of the root zone
- Assist in greenhouse and nursery activities

8.2.1.4 Soil Nutrient Sensors

These sensors measure some primary nutrients (N, P, K) and some micronutrients (Burton et al. 2020).

Applications

- Fertilization with precision to outstanding levels
- Early detection of nutrient deficiencies
- Improved control of fertilizer wastage and nutrient-rich runoff

8.2.2 Environmental Sensors

Farmers can track the microclimates of their fields with environmental sensors and weather stations. They are especially useful for measuring factors such as evapotranspiration and estimating irrigation needs, as well as for estimating pest activity.

8.2.2.1 Humidity Sensors

Measure relative humidity and the moisture content in the air (Huang et al. 2025).

Applications

- Disease forecasting, especially fungal infections
- Evapotranspiration determination
- Greenhouse management and ventilation

8.2.2.2 Rainfall Sensors

Include precipitation detectors and rain gauges that record the amount and intensity of rain in real-time (Kashyap and Kumar 2021a).

Applications

- Irrigation and flood risk management
- Water balance assessment in fields

8.2.2.3 Sunlight (Solar Radiation) Sensors

Measure the solar energy in the environment along with photosynthetically active radiation (PAR) (Belkher et al. 2025).

Applications

- Estimates potential photosynthesis
- Determines optimal greenhouse shading
- Predicts the growth rate of the crops

8.2.2.4 CO_2 Sensors

Measure the concentration of CO_2 gas in the air, which is especially important in closed environments like greenhouses.

Applications

- Improves the ventilation system
- Increases photosynthesis
- Greater yield in protected farming systems

8.2.3 Crop Health Sensors

Sensor technologies can identify plant stress before symptoms appear and capture patterns in growth and disease. Optical, spectral, and thermal technologies are used for these types of sensors.

8.2.3.1 Growth Sensors

These sensors use ultrasonic and LiDAR technologies to capture plant height, canopy area, and biomass.

Applications

- Monitoring growth rates
- Estimating yields
- Targeted variable rate spraying and harvesting

8.2.3.2 Stress Detection Sensors

These sensors capture temperature, water stress, and nutrient deficiency imbalances using infrared and hyperspectral imaging (Kashyap and Kumar 2021b).

Applications

- Detecting disease at the hierarchy of the system
- Detecting stress from drought
- Modifying controlled parameters of irrigation and fertilization

8.2.3.3 Leaf Wetness Sensors

These sensors capture moisture from the surfaces of foliage and are predictive of disease outbreaks such as blight and mildew (Huang et al. 2025).

Applications

- Forecasting the risk of fungal infections
- Timing the application of fungicides
- Supporting decision support systems

8.2.3.4 Chlorophyll Sensors

These sensors are chlorophyll meters and NDVI sensors used to estimate plant chlorophyll content. This indicates the overall vigor of the plant and the nutrient status (Huang et al. 2025).

Applications

- Estimating nitrogen needs
- Mapping the spatial variability of crop health
- Precision nutrient application improvement

8.2.4 Livestock and Machinery Monitoring Sensors

Modern farms comprise more than just fields of crops—they also include livestock and machinery, which also require constant oversight (Bernabucci et al. 2025).

8.2.4.1 Livestock Monitoring Technology

Wearable livestock monitoring technology (RFID tags, accelerometers, GPS collars, temperature biosensors) tracks the livestock's movement, behavioral patterns, and overall health (Bernabucci et al. 2025).

Applications

- Early detection of illness and heat stress
- Monitoring individual feeding
- Locating livestock on extensive farms
- Assistance with automated milking and feeding systems

8.2.4.2 Machinery Monitoring Technology

Tractors, harvesters, and sprayers are equipped with technology to monitor machinery performance, fuel consumption, engine status, and overall functioning. Common examples are GPS modules, vibration sensors, fuel sensors, and implement load sensors (Colaço et al. 2021).

Applications

- Preventive maintenance
- Improved field coverage and route optimization
- Reduced fuel wastage
- Better automation with more intelligent machinery

8.3 System Architecture for Real-Time Data Collection

A system architecture has been developed to enable real-time soil monitoring and artificial intelligence (AI) or machine learning (ML)-based decision-making for crop, fertilizer, and irrigation management. Smart sensors continuously monitor soil and acquire real-time data. The collected data is stored in the cloud, where it undergoes preprocessing and analysis to support ML applications and optimize the deployment of agricultural resources (Marjiya et al. 2025).

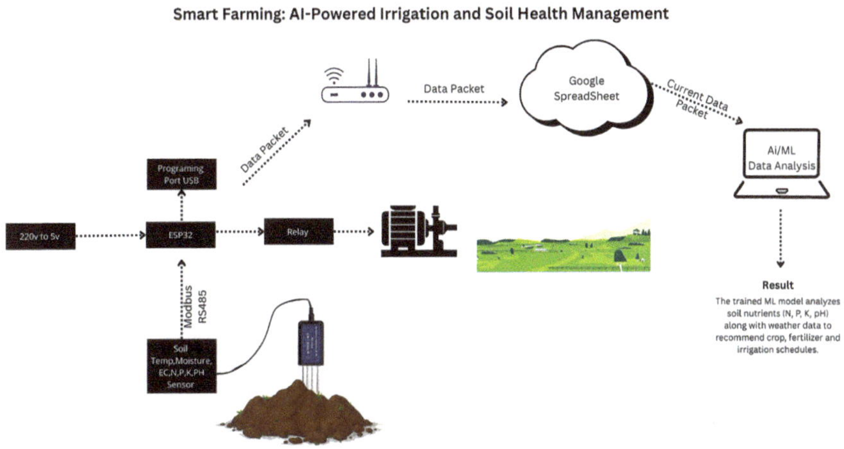

Fig. 8.2 IoT-integrated soil health and automated irrigation system architecture

8.3.1 Sensor Node Design and Deployment Strategies

The proposed architecture utilizes a layered approach. Within the soil monitoring layer, the system comprises three primary steps: architecture deployment, data acquisition, and crop recommendation. A multiparameter soil sensor is used to monitor soil conditions. The sensor records temperature, moisture content, electrical conductivity (EC), macronutrients (N, P, K), and pH, which collectively indicate soil status and hydric balance. The sensor connects to a computer using the Modbus RS-485 protocol, which is suitable for long-distance communication and noisy agricultural environments (Chourlias et al. 2025). Figure 8.2 presents the "IoT-Integrated Soil Health and Automated Irrigation System Architecture," illustrating the complete workflow from soil sensing and data transmission to AI-driven analysis and automated irrigation control.

The ESP32 microcontroller collects sensor data at predetermined intervals and serves as an edge-computing node. To ensure continuous operation and reliability, a relay module receives both AC 220 V and DC 5.0 V power supplies. A designated GPIO pin on the ESP32 controls the activation of a water pump according to preset instructions or AI-generated recommendations.

The ESP32 transmits data packets to a nearby wireless router using its onboard Wi-Fi module. The router forwards the data to the cloud, where it is stored in a Google Spreadsheet. This platform serves as a lightweight, cloud-based data logging solution that maintains a sequential record of sensor data and system actions. The tabular data format facilitates interoperability with other applications, enhances system visualization, and enables straightforward inspection by farmers and researchers (Ilyasu et al. 2025).

The AI and ML data analysis module runs on a computer or server and processes information stored in the cloud. The most recent sensor readings are integrated with soil and weather data, including rainfall, temperature, humidity, and their forecasts.

This combined dataset is analyzed using an optimized ML algorithm to generate soil condition-adaptive crop recommendations, appropriate fertilizer types and dosages (N, P, K, and pH balance), irrigation requirements based on soil moisture, and a calibrated weather forecast for irrigation (Chourlias et al. 2025; Kashyap and Kumar 2021a).

The results of the AI and ML analysis are presented to the user via a dashboard or application. The interface displays recommended adjustments, such as irrigation schedules (e.g., irrigate for 20 min), alerts (e.g., apply fertilizer due to low nitrogen levels), and suggestions for optimal crop selection for the season. These recommendations can be implemented manually by the farmer or automatically executed by the ESP32, thereby completing the data-driven smart farming system loop.

This architecture integrates low-cost sensors, an ESP32-based edge device, field communication, cloud-based data logging using Google Spreadsheet, and AI/ML analysis. Together, these components provide a comprehensive solution for real-time soil data collection and advanced farm management.

8.4 Data Acquisition Process

Data collection in smart farming begins with determining the appropriate sampling frequency and synchronizing sensor nodes distributed throughout the field. Sampling frequency specifies how often a sensor records data and must be optimized for each data stream. For instance, soil moisture and temperature require more frequent sampling because they are significant for precision irrigation, whereas parameters such as rainfall or less critical nutrient data can be sampled less often. Constructing meaningful data sets depends on achieving a degree of synchronization among sensors. Misalignment in sensor data can lead to incorrect interpretations. Therefore, real-time sampling and synchronized timestamps are essential for accurate agricultural data collection (Musanase et al. 2023). The data acquisition and processing architecture is illustrated in Fig. 8.3.

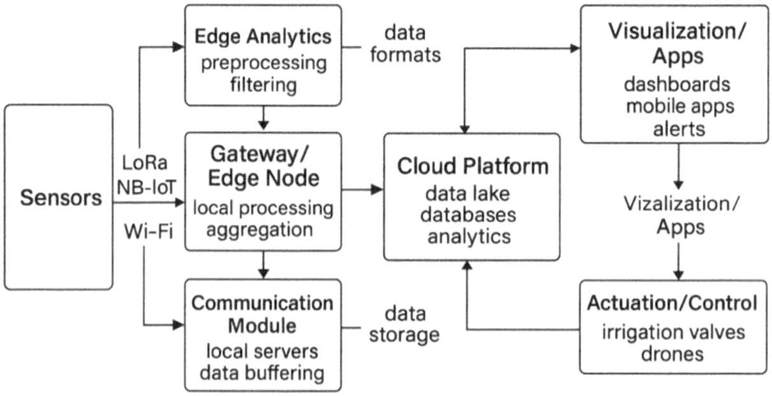

Fig. 8.3 Schematic representation of the data acquisition and processing architecture

8.4.1 Data Sampling Frequency and Synchronization

Data sampling frequency refers to the intervals at which a sensor collects and records information from its environment. This frequency is critical to the operational efficiency of sensors in real-time agricultural monitoring systems. In contemporary smart agriculture, parameters such as temperature, humidity, light, and soil moisture can fluctuate over varying durations, necessitating adjustments to the sampling interval. For example, during irrigation events, soil moisture may change significantly within hours, requiring a higher sampling frequency. In contrast, parameters like soil nutrient concentration typically change over longer periods, making less frequent sampling, such as weekly or monthly, sufficient. Selecting an appropriate sampling frequency is essential to ensure the system captures relevant changes without collecting redundant data. Overlapping samples should be minimized to avoid unnecessary information. Synchronization is also crucial for system efficiency. When collecting data from multiple sensors positioned at different locations, temporal alignment is necessary to preserve its value. Without synchronization, multisensor datasets may become inconsistent, potentially leading to erroneous decisions in irrigation scheduling, disease pattern prediction, or environmental analysis. Time synchronization is commonly achieved using Network Time Protocol (NTP) or timestamping gateways that manage incoming data (Musanase et al. 2023).

8.4.2 Real-Time Transmission and Streaming

Real-time transmission and data streaming transfer sensor information from the field to a processing unit, typically a computer or server, using Wi-Fi, LoRaWAN, Zigbee, cellular, or satellite connections. This continuous transfer enables farmers or automated systems to evaluate key metrics, identify issues, and implement corrective actions promptly. Real-time streaming is essential for applications requiring immediate responses, such as detecting sudden decreases in soil moisture, increases in temperature, or abnormal livestock movements. In contrast to systems that store data locally for later access, streaming systems eliminate record delays, thereby reducing the interval between sensor data collection and responsive decision-making. The selection of communication protocols depends on factors such as distance, bandwidth, power consumption, and land characteristics. For instance, low-power wide-area networks (LPWAN) are suitable for large farms where sensor nodes face energy and spatial constraints. Conversely, networks such as 4G and 5G are preferable for real-time crop monitoring because they can transmit large datasets, including images and videos. Effective real-time transmission also requires buffering and queueing mechanisms to prevent data loss during network downtime in loss-critical applications (Shahab et al. 2025).

8.4.3 Edge Versus Fog Versus Cloud Data Collection Approaches

Modern smart agriculture utilizes data collection across multiple computing levels: edge, fog, and cloud, each offering distinct advantages. Edge computing enables data collection and processing in close proximity to sensor nodes, resulting in negligible delay (Tandon and Gupta 2023). This approach is particularly beneficial for applications requiring immediate automated responses, such as irrigation, machine controllers, and safety systems. By performing local data computation, edge devices reduce transmission demands on communication networks and conserve energy, as only filtered or summarized data are forwarded to higher computing levels (Tandon and Gupta 2025).

Fog computing operates between edge devices and the cloud. Fog nodes offer greater computational, storage, and decision-making capabilities than farm-level sensors. This configuration reduces latency compared to cloud-only systems and provides more processing power than individual sensors. Fog computing is particularly suitable for medium- to high-complexity tasks, such as aggregating data from multiple fields, processing disease-risk predictive models, and coordinating large-scale sensor networks in real time (Wadhwa et al. 2024).

Cloud computing, positioned at the top tier of the structure, provides advanced analytics tools, scalable machine-learning capabilities, and extensive storage resources. The cloud is essential for long-term analysis, model training, identifying historical trends, and integrating with complex AI systems. Although cloud computing introduces increased latency due to processing distance, it excels at managing large volumes of data and delivering valuable insights at both farm and regional levels (Tandon et al. 2022). Together, edge, fog, and cloud computing create a layered architecture that balances speed, processing power, and scalability, enabling efficient and flexible data systems for diverse agricultural applications.

8.5 Data Processing and Storage

In real-time agricultural systems, raw data management begins at the sensors, also referred to as the edge, where initial readings are processed, formatted, and cleaned before being transmitted to the cloud or a central storage system for advanced computation. This preliminary filtering reduces the computational burden on the cloud, optimizes the quality of information transmitted for further analysis, and enhances system responsiveness to rapid atmospheric changes. At this stage, essential processes are performed, such as removing significant anomalies, data compaction, and converting analog signals to discrete binary form.

8.5.1 Preprocessing at Sensor/Edge Level

At the sensor or edge level, preprocessing is the initial stage for handling unfiltered data and occurs before uploading the data to higher-level processing layers. In modern agricultural environments, sensors collect data on soil composition, temperature, humidity, animal movement, and moisture content. Edge processing reduces communication costs by filtering data, streamlining values, and performing basic validations to eliminate irrelevant information. Consequently, real-time agricultural applications transmit only relevant, preprocessed data across the network, thereby enabling efficient, rapid processing.

8.5.2 Noise Reduction and Calibration

The verification and validation of sensor-based agricultural data require both filtering and calibration processes. Environmental factors such as wind, dust, variable sunlight, and electrical interference can compromise the reliability of sensor readings. Calibration aligns sensor outputs with established reference values, eliminating bias and systematic errors. Noise reduction, achieved through filtering and smoothing techniques, enhances data accuracy, precision, and overall quality. These processes ensure that decision-making in precision agriculture, including water scheduling and disease detection, relies on high-quality information.

8.5.3 Real-Time Databases and Data Pipelines

Sensor-based farm systems fundamentally depend on continuous, high-speed data storage and transfer. Real-time databases and data pipelines are essential for supporting analytical functions, live dashboards, and the collection of input from thousands of distributed sensors. Data streams incorporate retention endpoints and utilize edge devices to facilitate data mobility, while also ensuring data validation, transformation, and routing. Together, these systems enable automation and provide real-time access to current field information, supporting timely and informed interventions by farmers and their decision support systems throughout the farm ecosystem.

8.6 Case Studies

8.6.1 Precision Irrigation Systems

Precision irrigation systems analyze data in real time to optimize water utilization and enhance crop health. Soil moisture sensors monitor the availability of water to plants, enabling the system to autonomously determine the timing and quantity of irrigation required. Data-driven irrigation prevents both under-watering and over-watering. As a result, water retention and yield quality are improved. Numerous farms have reported superior crop performance and significant resource savings, illustrating the value of sensor- and data-driven agriculture (Lakhiar et al. 2024).

8.6.2 Greenhouse Monitoring

The integration of controlled environment greenhouses is essential for effective monitoring and management. Instrumentation that measures temperature, relative humidity, CO_2 concentration, and light intensity supports maintaining optimal growing conditions. Automated system adjustments are achieved through continuous monitoring, utilizing ventilation, heating, cooling, and shading mechanisms. With comprehensive data integration, controlled-environment greenhouses enable faster, healthier, and higher-quality plant growth. Data collected from these systems can be used to model and predict desired environmental conditions, thereby enhancing and refining greenhouse operations (Simo et al. 2022; Mansoor et al. 2025).

8.6.3 Large-Scale Farm Data Acquisition Projects

Large-scale farms monitor extensive tracts of land in real time by deploying thousands of multi-layered communication systems and distributed multi-sensors. These operations integrate drones, soil probes, satellite imagery, weather stations, and communication technologies for machinery into a unified data collection ecosystem. The resulting data support yield prediction and the scheduling of irrigation for vast agricultural areas. Additionally, the data is optimized to enable remote pest detection and resource management. Collectively, these practices illustrate the significance of real-time data integration and advanced system development in achieving high efficiency in commercial and national agriculture (Kunt 2025; Stamatescu et al. 2019).

8.7 Challenges

8.7.1 Network Coverage and Connectivity Issues

The implementation of online data systems in modern farming encounters challenges in network communications, particularly in rural and remote areas with inconsistent data coverage. Consequently, some farming locations experience interruptions in the data and communications required for precision agriculture. Data gaps frequently occur in crop location systems due to unreliable weather satellite data. Automated farming systems rely on continuous data streams to support precise, time-based decision-making (Miller et al. 2025).

8.7.2 Sensor Durability and Maintenance

Sensor accuracy declines over time due to malfunctions caused by extreme conditions such as high temperature, moisture, mud, and mechanical stress, which are prevalent in farm environments. These adverse conditions degrade sensor components and reduce battery lifespan. To maintain long-term functionality and withstand these challenges, sensors require regular calibration, maintenance, and cleaning. Additionally, the inherent operational complexity of large-scale farms, combined with the complexity and operational costs of sensor networks, renders scaling deployments inefficient for most businesses (Bouni et al. 2024).

8.7.3 Data Security and Privacy Concerns

The integration of connected systems and cloud-enabled data analytics in agriculture raises significant concerns regarding data security and privacy. Preventing unauthorized access, misuse, or abuse of sensitive information, such as yield predictions and resource consumption data, is essential. Protecting this data requires robust encryption protocols as well as strengthened access control and authentication mechanisms. These security and privacy considerations become particularly critical when external service providers or cloud computing vendors are involved (Rahaman et al. 2024).

8.7.4 Cost and Scalability Constraints

Sensor networks are expensive, and their deployment and scaling are often not cost-effective for most farmers, particularly those with small or medium-sized holdings. The high costs are attributed to purchasing, integrating, and maintaining hardware, as well as communication infrastructure and cloud platform fees. As hardware is updated and farm operations expand, integrating additional devices and infrastructure becomes necessary. The slow adoption of real-time agricultural technologies indicates that maintaining scalable and cost-effective solutions remains a significant challenge.

8.8 Conclusion

The integration of sensor networks with real-time data acquisition technologies constitutes a significant advancement in digital and precision agriculture, facilitating continuous observation and intelligent management of complex farm ecosystems. The combination of distributed sensing devices, Internet of Things (IoT) communication protocols, edge–fog–cloud computing frameworks, and artificial intelligence (AI)-driven analytics enables the transformation of raw environmental and crop data into actionable insights. These insights support precise irrigation, nutrient management, disease prediction, and resource optimization. The deployment of soil, environmental, crop-health, livestock, and machinery sensors establishes a comprehensive monitoring infrastructure that enhances productivity, sustainability, and resilience within farming systems. Despite these benefits, widespread adoption is hindered by challenges such as limited rural connectivity, insufficient sensor robustness, demanding maintenance requirements, cybersecurity risks, and high implementation costs, particularly for smallholder farmers. Overcoming these barriers will require the development of low-cost sensing technologies, energy-efficient communication networks, scalable data architectures, and secure digital platforms. Future smart farming systems are anticipated to increasingly integrate real-time sensing with advanced predictive analytics, autonomous agricultural equipment, and climate-responsive decision-support tools. This integration will enable highly adaptive agricultural systems that can address the growing global demand for food while promoting sustainable resource management and environmental stewardship.

References

Belkher S, Latrach B, Bekri W et al (2025) Improving the conversion efficiency of photosynthetically active radiation (PAR) absorbed by winter wheat under deficit irrigation using cost-effective autonomous IoT devices. J Agric Sci 17(4):52. https://doi.org/10.5539/jas.v17n4p52

Bernabucci G, Evangelista C, Girotti P et al (2025) Precision livestock farming: an overview on the application in extensive systems. Ital J Anim Sci 24(1):859–884. https://doi.org/10.1080/1828051X.2025.2480821

Bouni M, Hssina B, Douzi K et al (2024) Integrated IoT approaches for crop recommendation and yield-prediction using machine-learning. Internet Things 5(4):634–649. https://doi.org/10.3390/iot5040028

Burton L, Jayachandran K, Bhansali S (2020) Review—the "real-time" revolution for in situ soil nutrient sensing. J Electrochem Soc 167(3):037569. https://doi.org/10.1149/1945-7111/ab6f5d

Cheema SM, Pires IM (2025) AIoT based soil nutrient analysis and recommendation system for crops using machine learning. Smart Agric Technol 11:100924. https://doi.org/10.1016/j.atech.2025.100924

Chourlias A, Violos J, Leivadeas A (2025) Virtual sensors for smart farming: an IoT- and AI-enabled approach. Internet Things 32:101611. https://doi.org/10.1016/j.iot.2025.101611

Colaço A, Richetti J, Bramley R et al (2021) How will the next-generation of sensor-based decision systems look in the context of intelligent agriculture? A case-study. Field Crops Res 270:108205. https://doi.org/10.1016/j.fcr.2021.108205

Dahiphale D, Shinde P, Patil K et al (2025) Smart farming: crop recommendation using machine learning with challenges and future ideas. Artif Intell Appl. https://doi.org/10.47852/bonviewaia52026214

Huang F, Ryan G, Mustafa Z et al (2025) Flexible leaf wetness sensor based on laser-induced graphene for precision agriculture. Sens Actuators A Phys 388:116493. https://doi.org/10.1016/j.sna.2025.116493

Ilyasu U, Sani Z, Suleiman T (2025) Internet of things-based smart fish farming: application of smart sensors and computer vision to provide real-time monitoring and diagnosis in aquaculture. J Basics Appl Sci Res 3(2):70–77. https://doi.org/10.4314/jobasr.v3i2.8

Kashyap B, Kumar R (2021a) Sensing methodologies in agriculture for soil moisture and nutrient monitoring. IEEE Access 9:14095–14121. https://doi.org/10.1109/ACCESS.2021.3052478

Kashyap B, Kumar R (2021b) Sensing methodologies in agriculture for monitoring biotic stress in plants due to pathogens and pests. Inventions 6(2):29. https://doi.org/10.3390/inventions6020029

Kunt YE (2025) Development of a smart autonomous irrigation system using IoT and AI. arXiv. Available at: http://arxiv.org/abs/2506.11835

Lakhiar IA, Yan H, Zhang C et al (2024) A review of precision irrigation water-saving technology under changing climate for enhancing water use efficiency, crop yield, and environmental footprints. Agriculture 14:1141. https://doi.org/10.3390/agriculture14071141
Mansoor S, Iqbal S, Popescu SM et al (2025) Integration of smart sensors and IoT in precision agriculture: trends, challenges and future prospectives. Front Plant Sci 16:1587869. https://doi.org/10.3389/fpls.2025.1587869
Marjiya SA, Divya K, Latha K et al (2025) Real time smart farming analytics. Available at: www.ijedr.org
Miller T, Mikiciuk G, Durlik I et al (2025) The IoT and AI in agriculture: the time is now—a systematic review of smart sensing technologies. Sensors 25:3583. https://doi.org/10.3390/s25123583
Musanase C, Vodacek A, Hanyurwimfura D et al (2023) Data-driven analysis and machine learning-based crop and fertilizer recommendation system for revolutionizing farming practices. Agriculture 13(11):2141. https://doi.org/10.3390/agriculture13112141
Niranjan SK (2018) Proceedings of the second international conference on green computing and internet of things (ICGCIoT 2018), 16–18 August 2018, Karnataka, India. IEEE
Prasanna R, Saravanan P, Jenath M et al (2024) IoT based soil moisture monitoring and nutrients suggestion based on the soil pH value. In: Proceedings of the 4th international conference on power, energy, control and transmission systems (ICPECTS 2024). IEEE. https://doi.org/10.1109/ICPECTS62210.2024.10780246
Rahaman M, Lin CY, Pappachan P et al (2024) Privacy-centric AI and IoT solutions for smart rural farm monitoring and control. Sensors 24(13):4157. https://doi.org/10.3390/s24134157
Shahab H, Naeem M, Iqbal M et al (2025) IoT-driven smart agricultural technology for real-time soil and crop optimization. Smart Agric Technol 10:100847. https://doi.org/10.1016/j.atech.2025.100847
Simo A, Dzitac S, Badea GE et al (2022) Smart agriculture: IoT-based greenhouse monitoring system. Int J Comput Commun Control 17(6):5039. https://doi.org/10.15837/ijccc.2022.6.5039
Stamatescu G, Dragana C, Stamatescu I et al (2019) IoT-enabled distributed data processing for precision agriculture. arXiv. Available at: http://arxiv.org/abs/1906.02678
Tandon R, Gupta PK (2023) A hybrid security scheme for inter-vehicle communication in content centric vehicular networks. Wirel Pers Commun 129(2):1083–1096. https://doi.org/10.1007/s11277-023-10175-z
Tandon R, Gupta PK (2025) PALCT: vehicle-to-vehicle communication based on pseudonym assignment and encryption scheme using delay minimisation cover tree algorithm. Int J Vehicle Inform Commun Syst 10(2):145793
Tandon R, Verma A, Gupta PK (2022) A secure framework based on nature-inspired optimization for vehicle routing. In: Communications in computer and information science. Springer, Cham, pp 74–85. https://doi.org/10.1007/978-3-031-12638-3_7
Wadhwa H, Neoo H, Kaur M et al (2024) Analysis of IoT devices data using Bayesian learning on fog computing. In: Proceedings of the international conference artificial intelligence, blockchain, computing and security (ICABCS 2023). CRC Press/Balkema, London, pp 194–199. https://doi.org/10.1201/9781032684994-29

Smart Greenhouses and Controlled Environment Agriculture

9

G. S. Pradeep Ghantasala, Pellakuri Vidyullatha, and Iacovos Ioannou

Abstract

Smart greenhouses and controlled environment agriculture (CEA) represent innovations poised to transform agricultural practices. By integrating advanced technologies with traditional farming, these systems create highly efficient production environments with precise control over key variables. The use of sensors, Internet of Things (IoT) devices, automation, and artificial intelligence (AI) enables real-time monitoring and adjustment of environmental factors essential for plant growth, such as temperature, humidity, light intensity, CO_2 concentration, and nutrient levels. As a result, smart greenhouses can maintain optimal conditions throughout the entire plant growth cycle. These advancements lead to increased yields, improved product quality, and greater uniformity, while significantly reducing dependence on external weather conditions. CEA has been shown to facilitate year-round crop production and can potentially double yields on the same land area. Furthermore, compared to conventional field farming, water, fertilizer, and pesticide usage can be substantially reduced. Data-driven decision-making, powered by ML and predictive analytics, allows for early detection of plant stress, diseases, and resource inefficiencies, thereby minimizing losses and reducing operational costs. The integration of renewable energy sources with precision resource management further lowers the carbon footprint and enhances sustainability. Smart greenhouses offer significant advantages for

G. S. Pradeep Ghantasala (✉)
Department of Computer Science and Engineering, Alliance School of Advanced Computing, Alliance University, Bengaluru, Karnataka, India

P. Vidyullatha
Department of Computer Science and Engineering, Koneru Lakshmaiah Education Foundation, Vaddeswaram, Guntur, Andhra Pradesh, India

I. Ioannou
Department of Computer Science, Philips University, Nicosia, Cyprus

H. N. Oo et al. (eds.), *Transforming Agriculture with AI, IoT, and Blockchain*,
https://doi.org/10.1007/978-981-92-1491-4_9

urban agriculture, arid regions, and areas with unpredictable climates. These systems address global challenges related to food security and environmental sustainability. Although initial capital investment and complex engineering requirements present challenges, advancements in digital agriculture and declining sensor costs are driving broader adoption. Taken together, smart greenhouses and CEA provide a technologically advanced, environmentally sustainable solution to the future challenges of agriculture.

Keywords

Irrigation management · IoT · AI · Real-time monitoring · Controlled environment agriculture · Smart greenhouse

9.1 Introduction

Rapid population growth, climate change, limited cultivable land, and increasing demand for high-quality food are driving significant transformations in agriculture (Wijerathna-Yapa and Pathirana 2022). Traditional farming methods remain vulnerable to unpredictable weather, pest infestations, and inefficient resource use. Consequently, contemporary agricultural practices are increasingly adopting technology-based approaches to enhance productivity and promote environmental sustainability. Among these, smart greenhouses and controlled environment agriculture (CEA) have shown considerable potential to improve both the resilience and efficiency of food systems (Marín-Garcia et al. 2023).

Smart greenhouses are advanced agricultural facilities equipped with sensors, automated systems, and digital control mechanisms to regulate key environmental factors influencing plant growth (Bicamumakuba et al. 2025). Unlike traditional greenhouses, smart greenhouses utilize real-time data and intelligent management strategies to optimize growing conditions. Parameters such as temperature, humidity, light intensity, CO_2 concentration, and soil or nutrient solution properties are continuously monitored and adjusted to enable precise environmental control (Sharma and Kumar 2023).

CEA advances this concept by establishing fully managed cultivation systems that are often independent of external climatic conditions (Sharma et al. 2024). CEA incorporates technologies such as hydroponics, aeroponics, aquaponics, and vertical farming, which facilitate crop production in closed or semi-closed environments (Nwanojuo et al. 2025). These systems are engineered to maximize yield per unit area while significantly reducing water and chemical usage, as well as minimizing the overall environmental footprint. As a result, CEA is particularly suitable for urban and resource-constrained settings (Frimpong et al. 2023).

The integration of emerging technologies such as Internet of Things (IoT), artificial intelligence (AI), machine learning (ML), and data analytics into greenhouses and CEA systems has significantly expanded their capabilities (Gürsu 2024). These technologies enable predictive modeling, automated control, and early detection of plant stress, diseases, and nutrient deficiencies. Consequently, agricultural producers and farm managers can achieve higher yields, optimize resource allocation, minimize operational costs, and consistently produce crops of superior and uniform quality (Gupta et al. 2023).

Despite the numerous advantages of smart greenhouses and CEA, these approaches face challenges such as high initial investment costs, technological complexity, and the need for specialized expertise in operation and maintenance (Maraveas 2022). However, ongoing research, technological advancements, and decreasing costs of sensors and automation tools are facilitating the global adoption of these solutions. In the pursuit of sustainable and resilient global food systems, smart greenhouses and CEA are poised to play a key role in the future of agriculture (Mešić et al. 2024).

9.2 CEA Core Concepts

CEA refers to the management of key parameters within greenhouses, such as temperature, humidity, light, CO_2, nutrients, and airflow, to ensure consistent crop quality regardless of season (Ragaveena et al. 2021; Neo et al. 2022). In these systems, plants are cultivated in greenhouses using hydroponic or other soilless methods under artificial lighting and automated control systems, which together optimize resource utilization and minimize external disturbances (Prasetia et al. 2021). Figure 9.1 demonstrates that integrating sensors, IoT technologies, data analytics, and AI enables food production that is both efficient and environmentally sustainable, and can effectively withstand various stresses, particularly in urban or resource-limited settings (Sharma and Shivandu 2023).

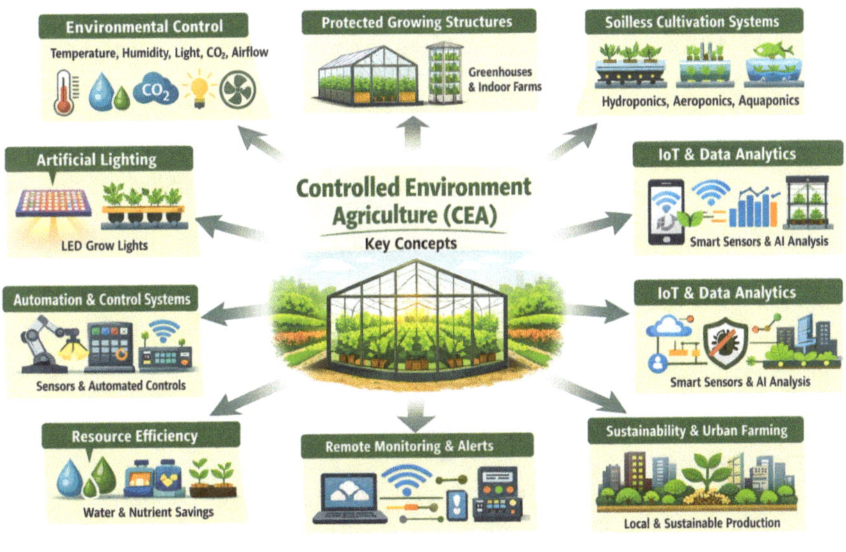

Fig. 9.1 Key concepts of the controlled agriculture system

Environmental Control

CEA relies on precise regulation of environmental factors such as temperature, humidity, light intensity, CO_2 concentration, and airflow (Hu et al. 2025; Baghalian et al. 2023). Maintaining these parameters at optimal levels throughout the plant life cycle ensures consistent yields, accelerated growth, and improved crop quality, independent of external weather conditions.

Protected Growing Structures

Crops are cultivated in tightly controlled environments, including greenhouses, indoor farms, and vertical farming units (Akpenpuun et al. 2025). These structures serve as physical barriers to extreme weather, pests, and diseases, while also enabling precise environmental regulation.

Soilless Cultivation Systems

CEA often employs hydroponics, aeroponics, and aquaponics, methods of cultivating plants without soil (Alnaqbi and Araci 2024). These systems deliver nutrients directly to the root zone, resulting in increased nutrient-use efficiency, reduced water use, and accelerated plant growth.

Artificial and Supplemental Lighting

LED grow lights provide plants with optimal light spectra and photoperiods required for efficient photosynthesis. Additionally, controlled lighting supports morphological development, enhances yield, and improves nutritional value. This technology also enables year-round crop production.

Automation and Control Systems

Automated systems equipped with sensors, controllers, and actuators continuously monitor and adjust environmental conditions to optimize suitability (Hebala and Ashour 2024). Automation reduces labor requirements, minimizes human error, and maintains stable conditions for crop growth.

IoT Integration

IoT-enabled sensors collect real-time data on environmental and plant parameters. These data support remote monitoring, automated alert notifications, and data-driven decision-making in precision agriculture.

Data Analytics and AI

Advanced analytics and AI models utilize historical and current data to forecast crop demand, detect early signs of stress or disease, and optimize resource allocation to maximize yield.

Resource Efficiency

CEA systems significantly reduce the consumption of water, fertilizer, and pesticides through precise control and recycling mechanisms (Cowan et al. 2022).

Consequently, CEA demonstrates both environmental sustainability and long-term economic viability.

Biosecurity and Crop Health
Controlled environments limit exposure to pests and pathogens, reducing the need for chemical pesticides and, consequently, improving both food safety and crop reliability.

Sustainability and Urban Agriculture
CEA enables local, self-sufficient food production in urban settings and in regions with limited natural resources. By maximizing crop yield per unit area and minimizing environmental impact, CEA supports the development of resilient and future-oriented crop production systems.

9.3 Smart Greenhouse Design and Components

A smart greenhouse refers to a controlled environment in which plant conditions are continuously monitored, analyzed, and regulated using modern technologies such as the IoT, automation, and AI (Riskiawan et al. 2024; Ghiasi et al. 2024). The physical design integrates smart features within the structure to optimize plant growth conditions and resource efficiency, as illustrated in Fig. 9.2.

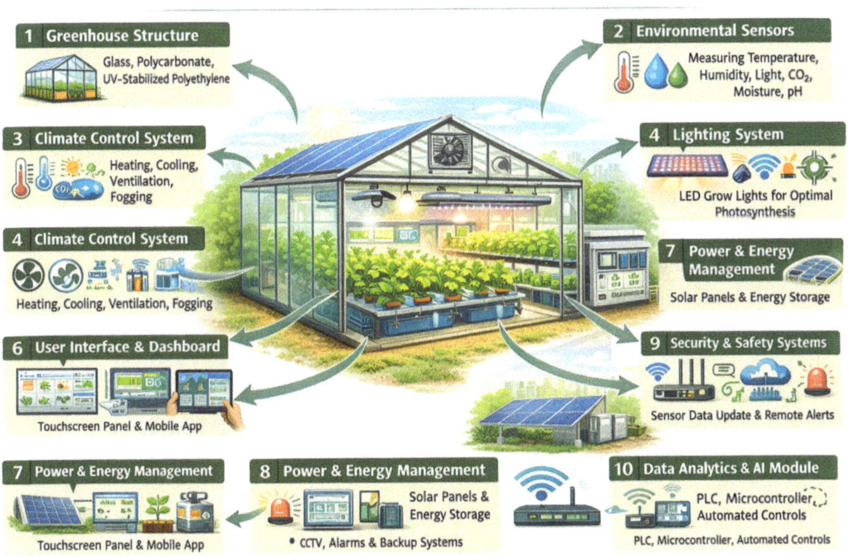

Fig. 9.2 Smart greenhouse design and components

Greenhouse Structure

The primary function of the physical structure is to provide crops with a protected environment. Typically, it is constructed from materials such as glass, polycarbonate, or polyethylene, which allow sunlight to pass through while insulating against external weather conditions (Zamani et al. 2025). In addition to sheltering crops, the structure also contains ventilation systems, sensors, lighting, and irrigation equipment.

Environmental Sensors

Sensors constitute the foundation of smart greenhouse systems. These devices collect data on parameters including temperature, humidity, soil or nutrient moisture, light intensity, CO_2 concentration, and pH. Accurate sensing enables real-time monitoring and facilitates precise environmental control.

Climate Control System

Heating, cooling, exhaust forcing, fogging, and ventilation are regulated through climate control systems to maintain optimal plant temperature and humidity levels (Yusuf et al. 2025). Automated climate control prevents plant stress and ensures stable growth conditions throughout both day and night.

Lighting System

Artificial lighting, particularly LED grow lights, is used to supplement or fully replace natural sunlight. Smart lighting systems control light intensity, spectrum, and photoperiod to optimize photosynthesis, flowering, and crop yield.

Irrigation and Nutrient Delivery System

Automated irrigation systems such as drip irrigation, hydroponics, and fertigation deliver water and nutrients directly to plant roots (Savvas et al. 2023). These systems reduce water wastage and ensure precise nutrient delivery to plants.

Control Unit (Controller)

The controller functions as the central processing unit of the smart greenhouse. It analyzes sensor data and initiates actions based on predefined thresholds or algorithms. Controllers may include microcontrollers, programmable logic controllers (PLCs), or embedded systems.

IoT Connectivity

The IoT enables sensors and controllers to connect via the Internet. Through mobile applications or web dashboards, farmers can remotely monitor greenhouse conditions and receive notifications if parameters exceed safe thresholds (Farooq et al. 2022a).

Automation and Actuators

Actuators, including motors, valves, pumps, and switches, execute control actions automatically. For example, these devices can open vents, activate fans, initiate irrigation, or adjust lighting in response to commands from the controller.

Data Analytics and AI Module

The collected data is stored and analyzed to identify patterns, predict crop requirements, detect diseases at early stages, and optimize energy and resource utilization. AI-based systems continually enhance their decision-making capabilities.

Power and Energy Management System

Smart greenhouse energy supply systems are typically highly efficient and often incorporate renewable energy sources, such as solar panels (Maraveas et al. 2021). Effective energy management ensures the continuous operation of sensors, controllers, and actuators.

User Interface and Monitoring Dashboard

The dashboards provide visual representations of both real-time and historical data. Users can configure limits, schedule operations, and manually override automated controls when necessary.

Security and Safety Systems

Security measures safeguard both data and physical facilities through restricted access, alarm systems, and data backup. Safety protocols are implemented to prevent disruptions caused by power outages or system failures.

9.4 Environmental Monitoring and Control Mechanisms

Both smart greenhouses and CEA depend heavily on environmental monitoring and control systems (Srimal et al. 2024). These systems provide optimal conditions for plant growth by continuously monitoring, analyzing, and adjusting environmental factors.

Environmental Monitoring

Environmental monitoring involves the continuous recording of key environmental factors in the greenhouse using various sensors (Maraveas and Bartzanas 2021).

- Temperature sensors monitor both air and soil temperatures to prevent plant heat or cold stress
- Humidity sensors measure atmospheric moisture levels to reduce the risk of fungal diseases and prevent plant water deficiency
- Light sensors assess both the intensity and duration of light necessary for photosynthesis
- CO_2 sensors monitor CO_2 concentrations to support optimal plant growth

- Soil and nutrient sensors measure soil moisture, pH, and electrical conductivity (EC) to ensure appropriate nutrient delivery

Continuous data collection enables early detection and mitigation of adverse conditions.

Data Acquisition and Processing

Sensor data is processed by data acquisition systems before being transmitted to controllers. Initially, the data undergoes filtering and digital conversion, followed by calculation of the deviation between actual and optimal setpoints.

Control Strategies

Control strategies are formulated using processed data:

- Threshold control, also known as ON/OFF control
- Rule-based control, which utilizes IF-THEN logic
- Intelligent control employing AI and ML algorithms for predictive regulation

These techniques facilitate rapid and precise adaptation to environmental changes.

Environmental Control Mechanisms

Control mechanisms regulate greenhouse environments by employing various actuators.

- Heating systems, such as heaters and boilers, increase the ambient temperature
- Cooling systems, including fans, cooling pads, and evaporative cooling units, decrease the temperature
- Ventilation systems regulate airflow and maintain appropriate humidity levels
- Lighting systems manage light intensity and photoperiod duration
- Irrigation and fertigation systems manage the delivery of water and nutrients

Feedback and Closed-Loop Control

A smart greenhouse functions as a closed-loop feedback system, in which controlled actions directly influence environmental conditions that are continuously monitored by sensors to maintain stability (Walczuch et al. 2022).

Automation and IoT Integration

Smart greenhouses utilize a closed-loop feedback system in which controlled actions directly affect environmental conditions, and sensors continuously monitor outcomes to maintain stability.

Benefits of Monitoring and Control Mechanisms

- Enhanced crop yield and improved product quality
- Reduced consumption of water, energy, and fertilizers
- Early detection of plant stress and diseases
- Continuous, climate-independent agricultural production throughout the year

Environmental monitoring and control systems convert conventional greenhouses into intelligent, self-regulating environments. By integrating sensors, automation, and data-driven management, smart greenhouses enable sustainable, efficient, and high-quality agricultural production.

9.5 IoT-Based Greenhouse Management

IoT-based greenhouse management integrates sensors, connectivity, and automation to enable continuous monitoring and real-time control of the greenhouse environment (Farooq et al. 2022b). In this management approach, IoT devices are equipped with sensors that measure key parameters, including temperature, humidity, light intensity, soil or nutrient moisture, pH, and CO_2 concentration. Data from these sensors is transmitted to a central controller or cloud server via wireless communication. Consequently, plant growth conditions can be monitored remotely at any time. Regular monitoring facilitates early detection of plant stress, disease, or environmental imbalances. Appropriate adjustments can then be implemented to optimize crop health and maximize yield.

Analysis of the collected data indicates that various greenhouse control systems, including irrigation, ventilation, heating, cooling, and lighting, are automatically managed (Nemali 2022). Cloud-based platforms and mobile applications enable remote monitoring and adjustment of greenhouse conditions. Additionally, big data analytics and AI support decision-making by forecasting crop requirements and maximizing yield. Overall, the implementation of IoT technology in greenhouse management enhances resource efficiency, significantly reduces operational costs (Titirmare et al. 2024), and promotes environmentally sustainable farming practices.

9.6 AI and Optimization Techniques

AI in Greenhouse Systems

AI enables smart greenhouses to move beyond conventional automation by facilitating intelligent decision-making. Deep learning models and ML algorithms analyze both real-time and historical sensor data to identify complex relationships between environmental variables and crop growth. AI applications include monitoring crop development, predicting yields, identifying diseases and pests through image recognition, and detecting plant stress (Bayar et al. 2025). By analyzing historical data, AI systems can anticipate plant needs and recommend optimal interventions before issues arise, thereby improving productivity and crop quality.

Optimization Techniques

Optimization techniques are designed to maximize crop yield and quality while minimizing resource consumption, including water, energy, and nutrients. Variables such as irrigation timing, nutrient concentration, temperature settings, and lighting duration are optimized using biologically inspired algorithms such as Genetic

Algorithms (GA), Particle Swarm Optimization (PSO), and Ant Colony Optimization (ACO) (Anukeerthana et al. 2024). The primary objective of these methods is to achieve optimal greenhouse operation by identifying the most effective values for control variables within specified constraints.

AI–Optimization Integration
Greenhouse systems demonstrate significant adaptability and efficiency when AI models are integrated with optimization methods (Kaur et al. 2024). AI models predict optimal environmental conditions, while optimization methods identify the control parameters that should be adjusted to achieve these conditions at minimal cost. This integration enables real-time control, reduces reliance on human intervention, enhances sustainability, and facilitates precision agriculture under rapidly changing climatic conditions. The key benefits of the AI-optimization integration are as follows:

- Increased crop yield and quality
- Reduced consumption of water, energy, and fertilizers
- Early detection of plant diseases and stress factors
- Autonomous and intelligent control of greenhouse environments

9.7 Applications and Case Studies of Smart Greenhouses and CEA

Smart greenhouses and CEA have emerged as effective approaches for achieving efficient, high-yield crop production in commercial settings, urban and vertical farming, and precision irrigation. Advanced sensor technologies, automation, the IoT, and AI collectively enable early disease detection, climate-resilient farming practices in extreme environments, and significant reductions in water, energy, and pesticide usage (Delfani et al. 2024). Several real-world case studies, as shown in Fig. 9.3, demonstrate that smart greenhouses and CEA technologies enhance not only productivity but also the sustainability and resilience of agricultural systems, thereby addressing global challenges such as food security, climate change, and resource scarcity.

Commercial Crop Production
Smart greenhouses are commonly utilized for cultivating high-value crops such as vegetables, fruits, flowers, and herbs. Precise regulation of temperature, humidity, light, and nutrient levels enables commercial farmers to achieve higher yields of superior quality within shorter timeframes. Additionally, automation reduces labor costs and mitigates crop losses due to pests, diseases, and weather fluctuations, thereby enhancing the sustainability of commercial farming.

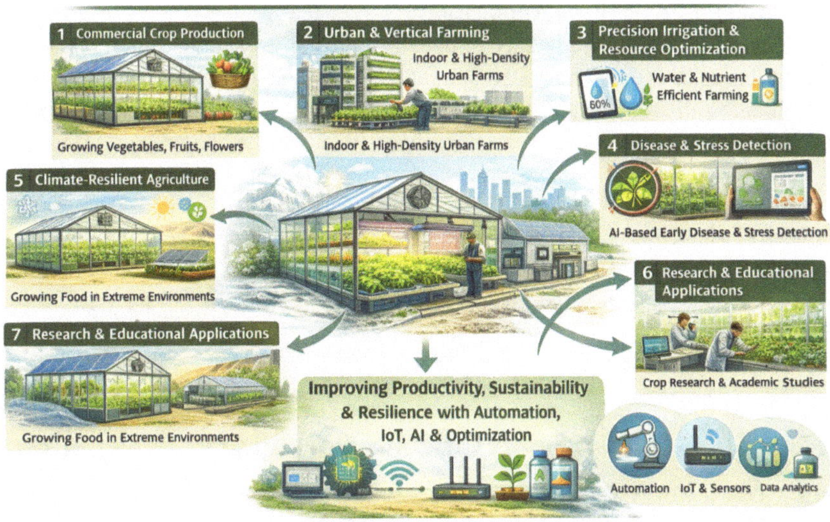

Fig. 9.3 Smart greenhouses and CEA applications

Urban and Vertical Farming

In urban areas with limited land, smart greenhouses and vertical farms enable indoor food production in controlled environments, such as on rooftops, in warehouses, and in shipping containers. CEA facilitates year-round cultivation near consumers, thereby reducing transportation costs, postharvest losses, and carbon emissions. This approach is especially effective for producing leafy greens and microgreens.

Precision Irrigation and Resource Optimization

Smart greenhouse systems utilize sensors and optimization algorithms to deliver water and nutrients precisely when required. Case studies indicate a significant reduction in water use, ranging from 40% to 60% compared with traditional farming methods, while maintaining or enhancing crop yields. These systems are particularly advantageous in regions experiencing water scarcity.

Disease and Stress Detection

Automatic image processing using AI programs and sensor analysis has been employed to detect early symptoms of plant stress, nutrient deficiencies, or diseases. Timely interventions informed by intelligent systems have reduced pesticide use and prevented crop losses.

Climate-Resilient Agriculture

The smart greenhouse system has been successfully implemented in regions characterized by harsh environmental conditions, such as deserts, frozen areas, and zones with highly variable weather. The case study demonstrates that controlled environments enable food production in extreme-climate regions.

Research and Educational Applications
Smart greenhouses are frequently utilized at universities and agricultural research institutions to facilitate controlled studies of crop genetics, nutrient optimization, and climate impact. The ability to repeatedly control and measure environmental conditions in smart greenhouses significantly advances innovation in modern agriculture.

Applications and case studies indicate that smart greenhouses and CEA can substantially enhance productivity, sustainability, and resilience. Through the integration of automation, IoT, AI, and optimization techniques, smart greenhouses and CEA address critical global challenges, such as food security, climate change, and resource scarcity.

9.8 Challenges in Smart Greenhouses and CEA

Smart greenhouses and CEA face several challenges, including significant upfront investment, high energy consumption, technical complexity, and dependence on reliable sensors, data management, and skilled personnel (Elakrouch et al. 2025). Additional issues, such as cybersecurity threats, limited crop diversity, scalability constraints, and frequent power or system failures, can disrupt habitat stability and profitability. As shown in Fig. 9.4, addressing these challenges by adopting renewable energy, robust system architecture, comprehensive workforce training, and cost-effective technologies is essential for the sustained ecological and economic viability of CEA systems.

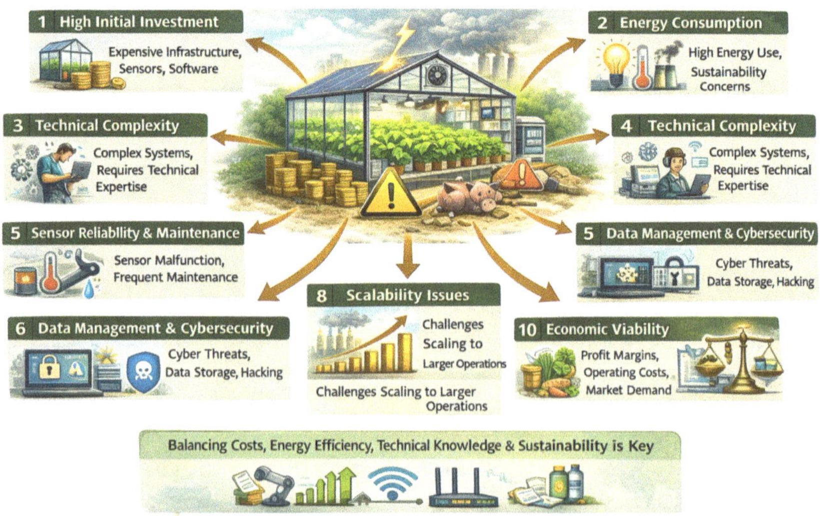

Fig. 9.4 Challenges in smart greenhouses

High Initial Investment

The capital requirements for smart greenhouse technology are substantial, encompassing infrastructure, sensors, automation, lighting, and control software. These significant initial investments often present a considerable barrier, particularly for small and medium-sized farmers.

Energy Consumption

These systems primarily rely on controlled environments, including lighting, heating, cooling, and ventilation systems. For example, indoor and vertical farming consume significant amounts of energy, making sustainability a critical concern in the absence of renewable energy sources.

Technical Complexity

Smart greenhouses represent sophisticated systems that integrate hardware, software, IoT networks, and data analytics. The operation and maintenance of these systems typically require a technician; however, such personnel may not always be available.

Sensor Reliability and Maintenance

Most sensors consist of electronic components. Exposure to moisture, dust, or chemicals can cause deterioration. Under these conditions, sensors may malfunction or provide inaccurate readings, leading to incorrect control decisions that adversely affect plant health. Regular recalibration and maintenance of sensors are essential, as they are part of the ongoing operational requirements of smart greenhouses.

Data Management and Cybersecurity

The substantial volume of data generated by IoT sensors requires secure storage and processing. Cybersecurity threats, including data theft, hacking, and unauthorized access, present significant challenges within greenhouse networks.

Limited Crop Variety

CEA is primarily profitable for high-value crops such as leafy vegetables and herbs. Its applicability to staple crops, including cereals, remains limited.

Dependence on Technology

System breakdowns resulting from power failures, network disruptions, or software errors may significantly disrupt greenhouse operations. Without a backup system, such failures can result in considerable crop losses.

Scalability Issues

Scaling smart greenhouse systems from pilot projects to commercial operations presents challenges, including increased costs, higher energy requirements, and greater complexity in system coordination.

Environmental Footprint
Although CEA reduces water and pesticide usage, its carbon footprint can be substantial when energy-intensive processes rely primarily on fossil fuels.

Economic Viability
Profitability depends on market demand, energy costs, and crop selection. In some regions, the payback period may extend over several years, discouraging the adoption of greenhouse technologies.

Although smart greenhouses and CEA offer significant advantages, challenges related to cost, energy efficiency, technical expertise, and sustainability remain unresolved. Advances in AI, renewable energy, and low-cost sensors are expected to mitigate these challenges.

9.9 Conclusion

Smart greenhouses and CEA represent major advancements in contemporary agricultural practices, integrating advanced technologies with the foundational principles of precision agriculture. Through the use of sensors, the IoT, automation, AI, and optimization techniques, CEA enables precise regulation of environmental parameters such as temperature, humidity, light, CO_2 concentration, and nutrient delivery. This level of control promotes optimal crop growth, higher yields, improved quality, and consistent year-round production regardless of external weather conditions. A primary advantage of smart greenhouse technology is the optimization of resource utilization. CEA consumes approximately three times less water, fertilizers, and pesticides compared to traditional farming methods. Additionally, CEA maximizes resource efficiency by achieving the highest possible productivity per unit area. This efficiency makes CEA particularly suitable for urban agriculture, regions experiencing water scarcity, and areas affected by climate change. Furthermore, the technology facilitates early detection of plant stress, diseases, and system inefficiencies, thereby reducing losses and improving overall operational efficiency. Despite these benefits, factors such as high initial investment, significant energy consumption, system complexity, and data security concerns have limited the widespread adoption of smart greenhouses. Ongoing advancements in integrating renewable energy sources, affordable sensors, cloud computing, and intelligent control algorithms are addressing these challenges. As technology progresses and economies of scale are realized, smart greenhouses are expected to become increasingly cost-effective. The continued innovation and adoption of these systems are likely to shape the future of efficient and sustainable agricultural practices.

References

Akpenpuun T, Sanusi H, Ogundele O et al (2025) Advancements in vertical farming: a review of potentials, challenges, and prospects. Korean J Agric Sci 52:619–638. https://doi.org/10.7744/kjoas.520418

Alnaqbi SK, Araci ZC (2024) Case study: identifying comparison criteria for AHP analysis of controlled environment agriculture systems. In: Proceedings of the IEEE international conference on technology management, operations and decisions (ICTMOD). IEEE, pp 1–5. https://doi.org/10.1109/ICTMOD

Anukeerthana BS, Lavanya DS, Gurucharran V et al (2024) Improving agricultural productivity and water usage through advanced ACO technique. In: Proceedings of the international conference on communication and signal processing (ICCSP). IEEE, pp 93–97

Baghalian K, Hajirezaei MR, Lawson T (2023) Editorial: Current and future perspectives for controlled environment agriculture (CEA) in the 21st century. Front Plant Sci 14:1334641. https://doi.org/10.3389/fpls.2023.1334641

Bayar J, Ali N, Cao Z et al (2025) Artificial intelligence of things (AIoT) for precision agriculture: applications in smart irrigation, nutrient and disease management. Smart Agric Technol 12:101629. https://doi.org/10.1016/j.atech.2025.101629

Bicamumakuba E, Reza MN, Jin H et al (2025) Multi-sensor monitoring, intelligent control, and data processing for smart greenhouse environment management. Sensors 25(19):6134. https://doi.org/10.3390/s25196134

Cowan N, Ferrier L, Spears B et al (2022) CEA systems: the means to achieve future food security and environmental sustainability? Front Sustain Food Syst 6:891256. https://doi.org/10.3389/fsufs.2022.891256

Delfani P, Thuraga V, Banerjee B et al (2024) Integrative approaches in modern agriculture: IoT, ML and AI for disease forecasting amidst climate change. Precis Agric 25:2589–2613. https://doi.org/10.1007/s11119-024-10164-7

Elakrouch M, Mohaddab M, Rachid A et al (2025) Soilless agriculture at a crossroads: strengths, challenges, and prospects of hydroponics, aquaponics, and bioponics in relation to precision farming. Sustainability 18(1):64. https://doi.org/10.3390/su18010064

Farooq MS, Riaz S, Helou MA et al (2022a) Internet of things in greenhouse agriculture: a survey on enabling technologies, applications, and protocols. IEEE Access 10:53374–53397. https://doi.org/10.1109/ACCESS

Farooq MS, Javid R, Riaz S et al (2022b) IoT based smart greenhouse framework and control strategies for sustainable agriculture. IEEE Access 10:99394–99420. https://doi.org/10.1109/ACCESS.2022.3204066

Frimpong F, Asante MD, Peprah CO et al (2023) Water-smart farming: review of strategies, technologies, and practices for sustainable agricultural water management in a changing climate in West Africa. Front Sustain Food Syst 7:1110179. https://doi.org/10.3389/fsufs.2023.1110179

Ghiasi M, Wang Z, Mehrandezh M et al (2024) A systematic review of optimal and practical methods in design, construction, control, energy management and operation of smart greenhouses. IEEE Access 12:2830–2853. https://doi.org/10.1109/ACCESS.2023.3346436

Gupta S, Berenji HR, Shukla M et al (2023) Opportunities in farming research from an operations management perspective. Prod Oper Manag 32(6):1577–1596. https://doi.org/10.1111/poms.13967

Gürsu H (2024) Waste-based vertical planting system proposal to increase productivity in sustainable horticulture; “PETREE”. Sustainability 16(8):3125. https://doi.org/10.3390/su16083125

Hebala A, Ashour HA (2024) Realization of cooperative conveyor-AGV smart luggage handling system through AI-based object detection and IoT. In: Proceedings of the international conference on computer theory and applications (ICCTA). IEEE, pp 22–27

Hu J, Wang X, Tan S (2025) Energy management model in controlled environment agriculture: a review. Energies 18(17):4544. https://doi.org/10.3390/en18174544

Kaur A, Sonawane V, Rosha P (2024) Energy efficiency optimization strategies for greenhouse-based crop cultivation: a review. Can J Chem Eng 102(3):1051–1065. https://doi.org/10.1002/cjce.25131

Maraveas C (2022) Incorporating artificial intelligence technology in smart greenhouses: current state of the art. Appl Sci 13(1):14. https://doi.org/10.3390/app13010014

Maraveas C, Bartzanas T (2021) Application of internet of things (IoT) for optimized greenhouse environments. AgriEngineering 3(4):954–970. https://doi.org/10.3390/agriengineering3040060

Maraveas C, Loukatos D, Bartzanas T et al (2021) Smart and solar greenhouse covers: recent developments and future perspectives. Front Energy Res 9:783587. https://doi.org/10.3389/fenrg.2021.783587

Marín-Garcia E, Torres-Marín JN, Chaverra-Lasso A (2023) Smart greenhouse and agriculture 4.0. Rev Cient 46:37–50. https://doi.org/10.14483/23448350.19816

Mešić A, Jurić M, Donsì F et al (2024) Advancing climate resilience: technological innovations in plant-based, alternative and sustainable food production systems. Discov Sustain 5:423. https://doi.org/10.1007/s43621-024-00581-z

Nemali K (2022) History of controlled environment horticulture: greenhouses. HortScience 57(2):239–246. https://doi.org/10.21273/HORTSCI16160-21

Neo DCJ, Ong MMX, Lee YY et al (2022) Shaping and tuning lighting conditions in controlled environment agriculture: a review. ACS Agric Sci Technol 2(1):3–16. https://doi.org/10.1021/acsagscitech.1c00241

Nwanojuo MA, Anumudu CK, Onyeaka H (2025) Impact of controlled environment agriculture (CEA) in Nigeria, a review of the future of farming in Africa. Agri 15(2):117. https://doi.org/10.3390/agriculture15020117

Prasetia Y, Putrada AG, Rakhmatsyah A (2021) Evaluation of IoT-based grow light automation on hydroponic plant growth. J Ilm Tek Elektro Komput Inform 7(2):314–325. https://doi.org/10.26555/jiteki.v7i2.21424

Ragaveena S, Shirly Edward A, Surendran U (2021) Smart controlled environment agriculture methods: a holistic review. Rev Environ Sci Biotechnol 20:887–913. https://doi.org/10.1007/s11157-021-09591-z

Riskiawan HY, Gupta N, Setyohadi DPS et al (2024) Artificial intelligence enabled smart monitoring and controlling of IoT-green house. Arab J Sci Eng 49:3043–3061. https://doi.org/10.1007/s13369-023-07887-6

Savvas D, Giannothanasis E, Ntanasi T et al (2023) State of the art and new technologies to recycle the fertigation effluents in closed soilless cropping systems aiming to maximise water and nutrient use efficiency in greenhouse crops. Agronomy 14(1):61. https://doi.org/10.3390/agronomy14010061

Sharma PK, Kumar S (2023) Soil temperature and plant growth. In: Soil physical environment and plant growth. Springer, Cham. https://doi.org/10.1007/978-3-031-28057-3_7

Sharma K, Shivandu SK (2023) Integrating artificial intelligence and internet of things (IoT) for enhanced crop monitoring and management in precision agriculture. Sensors Int 5:100292. https://doi.org/10.1016/j.sintl.2024.100292

Sharma A, Hazarika M, Heisnam P et al (2024) Controlled environment ecosystem: a plant growth system to combat climate change through soilless culture. Crop Des 3(1):100044. https://doi.org/10.1016/j.cropd.2023.100044

Srimal MDW, Aboobucker MSM, Sha HH et al (2024) Emerging frontiers in CEA: an IoT-based system architecture for smart, sustainable, and high-yielding controlled environment agriculture. IOP Conf Ser Earth Environ Sci 1401(1):012004. https://doi.org/10.1088/1755-1315/1401/1/012004

Titirmare S, Margal P, Gupta S et al (2024) AI-powered predictive analytics for crop yield optimization. In: Agriculture 4.0: smart farming with iot and artificial intelligence. CRC Press, London, pp 1–15. https://doi.org/10.1201/9781003570219-5

Walczuch D, Nitzsche T, Seidel T et al (2022) Overview of closed-loop control systems and artificial intelligence utilization in greenhouse farming. In: Proceedings of the IEEE international conference on omni-layer intelligent systems (COINS). IEEE, pp 1–6

Wijerathna-Yapa A, Pathirana R (2022) Sustainable agro-food systems for addressing climate change and food security. Agriculture 12(10):1554. https://doi.org/10.3390/agriculture12101554

Yusuf AG, Al-Yahya FA, Saleh AA et al (2025) Optimizing greenhouse microclimate for plant pathology: challenges and cooling solutions for pathogen control in arid regions. Front Plant Sci 16:1492760. https://doi.org/10.3389/fpls.2025.1492760

Zamani F, Duri LG, Mori M et al (2025) Advances in light manipulation in greenhouse horticulture: the innovative smart covers. Front Plant Sci 16:1640530. https://doi.org/10.3389/fpls.2025.1640530

Blockchain for Agri-Supply Chain Traceability

10

Shruti Mishra and Diksha Bali

Abstract

In the context of globalization, consumers are becoming highly attentive to the authenticity, safety, and legitimacy of their food. Policymakers and governments are similarly prioritizing nutritional security alongside food availability for the general population. The global agri-food supply chain faces persistent challenges, including food counterfeiting, contamination risks, and inefficient product recall procedures, highlighting the need for improved traceability and transparency. To address these issues, the agriculture sector is adopting blockchain technology, a distributed and decentralized digital ledger, to create secure, transparent, and trustworthy food systems. This chapter examines the role of blockchain in strengthening traceability by overcoming the limitations of traditional systems, which often rely on centralized databases, fragmented data, limited stakeholder transparency, and manual, paper-based processes. Blockchain provides decentralization, immutability, real-time transparency, and enhanced security, enabling comprehensive end-to-end tracking from farm to consumer and offering verifiable records at each stage of a product's lifecycle. This chapter further emphasizes the transformative potential of blockchain technology through real-world applications in the agri-supply chain. For instance, Walmart, in partnership with IBM and Tsinghua University in China, successfully reduced pork product traceability time from days to seconds. The Coffee Board of India and Eka developed a blockchain-based marketplace that offers real-time pricing and transparent trading to over 350,000 coffee growers. Starbucks collaborated with Microsoft to develop a digital traceability tool that allows users to follow coffee beans from farm to cup, thereby promoting transparency and improving consumer trust. By exploring these cases, this chapter demonstrates both the

S. Mishra · D. Bali (✉)
ICAR-National Institute of Agricultural Economics and Policy Research, New Delhi, India

H. N. Oo et al. (eds.), *Transforming Agriculture with AI, IoT, and Blockchain*,
https://doi.org/10.1007/978-981-92-1491-4_10

advantages and challenges of integrating blockchain technology into agri-food supply chains. While the technology enhances food safety, operational efficiency, and trust, issues such as high implementation costs, scalability, data integrity, and regulatory uncertainty need to be addressed to realize its full potential. In conclusion, blockchain offers a transformative approach to enhancing traceability and ensuring secure, transparent, and resilient global agri-food supply chains.

Keywords

Blockchain · Agri-supply chain · Globalization · Traceability

10.1 Introduction

Global agri-food supply chains, encompassing production, processing, transportation, certification, and retailing across distributed geographic networks, have expanded significantly in both size and complexity. While this integration has enhanced productivity and market reach, it has also exposed vulnerabilities in food safety, authenticity, and traceability. Conventional traceability systems often struggle to ensure data accuracy, interoperability, and real-time responsiveness due to reliance on paper records, isolated digital files, or centralized databases. Fragmented documentation, delays in accessing critical information, and susceptibility to data manipulation are persistent issues within these systems. Such deficiencies have contributed to several food safety incidents and fraud cases, where inadequate traceability impeded timely recalls and effective risk management (Aung and Chang 2014; Charlebois et al. 2014; Demestichas et al. 2020).

Consumer expectations have evolved, with a growing demand for transparency regarding product origin, environmental impact, ethical sourcing, and quality assurance (Pozelli Sabio and Spers 2022). Rather than relying solely on branding or certification labels, contemporary consumers increasingly seek verifiable information (Charlebois et al. 2014). Mislabeling and fraud remain significant challenges in global commodity sectors such as coffee, meat, seafood, and horticultural produce (Galvez et al. 2018). Consequently, researchers and industry stakeholders have pursued new technological frameworks capable of delivering end-to-end visibility, accuracy, and trust across agri-food supply chains (Lv et al. 2023).

The unique characteristics of agricultural products highlight the suitability of technological interventions. Agri-food products exhibit biological variability, perishability, and heightened sensitivity to environmental factors such as humidity, temperature, and storage duration (Bolívar et al. 2025). Agricultural supply chains typically involve farmers, brokers, processors, cold-chain operators, distributors, inspectors, and retailers, each maintaining separate data systems. This fragmentation often results in inconsistent or incomplete records, complicating traceability efforts (Behnke and Janssen 2020). These challenges are further exacerbated by vulnerabilities to fraud and adulteration, especially in high-value commodities or complex international supply chains (Bouzembrak et al. 2018).

Blockchain technology enables a unified and tamper-proof traceability framework that addresses several of these challenges. Transaction events such as harvesting, processing, and distribution can be recorded sequentially, with each product batch assigned a unique digital identity. Integration with Internet of Things (IoT) sensors enables real-time environmental data to be recorded directly on the ledger, while smart contracts automate verification processes and reduce administrative burdens (Leng et al. 2018). These capabilities significantly enhance traceability accuracy, minimize the risk of human manipulation, and expedite the retrieval of supply-chain data during product recalls (Yao and Zhang 2022).

However, current research indicates that blockchain is not a comprehensive solution despite its advantages. Persistent challenges include data entry reliability, digital literacy, deployment costs, interoperability, and governance structures (Behnke and Janssen 2020; Kamilaris et al. 2019). These limitations highlight the need for further research into regulatory frameworks, inclusive implementation models, and socio-technical integration to ensure that smallholders and marginalized stakeholders benefit from digital traceability innovations. The existing literature provides a foundation for understanding how blockchain technology could transform agri-food traceability and identifies the issues that must be addressed for effective implementation. This chapter contributes to the growing body of knowledge by analyzing how blockchain can enhance transparency, accountability, and resilience in agricultural systems through the integration of supply-chain, governance, and technological perspectives.

10.2 Methods and Framework

The approach adopted in this chapter is grounded in a qualitative, interpretive research design that reflects the multidisciplinary and dynamic nature of blockchain applications in agri-food traceability systems. A solely technical or descriptive methodology is insufficient, as blockchain adoption in agriculture spans institutional, technological, and socioeconomic domains. Instead, a triangulated methodology is employed, comprising a systematic literature review, comparative case-study analysis, and conceptual synthesis. This approach facilitates a comprehensive understanding of both the practical applications and theoretical foundations of blockchain-based traceability in agriculture.

A comprehensive literature review forms the analytical framework for this chapter. This review identifies the primary challenges associated with conventional traceability systems and highlights the novel features introduced by blockchain technology. Key studies addressing traceability bottlenecks, data integrity issues, and the capacity of distributed ledger technologies to enhance transparency, immutability, and multi-stakeholder trust include those by Kamilaris et al. (2019), Aung and Chang (2014), Tian (2017), Galvez et al. (2018), and Behnke and Janssen (2020). The integration of these findings provides a holistic perspective on the relevance of blockchain in agricultural contexts.

Beyond the literature review, this chapter investigates functional blockchain applications within various agri-food systems through comparative case-study analysis. Case selection was based on applicability, verifiability, and the capacity to support rigorous analytical generalization, following theoretical sampling principles (Ebneyamini and Sadeghi Moghadam 2018). Notable examples include Walmart's pork traceability system in China, Sahyadri Farms' farmer-driven digital traceability system in India, Starbucks' farm-to-cup transparency initiative, and the blockchain-enabled marketplace of the Coffee Board of India. Employing multiple cases enables cross-context comparison, facilitating the identification of both recurring patterns and context-specific challenges.

This chapter culminates in the development of a conceptual framework for blockchain-enabled traceability in agriculture, synthesizing insights from both the literature and case studies. This synthesis incorporates organizational factors such as governance, interoperability, and stakeholder incentives; technological components including digital identity, smart contracts, IoT-blockchain integration, and permissioned ledger structures; and policy considerations such as regulation, standardization, and digital infrastructure. The resulting framework aligns with established methodological traditions in information systems research, where theoretical constructs and empirical observations are integrated through conceptual modeling (Gregor 2006). This systematic synthesis provides a coherent structure for evaluating blockchain's transformative potential and identifying future research directions in agri-food traceability. Through this triangulated methodological approach, this chapter aims to deliver a comprehensive understanding of how blockchain can enhance resilience, accountability, and transparency in agricultural supply chains. A schematic representation of the triangulated research methodology is presented in Fig. 10.1.

10.3 Blockchain Systems: An Overview

A blockchain is a distributed and decentralized digital ledger that securely and transparently maintains an immutable audit trail of transactions. Rather than relying on a single central authority, blockchain operates as a network infrastructure that establishes trust through distributed verifiability, auditability, and consensus (Crosby et al. 2016). Transactions are stored as a sequence of blocks arranged in chronological order. Each subsequent block is cryptographically linked to its predecessor by incorporating its hash value, or digital fingerprint. This structure ensures immutability, rendering records tamper-proof because data, once added, cannot be altered or deleted (Tian 2017).

Distributed consensus ensures that all network nodes reach agreement on the validity of transactions before new data is added. Various consensus mechanisms exist, including Proof of Work (PoW), Proof of Stake (PoS), and Practical Byzantine Fault Tolerance (PBFT) (Kamilaris et al. 2019). In addition to record-keeping, blockchain technology enables smart contracts, which are digital protocols with terms directly encoded into code. These contracts automatically execute predefined

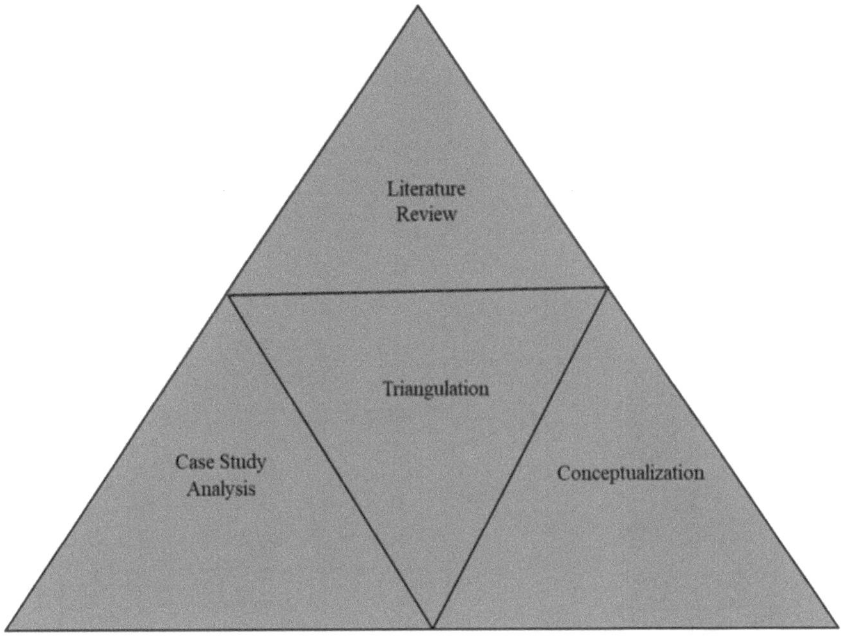

Fig. 10.1 Schematic representation of the triangulated research methodology

actions when specific conditions are met, streamlining processes such as payments, compliance checks, and contract enforcement without intermediaries (Leng et al. 2018). Most supply chain implementations utilize permissioned blockchains, such as Hyperledger Fabric, where participation is restricted to authorized entities. In contrast, public blockchains like Bitcoin are open to all users. Permissioned systems are generally more suitable for business applications in agri-food traceability, offering faster processing and enhanced privacy (Behnke and Janssen 2020). Figure 10.2 shows a schematic of stage-wise information capture and its transparent communication to the consumer.

10.4 Blockchain-Enabled Agri-Supply Chain Traceability

Blockchain represents a transformative digital infrastructure capable of addressing persistent traceability challenges in agricultural supply chains. Its core attributes—decentralization, immutability, transparency, and shared verification—render it particularly suitable for agricultural supply systems characterized by multiple intermediaries, fragmented recordkeeping, and heightened vulnerability to data manipulation, fraud, and adulteration. Stakeholders such as farmers, distributors, processors, logistics providers, and retailers participate in complex networks managing agricultural products. Traditional centralized systems frequently experience data silos, inconsistent reporting, and limited interoperability, which impede reliable

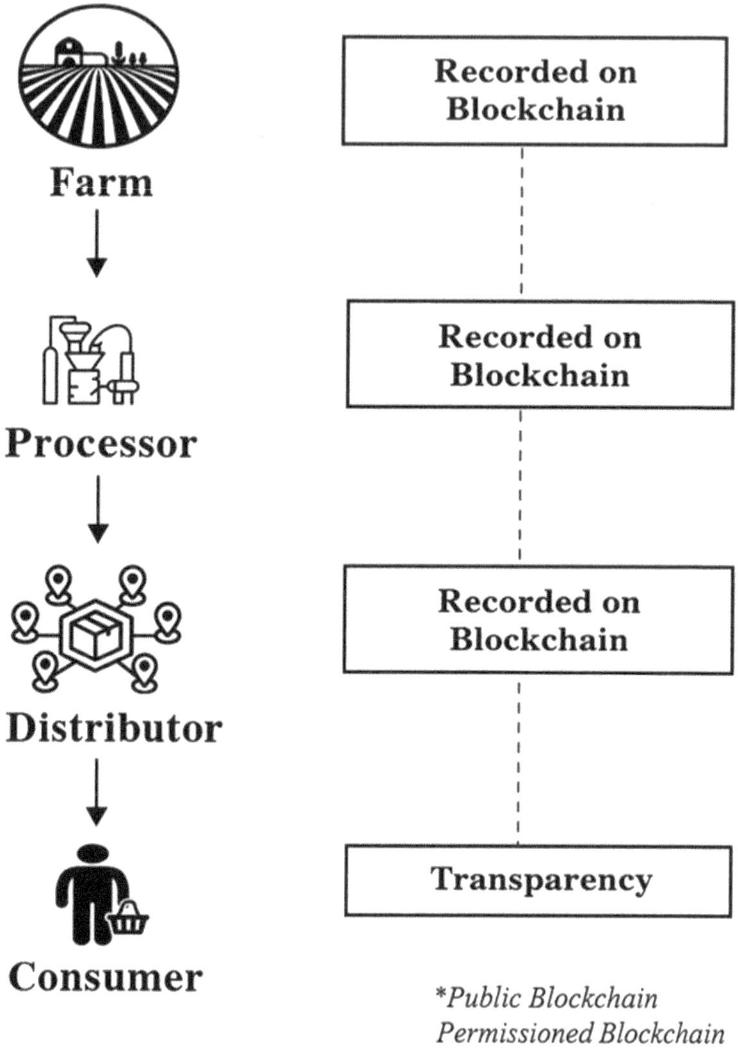

Fig. 10.2 Information capture at each stakeholder stage with transparent communication to the consumer

documentation across supply chain stages. In contrast, blockchain offers a distributed ledger architecture that enables a unified and tamper-resistant traceability framework by securely timestamping, cryptographically linking, and making each transaction accessible to authorized participants (Kamilaris et al. 2019; Tian 2017).

The unique characteristics of perishable and quality-sensitive commodities highlight the relevance of blockchain technology in agriculture. Ensuring safety and regulatory compliance requires careful handling, cold chain maintenance, and timely documentation for many food products. The immutability and auditability of blockchain records enhance accountability throughout the supply chain, reducing the risk of fraudulent relabeling and improving the efficiency of product recalls (Galvez et al. 2018). Furthermore, blockchain enables real-time data acquisition through integration with IoT devices and supports automated rule enforcement via smart contracts. This approach fosters a more reliable, data-driven process for verifying sustainability, safety, and authenticity in agri-food production. Several applications of blockchain in this context are outlined below.

10.4.1 Promoting Transparency in Production and Retail

Blockchain technology offers a significant means to enhance traceability and transparency across the entire agricultural supply chain. This technology enables verification and accountability regarding the origin, quality, and distribution of products throughout all stages. By establishing a decentralized and shared ledger, blockchain ensures that essential information about production, transportation, processing, and storage is securely recorded and readily accessible to all relevant stakeholders, including farmers, distributors, retailers, and consumers. The availability of consistent and immutable data fosters accountability and trust among participants. Enhanced transparency allows consumers to verify product authenticity and helps combat fraud, counterfeiting, and misrepresentation. For example, comprehensive product histories can be accessed by scanning a QR code on packaging. Walmart's IBM Food Trust pilot program significantly improved responsiveness and transparency by reducing traceability time from 7 days to approximately 2.2 s (Kamath 2018).

10.4.2 Agri-Food Systems: Blockchain Architecture and Ledger Design

Architectural design plays a critical role in agricultural blockchain systems by balancing privacy, scalability, and transparency. Most implementations utilize permissioned blockchains, such as Hyperledger Fabric, which restrict validation rights to trusted organizations and manage access. Consensus algorithms, including Raft and Practical Byzantine Fault Tolerance (PBFT), are employed to achieve higher throughput and lower latency, both of which are essential for processing high-volume agricultural data (Behnke and Janssen 2020). Ledger architecture often distinguishes between on-chain and off-chain data. Cryptographic hashes are stored on-chain to ensure data integrity, whereas large files or sensor outputs are maintained off-chain. This hybrid approach preserves immutability and supports scalability. Successful deployment further depends on interoperability with existing platforms and regulatory databases.

10.4.3 Digital Identity of Agricultural Products

Digital identity forms the foundation of blockchain-based traceability. Each product unit or batch receives a unique identifier, such as a QR code, RFID tag, or NFC token, which links physical goods to digital records. Throughout harvesting, processing, transportation, and retail, these identifiers ensure a continuous flow of information. In high-value commodities such as coffee and horticultural products, digital identity supports compliance with safety and quality standards. It also enables the verification of certifications, including organic, fair trade, and geographic indication (Galvez et al. 2018; Tian 2017).

10.4.4 Automated Compliance and Smart Contracts

Smart contracts within blockchain networks are self-executing protocols that automatically enforce predetermined rules. These contracts automate payments, verify certifications, expedite verification processes, and issue alerts when quality standards are compromised in agricultural supply chains. Leng et al. (2018) demonstrate that recording temperature variations and noncompliance incidents through smart contracts enhances cold-chain traceability. Automation supports timely decision-making for perishable goods, reduces human error, and decreases administrative workload.

10.4.5 Integration with IoT and Sensor Technologies

The integration of IoT devices with blockchain technology enhances traceability by eliminating the need for manual data entry. Sensors transmit data directly to blockchain networks, continuously recording variables such as location, temperature, and humidity. According to Leng et al. (2018), this integration strengthens cold-chain management, prevents unreported excursions, and enables real-time intervention. These capabilities facilitate proactive risk mitigation and improve data accuracy.

10.4.6 Real-Time Product Tracking and Quality Verification

Blockchain technology enables the integration of IoT devices for real-time tracking and quality assurance. Throughout planting, processing, and transportation, equipment such as GPS locators and environmental sensors collect data that is permanently recorded on the blockchain. Smart contracts automate responses to deviations, including halting production or issuing alerts. This approach allows for rapid identification and isolation of contaminated products, thereby enhancing food safety. For example, in cases of processing anomalies, blockchain-based egg traceability systems have facilitated the identification of contamination sources from small farms.

10.4.7 Consumer Transparency and Trust Mechanisms

Blockchain technology enhances transparency for consumers by providing access to comprehensive product histories. By scanning QR codes or RFID tags, consumers can obtain information regarding farm origin, production methods, certifications, and logistics. Galvez et al. (2018) found that consumers perceive blockchain-verified information as more credible than conventional labels. This level of transparency supports brand loyalty, enables premium pricing, and promotes ethical sourcing.

10.4.8 Governance, Interoperability, and Standards

The effectiveness of a blockchain system is largely determined by its governance, which encompasses decisions related to dispute resolution, validation authority, and data access. Behnke and Janssen (2020) argue that inadequate governance can result in the re-centralization of control, thereby undermining the foundational promise of decentralization associated with blockchain technology. Interoperability is also essential; blockchain platforms should be compatible with standards such as GS1 identifiers and comply with FAIR data principles. Recent developments in cross-chain protocols and multi-chain architectures provide innovative mechanisms for coordinating data flows across various industries and geographic regions.

10.5 Applications of Blockchain in Agri-Food Systems

The adoption of blockchain technology in agricultural supply chains has progressed beyond experimental pilots to encompass a growing number of operational deployments across diverse commodities and geographical regions. These implementations illustrate the potential of blockchain to enhance traceability, increase transparency, and enable more reliable information sharing among supply chain participants. While the maturity of blockchain applications varies significantly, documented real-world deployments provide valuable insights into blockchain performance under market conditions and highlight persistent challenges. The following subsections synthesize verified global examples sourced from reputable industry reports, academic literature, and publicly documented deployments.

10.5.1 Global Implementation

The adoption of blockchain technology in agri-food systems has increased significantly in sectors characterized by acute traceability challenges and heightened consumer transparency expectations. Early initiatives focused on commodities such as seafood, beef, pork, coffee, dairy, and high-value horticultural products, where risks of fraud, mislabeling, and safety violations can result in substantial economic and reputational losses. Kamilaris et al. (2019) report that initial blockchain pilots were predominantly implemented within multinational supply chains, particularly those involving major retailers such as Walmart, Carrefour, and Albertsons, with the objective of improving recall efficiency and restoring consumer confidence following food safety incidents.

With technological advancements, blockchain applications have extended beyond corporate-driven models to encompass government-led agricultural programs, national commodity boards, and farmer cooperatives. Prominent examples include digital marketplaces in India, fishery traceability projects in Southeast Asia, beef authentication systems in Australia, and cooperative-based traceability initiatives in Europe and Latin America. These developments illustrate the versatility of blockchain in enhancing both domestic food systems and export-oriented supply chains across diverse agricultural sectors experiencing increasing traceability requirements.

10.6 Case Studies

10.6.1 Walmart–IBM–Tsinghua University: Pork Traceability in China

The blockchain-enabled pork traceability project implemented by Walmart in China exemplifies the application of blockchain technology in agriculture. Utilizing IBM's Hyperledger Fabric, the system enables real-time tracking of pork from farm to retail. This initiative reduced the time required to retrieve product histories from days to seconds, thereby improving recall responsiveness (Kamilaris et al. 2019). Developed in collaboration with Tsinghua University, the project aimed to address fraud and contamination scandals that had previously affected China's pork supply chain. By ensuring that batch numbers, farm origins, certifications, and logistics data are immutable and verifiable, the project enhanced operational efficiency and regulatory confidence. The success of this pilot project subsequently influenced Walmart's blockchain traceability initiatives for leafy greens in the United States, demonstrating the scalability of the technology beyond a single commodity.

10.6.2 Coffee Board of India and Eka Software: A Blockchain Marketplace for Coffee

The Coffee Board of India introduced one of the earliest government-supported blockchain platforms for agricultural marketing, establishing a digital marketplace that directly connects growers and buyers. Developed in collaboration with Eka Software Solutions, the system assigns each coffee lot a unique digital identity and records farm-level production and processing data on a blockchain ledger. This approach aims to reduce information asymmetry and enhance the market visibility of small and marginal producers.

Academic and industry evaluations suggest that the platform enhances transparency and traceability in both domestic and export markets by verifying origin, varietal characteristics, processing methods, and quality attributes. This initiative facilitates more equitable price discovery and increases the global competitiveness of Indian coffee by fostering trust among supply-chain participants. However, the advantages are most evident in the branded premium segment, consistent with existing literature, while the largely unbranded domestic market in India may experience limited immediate benefits (Maity and Sharma 2020).

10.6.3 Starbucks: Farm-to-Cup Transparency

Starbucks implemented a blockchain-based transparency initiative that enables consumers to trace coffee beans from farms to retail outlets. This platform utilizes Microsoft Azure's Blockchain Service to provide data on supply-chain movements, roasting locations, farmer cooperatives, and farm origin. While Starbucks' internal data processes are not fully disclosed in academic literature, the initiative is frequently cited as a prominent example of blockchain's potential to enhance ethical sourcing and consumer engagement. The rationale for this implementation is substantiated by consumer behavior research. Rueda et al. (2017) found that supply-chain transparency significantly influences consumer perceptions of sustainability and ethical responsibility, particularly within specialty coffee markets. Starbucks' adoption of blockchain technology aligns with broader industry trends and illustrates how digital transparency can reinforce brand integrity for multinational corporations.

10.6.4 Sahyadri Farms (India): Blockchain for Farmer Cooperatives

Sahyadri Farms, one of the largest farmer-producer organizations (FPOs) in India, has implemented blockchain-supported traceability across various horticultural products. The cooperative systematically records production data, input usage, pesticide residue test results, logistics information, and financial transactions through QR-coded packages linked to blockchain records. This approach enhances consumer trust and supply-chain accountability, particularly for export-oriented fruits and vegetables. The model is distinguished by its inclusivity, as smallholder farmers collectively engage in the digital ecosystem, and blockchain data facilitates transparent revenue sharing. Research on cooperative-driven digitalization in India indicates that such models can improve growers' price realization and reduce dependence on intermediaries (Somwanshi et al. 2022). Sahyadri Farms demonstrates that blockchain technology can be applied beyond corporate governance to empower farmer-led organizations.

10.6.5 Additional Global Cases: Bumble Bee Foods, Provenance, and Beef Ledger

Beyond Asia, blockchain adoption has expanded into international fisheries, livestock, and specialty food systems. For instance, Bumble Bee Foods implemented blockchain technology to trace yellowfin tuna sourced from Indonesia, providing consumers with information on catch location, handling practices, and sustainability certifications (Galvez et al. 2018). The UK-based technology company Provenance piloted blockchain-based fish traceability in Southeast Asia to address illegal, unreported, and unregulated fishing. In Australia, BeefLedger utilized blockchain to authenticate beef exports to China, addressing concerns regarding product counterfeiting and misleading labeling in premium meat markets.

These implementations demonstrate the blockchain's adaptability in addressing transparency challenges across a range of commodity sectors and regulatory contexts. Several patterns emerge from the reviewed cases. Corporate-led initiatives, such as those by Walmart and Starbucks, prioritize consumer trust, recall efficiency, and brand reputation protection. In contrast, government-led and cooperative-led efforts, including the Coffee Board of India's platform and Sahyadri Farms' system, emphasize input quality assurance, equitable data sharing, and farmer empowerment.

Although the motivations for adoption vary, these cases collectively illustrate blockchain's capacity to establish secure, verifiable, and interoperable data trails that enhance traceability. However, consistent challenges persist, including integration with existing systems, effective governance design, affordability for smallholders, and the need for robust digital infrastructure. Successful deployments typically feature strong institutional leadership, well-defined governance structures, active stakeholder participation, and alignment with broader digitalization policies.

10.7 Implementation Challenges

10.7.1 Data Integrity at the Point of Entry

A primary challenge in blockchain-enabled agri-food traceability is ensuring data accuracy at the point of entry. Although blockchain technology guarantees immutability after data is recorded, it does not verify the correctness or authenticity of manually entered information. Consequently, agricultural systems remain highly susceptible to the "garbage in, garbage out" problem (Galvez et al. 2018; Rana 2020). Contributing factors include low digital literacy and fragmented documentation practices. To improve data integrity, the broader digital infrastructure should be enhanced through expanded mobile connectivity, reliable internet access, and the deployment of sensor networks that automate data collection and minimize human intervention (Salah et al. 2019). The adoption of digital identity systems, shared digital platforms, and IoT-enabled real-time monitoring further improves data accuracy and reduces opportunities for data manipulation. Implementing GS1 identifiers and interoperable data schemas standardizes data formats across stakeholders, while capacity-building initiatives and digital literacy training enable farmers and intermediaries to use digital tools effectively (Astill et al. 2019; Elkoraichi et al. 2025). Collectively, these measures support the creation of credible blockchain records and the maintenance of high-quality data throughout the traceability system.

10.7.2 Scalability and Technical Performance Constraints

Blockchain systems must process high-frequency data from sensors, logistics operations, and smart contracts, which can lead to performance bottlenecks as transaction volumes increase. Behnke and Janssen (2020) highlight that these technical limitations are particularly pronounced in rural areas, where intermittent connectivity, unreliable power supply, and outdated digital infrastructure further constrain system efficiency. To improve scalability and enable real-time operation, it is essential to concurrently develop rural broadband infrastructure, deploy IoT devices, and enhance computing power. The adoption of globally recognized interoperability standards, such as GS1 identifiers and harmonized data schemas, reduces processing redundancies and facilitates efficient data exchange across platforms. Astill et al. (2019) suggest that targeted investments in research on lightweight consensus algorithms and hybrid blockchain architectures can mitigate performance constraints. Furthermore, integrating blockchain into national digital agriculture platforms can provide shared infrastructure, significantly lowering technical barriers and supporting the development of scalable, interoperable blockchain ecosystems.

10.7.3 High Implementation Costs and Economic Barriers

A primary challenge associated with the adoption of blockchain technologies is the significant financial burden imposed on smallholder farmers, who must invest in tagging devices, smartphones, IoT sensors, and pay platform subscription fees (Treiblmaier 2019). Agribusinesses may also encounter additional barriers due to the high costs of integrating blockchain technology with existing systems or upgrading infrastructure. To address these economic challenges, targeted support such as credit assistance, digital tool subsidies, and publicly funded digital platforms that reduce individual adoption costs is necessary. Capacity-building programs, digital literacy training, and extension services can enhance farmers' understanding of blockchain technologies and increase their confidence in their financial benefits (Elkoraichi et al. 2025). Cooperative-led cost-sharing models, supported by shared IoT infrastructure and standardized tagging systems, can further distribute financial burdens among groups. Government investment in these measures can help ensure that high implementation costs do not exclude smallholders from participating in digital supply chains.

10.7.4 Governance Complexities and Power Asymmetries

Governance remains a significant challenge because blockchain networks rely on specific rules governing validator roles, data access permissions, and dispute resolution procedures. Dominant actors, typically large retailers or technology providers, may centralize ownership of the blockchain network, which undermines the decentralization ideal of blockchain and erodes institutional trust. According to Behnke and Janssen (2020), effective governance requires equitable access to validation processes, transparent regulations concerning data visibility, and representation of farming organizations and cooperatives in decision-making bodies. Policies aimed at preventing digital monopolies, promoting competition between platforms, and ensuring just principles in blockchain governance are essential for maintaining blockchain integrity. Inclusive governance frameworks help ensure that the benefits of blockchain are distributed equitably rather than perpetuating existing power imbalances.

10.7.5 Regulatory Uncertainty and Lack of Standards

The adoption of blockchain technology is significantly impeded by the absence of clear regulatory frameworks. Most countries have yet to define the legal status of blockchain records, smart contracts, and digital certifications. Furthermore, the lack of standardized identifiers such as GS1, interoperable data schemas, and compliance templates can result in fragmented blockchain systems that are incompatible with cross-border trade requirements (Astill et al. 2019). To address these challenges, governments should establish legal clarity by defining the enforceability of blockchain-generated data, developing regulations for smart contracts, and aligning blockchain systems with existing export, food safety, and inspection regulations. Additionally, international collaboration is essential for developing cross-border interoperability in export-oriented agricultural supply chains.

10.7.6 Socio-Institutional Resistance and Trust Barriers

Adoption of blockchain technology faces significant constraints, primarily due to socio-institutional resistance. Stakeholders such as farmers, processors, and intermediaries often express concerns that blockchain may increase surveillance, expose profit margins, or diminish their negotiating power. In many informal agricultural markets, blockchain-enabled transparency is viewed as undesirable because deliberate opacity is used to maintain control. Addressing these challenges requires robust privacy protections, equitable data-sharing agreements, and transparent benefit-sharing mechanisms to foster trust, especially among marginalized groups. Additionally, demonstration pilots and farmer-led trials can enhance confidence in blockchain systems by providing tangible evidence of improvements in traceability and livelihoods.

10.8 Conclusion and Way Forward

Blockchain-enabled traceability represents a significant advancement in agri-food governance, with the potential to address persistent challenges related to transparency, trust, and data integrity. Due to the inherent complexity of production networks and susceptibility to fraud, the decentralized architecture of blockchain offers distinct advantages for enhancing traceability within agricultural supply chains. This chapter analyzes the core characteristics of blockchain technology and its capacity to improve food production processes, logistics visibility, verification mechanisms, and documentation accuracy. Case studies, including Walmart's initiative in China and Starbucks' transparency efforts, demonstrate the successful application of blockchain in traceability. These examples indicate that blockchain has progressed from theoretical exploration to practical implementation, strengthening recall responsiveness, reducing fraudulent activities, and facilitating equitable market access.Nevertheless, several constraints persist, including quality assurance challenges, disparities in digital infrastructure, scalability limitations, prohibitive costs for smallholders, and complex governance issues related to data rights. Consequently, the successful deployment of blockchain requires a supportive institutional, regulatory, and socioeconomic context, as well as enabling policies and coordinated stakeholder efforts.

Looking ahead, integrating blockchain with complementary technologies such as the IoT, AI, and digital identity systems will be essential for optimizing traceability processes. Interoperability among different blockchain networks may provide advantages in compliance, cross-border trade, and sustainability reporting. Further research should investigate the socioeconomic impacts of blockchain, particularly regarding power dynamics and market inclusion among actors in agricultural supply chains. The extent to which blockchain benefits small-scale farmers will depend on targeted policy interventions that promote capacity building and equitable access.

Blockchain presents a transformative opportunity to enhance traceability and foster trust within agri-food systems. Its future effectiveness relies on careful integration with existing frameworks, inclusive governance structures, and sustained collaboration among researchers, policymakers, and industry stakeholders. As agriculture confronts globalization and sustainability challenges, blockchain emerges as an important tool for improving information sharing and advancing transparency in food systems.

References

Astill J, Dara RA, Campbell M et al (2019) Transparency in food supply chains: a review of enabling technology solutions. Trends Food Sci Technol 91:240–247. https://doi.org/10.1016/j.tifs.2019.07.024

Aung MM, Chang YS (2014) Traceability in a food supply chain: safety and quality perspectives. Food Control 39:172–184. https://doi.org/10.1016/j.foodcont.2013.11.007

Behnke K, Janssen M (2020) Boundary conditions for traceability in food supply chains using blockchain technology. Int J Inf Manag 52:101969. https://doi.org/10.1016/j.ijinfomgt.2019.05.025
Bolívar J, Cantillo V, Miranda P (2025) Agri–food supply chain design for perishable products: application to small-scale farmers. Oper Res Int J 25:26. https://doi.org/10.1007/s12351-024-00878-x
Bouzembrak Y, Klüche M, Gavai A et al (2018) Internet of things in food safety: literature review and a bibliometric analysis. Trends Food Sci Technol 79:21–32. https://doi.org/10.1016/j.tifs.2018.06.008
Charlebois S, Sterling B, Haratifar S et al (2014) Comparison of global food traceability regulations and requirements. Compr Rev Food Sci Food Saf 13:1104–1123. https://doi.org/10.1111/1541-4337.12101
Crosby M et al (2016) Blockchain technology: beyond bitcoin. Appl Innov Rev 2:6–19
Demestichas K, Peppes N, Alexakis T et al (2020) Blockchain in agriculture traceability systems: a review. Appl Sci 10:4113. https://doi.org/10.3390/app10124113
Ebneyamini S, Sadeghi Moghadam MR (2018) Toward developing a framework for conducting case study research. Int J Qual Methods 17:1–11. https://doi.org/10.1177/1609406918817954
Elkoraichi Y, Elfezazi S, Belhadi A (2025) Analysis of barriers to blockchain technology adoption in the African agri-food supply chain. Discov Sustain 6:289. https://doi.org/10.1007/s43621-025-00489-0
Galvez JF, Mejuto JC, Simal-Gandara J (2018) Future challenges on the use of blockchain for food traceability analysis. Trends Anal Chem 107:222–232. https://doi.org/10.1016/j.trac.2018.08.011
Gregor S (2006) The nature of theory in information systems. MIS Q 30:611–642. https://doi.org/10.2307/25148742
Kamath R (2018) Food traceability on blockchain: Walmart's pork and mango pilots with IBM. J Br Blockchain Assoc 1:1–12. https://doi.org/10.31585/jbba-1-1-(10)2018
Kamilaris A, Fonts A, Prenafeta-Boldú FX (2019) The rise of blockchain technology in agriculture and food supply chains. Trends Food Sci Technol 91:640–652. https://doi.org/10.1016/j.tifs.2019.07.034
Leng K, Bi Y, Jing L et al (2018) Research on agricultural products supply chain inspection system based on internet of things and blockchain technology. Futur Gener Comput Syst 108:1047–1055. https://doi.org/10.1016/j.future.2017.08.032
Lv G, Song C, Xu P et al (2023) Blockchain-based traceability for agricultural products: a systematic literature review. Agriculture 13:1757. https://doi.org/10.3390/agriculture13091757
Maity P, Sharma S (2020) Blockchain technology in coffee industry in India. PRAGATI J Indian Econ 7:61–75. https://doi.org/10.17492/pragati.v7i1.196024
Pozelli Sabio R, Spers EE (2022) Consumers' expectations on transparency of sustainable food chains. Front Sustain Food Syst 6:853692. https://doi.org/10.3389/fsufs.2022.853692
Rana S (2020) Blockchain-based traceability and transparency in agricultural supply chains: challenges and opportunities. Turk J Comput Math Educ 11:1948–1956. https://doi.org/10.17762/turcomat.v11i3.13591
Rueda X, Garrett RD, Lambin EF (2017) Corporate investments in supply chain sustainability: selecting instruments in the agri-food industry. J Clean Prod 142:2480–2492. https://doi.org/10.1016/j.jclepro.2016.11.026
Salah K, Nizamuddin N, Jayaraman R et al (2019) Blockchain-based soybean traceability in agricultural supply chain. IEEE Access 7:73295–73305. https://doi.org/10.1109/ACCESS.2019.2918000
Somwanshi G, Deshmukh P, Anantwar G et al (2022) Empowering farmers with blockchain-based transparency throughout the value chain. Indian J Plant Genet Resour 35:41–45. https://doi.org/10.5958/0976-1926.2022.00039.0
Tian F (2017) A supply chain traceability system for food safety based on HACCP, blockchain & IoT. ICSSSM, Dalian
Treiblmaier H (2019) Combining blockchain technology and the physical internet to achieve triple bottom line sustainability: a comprehensive research agenda for modern logistics and supply chain management. Logistics 3:1. https://doi.org/10.3390/logistics3010010
Yao Q, Zhang H (2022) Improving agricultural product traceability using blockchain. Sensors 22:3388. https://doi.org/10.3390/s22093388

Smart Contracts and Decentralized Marketplaces in Agriculture

11

Soumya Mohapatra, P. Sruthi, Ramadas Sendhil, and Anuj Kumar

Abstract

Disruptive technologies, including smart contracts, serve as automated, enforceable agreements. Decentralized marketplaces serve as self-executing platforms for transactions. Both innovations are transforming global agriculture and are widely regarded as promising tools for development. Blockchain-based solutions improve efficiency, transparency, and trust among stakeholders within agricultural supply chains. The adoption of digital technologies has accelerated in recent years due to their diverse benefits, including real-time trading, Internet of Things (IoT)-enabled logistics, subsidy disbursements, and insurance claims in agriculture. However, significant challenges persist, particularly for smallholders. These challenges include limited digital literacy, interoperability across platforms, scalability issues, and legal complexities. This chapter presents potential strategies, including capacity building, improving digital literacy, fostering public–private partnerships, and increasing investment in research. These approaches aim to enhance stakeholder engagement throughout the value chain. Coordinated efforts by multiple stakeholders, with government support, are essential to build and strengthen the digital ecosystem in agriculture. Such initiatives can align the sector with global objectives for inclusive growth and sustainable development.

S. Mohapatra
ICAR-Central Institute of Post-Harvest Engineering & Technology, Ludhiana, Punjab, India

P. Sruthi
ICAR-National Dairy Research Institute, Karnal, Haryana, India

R. Sendhil (✉)
Pondicherry University, Kalapet, Puducherry, India
e-mail: sendhil.eco@pondiuni.ac.in

A. Kumar
Indian Institute of Wheat and Barley Research, Karnal, Haryana, India

H. N. Oo et al. (eds.), *Transforming Agriculture with AI, IoT, and Blockchain*,
https://doi.org/10.1007/978-981-92-1491-4_11

Keywords

Blockchain · Disruptive technologies · Trade platforms · Legal contract

11.1 Introduction

The agriculture sector is undergoing a significant digital transformation. Among the most disruptive technologies in digital agriculture are smart contracts and decentralized marketplaces, both of which rely on blockchain technology and represent promising drivers of economic development. A smart contract is defined as "an automatable and enforceable agreement; automatable by a computer, although some parts may require human input and control; enforceable either by legal enforcement of rights and obligations or via tamper-proof execution of computer code" (Clack et al. 2016). Decentralized marketplaces are blockchain-based self-executing platforms that enable individuals to interact and transact across regions. Together, these innovations enhance transparency, efficiency, and trust among supply chain actors, and have experienced rapid adoption in recent years (Mohapatra et al. 2025; Tripoli and Schmidhuber 2020; Kamilaris et al. 2019). Smart contracts integrated with decentralized marketplaces are considered efficient because they enable transparent exchanges, with every transaction recorded in a shared ledger. This stands in contrast to conventional markets, which are characterized by numerous intermediaries, information asymmetry, delayed payments, and a lack of traceability and transparency (Kshetri 2021).

Agricultural value chains can be transformed through the adoption of disruptive innovations, resulting in a collaborative digital ecosystem that enhances fairness in trade, traceability, and inclusivity (Sander et al. 2018). Globally, such applications are emerging in areas including commodity trading, agricultural insurance, traceability, certification, land registration, and input delivery (Lin et al. 2019; Mao et al. 2018). For example, blockchain-enabled platforms for coffee trading, beef exports, and solid waste management are operating successfully. Nevertheless, adoption is constrained by inadequate infrastructure, insufficient regulatory frameworks, and institutional challenges, including limited digital literacy, platform interoperability, scalability concerns, and the need for legally binding contracts (Paliwal et al. 2020).

In this context, this chapter examines the conceptual foundations, applications, potential impacts, challenges, and opportunities associated with smart contracts and decentralized marketplaces in agriculture (Fig. 11.1). It further addresses the ongoing debate regarding digital transformation in the agricultural sector and the transition of emerging economies toward data-driven, participatory, and equitable agri-food systems.

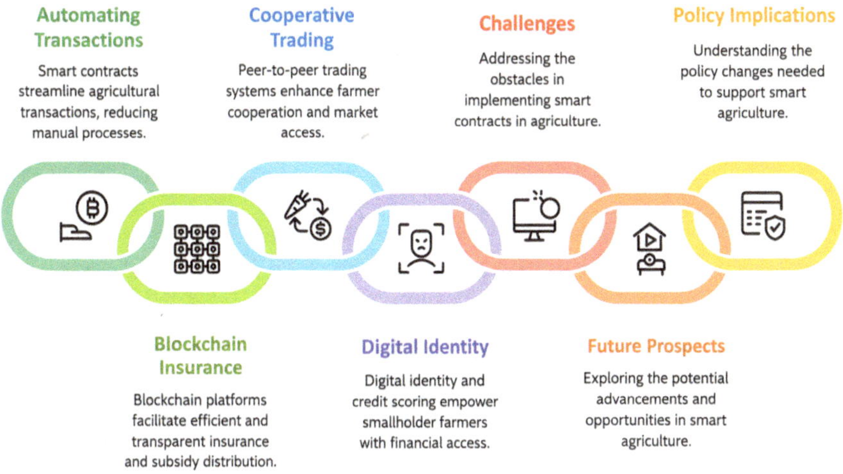

Fig. 11.1 Conceptual framework. (The above figure is generated from https://app.napkin.ai/)

11.2 Role of Smart Contracts in Automating Agricultural Transactions

Historically, agricultural transactions required multiple manual steps, often leading to errors and inefficiencies in the transaction system. These processes typically included multitiered approvals, which contributed to payment delays and other transaction-related challenges. Smart contracts have consequently emerged as an effective means to automate transactions through end-to-end encryption, thereby enhancing operational security and accuracy (Krithika 2022). Smart contracts are self-executing agreements designed to transform the management of agricultural transaction systems. These digital agreements incorporate conditional logic and are written as computer code on blockchain networks. They utilize distributed ledger technologies, such as Polygon, Ethereum, and Hyperledger, to automatically execute actions once predefined conditions are met. In the agricultural sector, where informal contracts are common, blockchain-based smart contracts promote trust and transparency along supply chains and increase market network efficiency by removing intermediaries (Kamilaris et al. 2019).

11.2.1 Key Roles of Smart Contracts in Agricultural Transactions

Smart contracts enhance farmers' bargaining power and reduce their reliance on intermediaries, enabling them to secure better prices for their produce. The application of smart contracts encompasses input supply, production, logistics, marketing, and quality regulation (Fig. 11.2). Decentralized ledgers, when integrated with smart sensors, automate logistics and facilitate real-time data storage to maintain

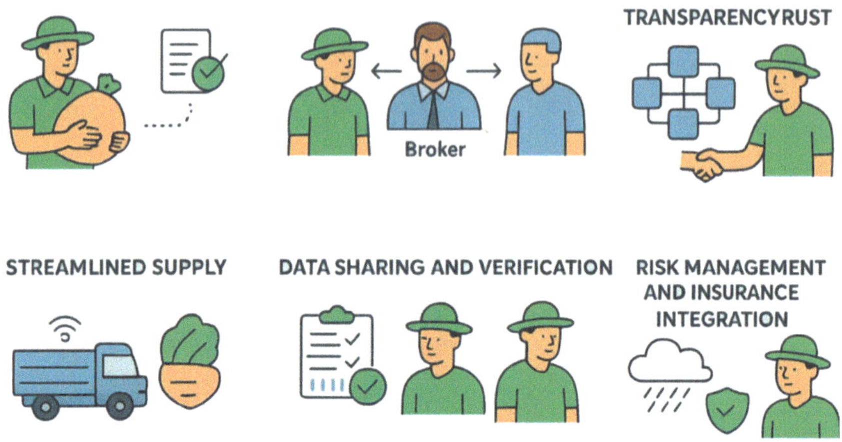

Fig. 11.2 Application of smart contracts in agriculture. (*Source:* Puthenveettil and Sappati (2024))

coordination among stakeholders throughout the supply chain. Upon delivery and verification of produce via IoT-based sensors, payments are executed automatically through the smart contract, removing the need for manual intervention. This automation accelerates payment settlements, reduces risks related to fraud, disputes, and manipulation, and extends to trade finance and warehouse receipt systems, where smart contracts enable instant collateral verification and loan disbursement.

11.3 Insurance and Subsidy Disbursement Through Blockchain Platforms

Agricultural insurance and subsidy programs in developing countries frequently experience systemic inefficiencies, such as delayed verification, lack of transparency, and misallocation of funds. The financial resources intended for marginal and smallholder farmers are often diverted through fraudulent claims and documentation, thereby diminishing the effectiveness of government assistance programs. The absence of robust regulatory frameworks, insufficient verification procedures, and reliance on paper-based processes further facilitate fraudulent activities within the agricultural finance system. These persistent challenges undermine the objectives of support schemes and erode farmers' trust.

11.3.1 Smart Contract-Enabled Solutions

The implementation of blockchain-integrated smart contracts enables comprehensive transaction recording and maintains a secure database of stakeholders. This approach reduces transaction time and enables immediate settlement of funds, thereby addressing inefficiencies in agricultural subsidy disbursements and insurance claims processes. Additionally, it streamlines on-farm verification for insurance claims by generating automated weather data and supporting index-based claim settlements. The system provides real-time verification data and securely stores farmers' digital records, ensuring the timely delivery of benefits to intended recipients. Furthermore, the adoption of smart contracts lowers transaction costs by minimizing administrative procedures and reducing verification and reporting expenses.

11.3.1.1 India's Fertilizer Subsidy Program: A Case Study

India operates one of the world's largest fertilizer subsidy programs, designed to reimburse manufacturers for the difference between production costs and sale prices. To enhance efficiency and transparency in fertilizer supply chains and ensure timely production, procurement, and distribution, NITI Aayog, in collaboration with Intel and PwC India, initiated a blockchain-based pilot project (Fig. 11.3).

The initiative integrated production, dispatch, and sales records into a shared blockchain ledger, granting stakeholders real-time access to verified data. This system substantially reduced reliance on paper-based operations, enabled instant

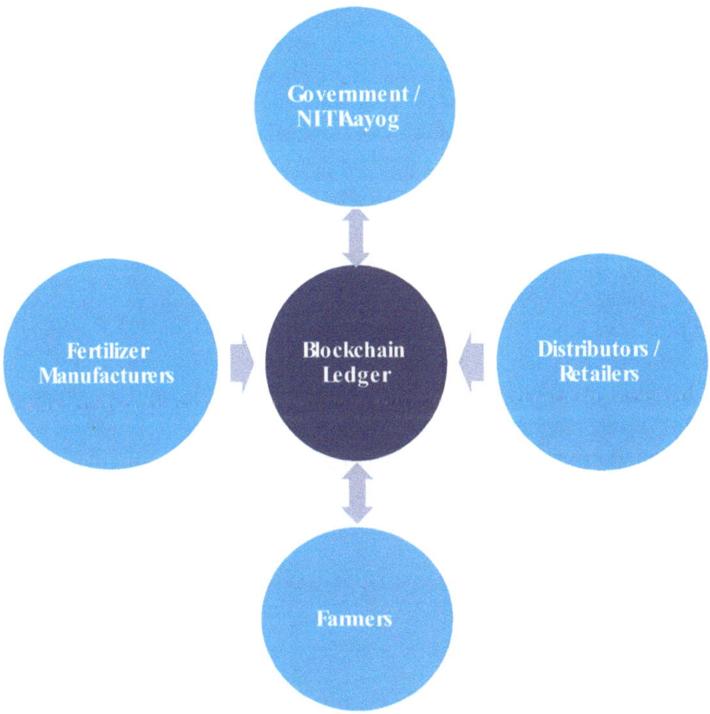

Fig. 11.3 Blockchain-enabled fertilizer subsidy management

dispatch certifications, and automated data entry into enterprise resource planning (ERP) systems. Synchronizing records across entities minimized settlement disputes, enhanced auditability, and demonstrated potential to streamline workflows within the agricultural finance system.

11.4 Farmer Cooperatives and Peer-to-Peer Trading Systems

Agricultural markets in developing nations face significant constraints, including reliance on intermediaries, high transaction costs, information asymmetry, inadequate infrastructure, and limited bargaining power for producers. The cooperative marketing model, when integrated with smart contract technology and blockchain systems, has regained relevance by facilitating a peer-to-peer (P2P) digital exchange ecosystem that reduces reliance on traditional middlemen (Leduc et al. 2021). Distributed ledger technologies enable cooperatives to communicate directly with farmers, foster trust among stakeholders, and verify digital transactions (Kumarathunga et al. 2022).

11.4.1 Evolution of Cooperative Digitalization

In developing countries such as India, traditional agricultural markets, including APMC mandis and e-NAM, as well as cooperatives such as farmer-producer companies and dairy cooperative societies, frequently encounter challenges related to centralized control, delayed payments, and labor-intensive manual accounting procedures. These entities also rely on centralized intermediaries for logistics management and price discovery. In contrast, smart-contract-enabled digital marketplaces facilitate automatic fund disbursement and ensure compliance with market regulations and cooperative rules. Decentralized smart contracts allow farmers to conduct direct digital transactions with other parties, enable instant fund settlement, and enhance farmers' profitability by reducing reliance on intermediaries (Daraghmi et al. 2024). AMUL, the largest dairy cooperative in India, utilizes smart contracts to maintain transparency and traceability within the milk procurement system throughout the dairy value chain (NDDB 2023). Table 11.1 presents a comparison between traditional and smart contract-enabled cooperative marketing models.

11.4.2 Peer-to-Peer (P2P) Trading Models in Agriculture

Blockchain-powered agricultural cooperatives are developing along three primary peer-to-peer (P2P) models:

1. Input exchange systems facilitate the sharing of community-based seeds, fertilizers, agrochemicals, and other inputs, with transactions digitally recorded on decentralized ledgers to enhance reliability.

Table 11.1 Traditional vis-à-vis smart contract-enabled cooperative marketing

Particulars	Traditional cooperatives	Smart contract-enabled cooperatives
Transparency	Low-to-moderate	Very high
Payment system	Time-consuming (payment cycle continues up to 10–15 days)	Instant (digital transaction)
Accounting	Manual- and paper-based record keeping	Decentralized digital ledger technologies
Price discovery	Centralized control by the intermediaries	Peer-to-peer negotiation
Governance	Elected members and committees	Digital and algorithmic

2. Commodity trading platforms establish direct connections between farmers and buyers by offering digital marketplaces where agricultural produce can be listed and auctioned (e.g., Ninjacart and Agribazaar).
3. Asset sharing networks allow cooperatives to rent machinery and equipment to members using verified smart contracts (e.g., Trringo).

Blockchain-enabled contracts implemented by cooperatives have the potential to increase farmers' net income by 12%–15% through real-time price discovery, enhanced bargaining power, and reduced intermediary commissions (Yadav et al. 2023).

11.4.3 Digital Grain Marketplace by TS-MARKFED: A Case Study

In 2022, the Telangana State Cooperative Marketing Federation (TS-MARKFED) piloted a digital grain marketplace initiative to establish a blockchain-integrated grain trading platform. The platform registered over 30,000 farmers and enabled buyers to bid through digital contracts. This initiative reduced transaction costs, enhanced collateral access via warehouse receipts, and facilitated the realization of instant payments for farmers. The key features of the scheme are summarised in Table 11.2.

This platform enables price fixation at the harvest stage and automatic settlement of payments when product quality meets predefined standards. Consequently, smart contracts demonstrate the potential and scalability of blockchain technology in cooperative marketing by promoting transparency, credibility, and efficiency throughout supply chains. Nevertheless, the adoption of smart contracts poses several challenges related to regulatory frameworks and governance that require careful monitoring mechanisms (Puthenveettil and Sappati 2024). Key constraints associated with the use of smart contracts and blockchain in cooperative marketing models include the following:

- *Cybersecurity risk*: potential data privacy breaches during the design of smart contracts

Table 11.2 Key features of the digital grain marketplace

Characteristics	Outcomes
Collectivization of farmers	More than 30,000 farmers were registered through cooperatives
Payment settlement	Payment cycle reduced from 12 days to 24 h through a smart contract-based instant payment
Quality control of the produce	Reduction in disputes by 12% through IoT-based digital verification
Warehouse receipts as digital tokens	Collateralization of loans increased by 22%

- *Digital literacy*: limited digital proficiency among stakeholders
- *Data compatibility*: challenges in harmonizing data across various regional platforms

India's Multi-State Cooperative Societies (Amendment) Act, 2023, emphasizes digital accounting and auditing and provides a robust policy framework for future integration into smart contract-enabled cooperatives.

11.5 Digital Identity and Credit Scoring for Smallholder Farmers

In developing nations, the agriculture sector is predominantly composed of marginal and smallholder farmers. Although financial inclusion has expanded significantly, smallholders often remain excluded from formal credit. Informal credit systems typically rely on land and collateral-based appraisals, which systematically exclude landless and tenant farmers. However, blockchain-integrated digital identities (DIDs) provide a secure and trusted database of farmers' transactions, production, and cooperative participation (Clark et al. 2025). This approach facilitates the development of a digital, decentralized credit scoring ecosystem, thereby supporting the establishment of a reliable and transparent formal credit system in agriculture.

A decentralized identity (DID) system comprises three primary components: credential issuers, such as government agencies and cooperatives that verify user-updated data; a blockchain registry, which stores information accessible exclusively with the farmer's consent; and financial institutions, which access anonymized data to estimate credible credit scores using artificial intelligence (AI) algorithms (Saurabh et al. 2023). Figure 11.4 illustrates the mechanism of digital identity and credit scoring in agriculture.

11.5.1 Cooperatives' Integration into Smart Contract-Based Digital Credit Systems

Cooperatives serve as essential verifiers of member data and ensure transparency in transaction processes. Blockchain-enabled cooperative databases frequently verify

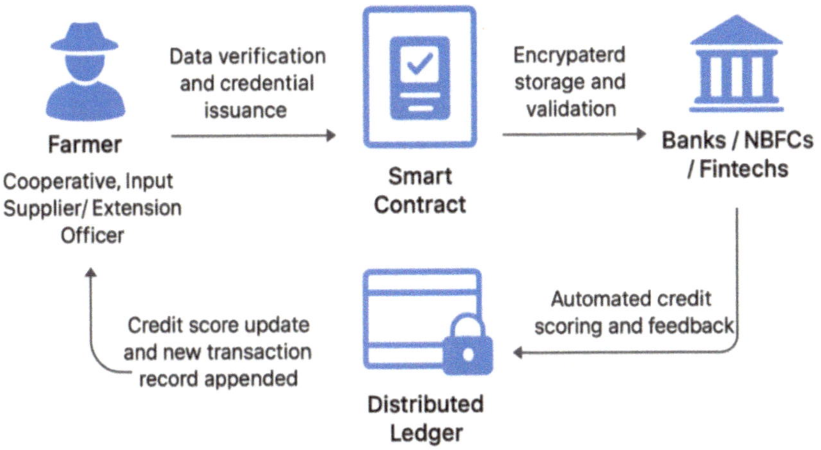

Fig. 11.4 Digital identity and credit scoring in agri-finance

Table 11.3 Impact of blockchain-based digital credit platforms

Platform	Country	Technology	Farmers' coverage	Impact on credit access
Agristack	India	AI + Blockchain	25 crores	Increased loan approvals by 22%
HARA	Indonesia	Blockchain	2 lakhs	Improved credit access for tenant farmers by 25%
Twiga foods	Kenya	Mobile ledger	4000	Increased working capital flow by 18%

Source: GoI (2023), Mulinge et al. (2021), Rijanto (2021)

yield, subsidy, and payment data submitted by members and incorporate this information into credit scoring mechanisms. This integration enhances smallholders' access to institutional credit (Noor et al. 2022). Table 11.3 summarizes the collective impact of various smart contract platforms globally on farmers' credit accessibility.

11.5.2 Agristack Initiative in India: A Case Study

Agristack is an initiative by the Government of India designed to establish a digital public infrastructure for the agriculture sector. Its objectives include enhancing farmers' access to credit, delivering real-time data-driven solutions, and streamlining the distribution of government benefits. The pilot phase focused on managing data on land, crops, and credit repayment for Indian farmers (GoI 2023).

The principal features of this initiative include the following:

- Assignment of unique digital farmer identification numbers to all farmers
- Reduction of loan approval duration from 10 days to 2 days
- Fifty percent reduction in the loan default rate
- Twenty-two percent increase in the loan approval rate.
- Estimation of digital credit scores utilizing yield and transaction data

Additionally, the initiative integrates transaction data from the e-NAM and PM-KISAN platforms into farmers' unique digital identification profiles, thereby enabling banks to conduct more reliable loan risk assessments.

11.6 Challenges in the Adoption of Smart Contracts and Decentralized Marketplaces

The adoption of blockchain-enabled decentralized smart contracts in agriculture faces several socioeconomic, institutional, and technical constraints, as shown in Fig. 11.5.

- *Scalability challenges*: High infrastructure and energy costs, ongoing system maintenance, and limited internet connectivity in remote areas restrict the scalability of digital financial systems.
- *Interoperability constraints*: Existing information systems and government databases are often incompatible with decentralized ledger technology, hindering market operations across platforms.
- *Digital illiteracy and accessibility barriers*: Smallholder farmers often lack awareness of and access to digital innovations, limiting adoption.

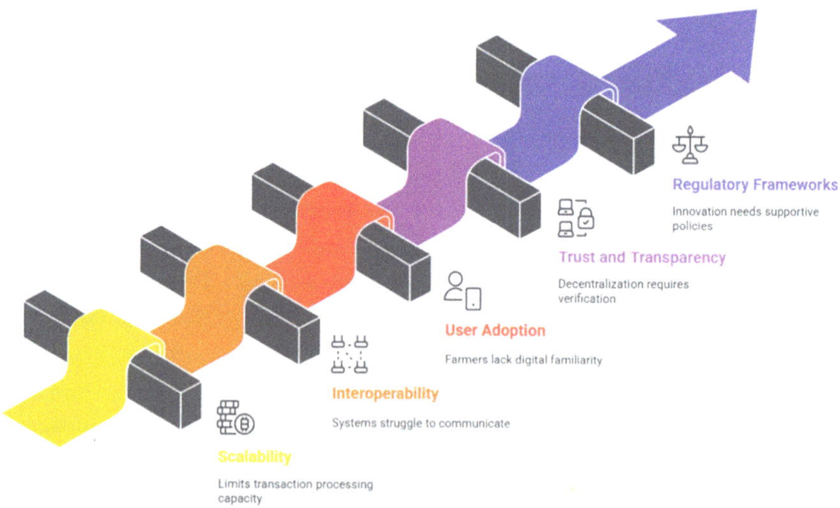

Fig. 11.5 Challenges of smart contracts in agriculture. (The above figure is generated from https://app.napkin.ai/)

- *Trust and transparency issues*: Ensuring privacy, trust, and transparency, as well as protecting farmers' personal data from breaches and unauthorized access, remains a significant challenge in developing nations.
- *Lack of standardized regulatory frameworks*: The absence of uniform protocols, supportive policies, and clear legal recognition of blockchain-enabled transactions hampers the effective implementation of smart contracts in agriculture.

11.7 Future Prospects of Smart Contracts and Decentralized Marketplaces in Agriculture

The future of agriculture is increasingly shaped by digital technologies, with smart contracts and decentralized marketplaces emerging as foundational elements for next-generation agricultural and economic development. As regulatory frameworks improve, the integration of these technologies within the agricultural sector is expected to become a mainstream mechanism for trade, value addition, and governance (Kamilaris et al. 2019). Blockchain-enabled validation can facilitate trust among participants by ensuring accountability, providing direct access to commodities, enabling fair pricing for all stakeholders, and supporting real-time traceability along the value chain. This approach also enhances trade transparency and promotes higher welfare outcomes. Automated processes enabled by IoT, spanning from input procurement to insurance claims, contribute to cost reduction and efficiency improvements at all levels, while addressing long-term sustainability challenges (Fig. 11.6). Aggregators such as Farmer-Producer Organizations (FPOs), cooperatives, and start-ups must take advantage of these innovations to help countries achieve policy objectives related to inclusive growth and sustainable development. Realizing the full potential of digital agriculture requires strengthened infrastructure, increased investment in value chains, simplified legal frameworks, and enhanced institutional capacity. Furthermore, public–private partnerships, supported by international organizations, are essential for establishing and reinforcing the digital ecosystem in agriculture.

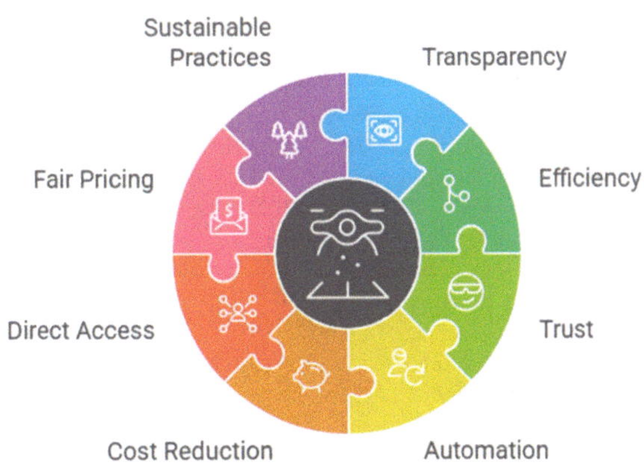

Fig. 11.6 Future prospects of digital agriculture. (The above figure is generated from https://app.napkin.ai/)

11.8 Conclusions

The agriculture sector has undergone continuous transformation since the advent of civilization and is currently progressing toward a digital ecosystem. Disruptive technologies, such as smart contracts and decentralized marketplaces, are regarded as promising innovations for agricultural and economic development. However, several challenges persist, including limited digital literacy, interoperability among transaction platforms, scalability concerns, and contract-related legal complexities. To ensure inclusive participation across the value chain, it is essential to implement capacity-building programs for chain actors and targeted digital literacy initiatives, particularly for smallholders, at all levels. Furthermore, governments, in collaboration with private entities, should invest in research to enhance platform security, protect data privacy, and promote farmer-centric trading mechanisms. Addressing these challenges requires multilevel policy support, strategic investment in digital infrastructure, and sustained efforts to strengthen overall capacity. Such coordinated actions will reinforce the digital ecosystem and advance global objectives of inclusive growth and sustainable development.

References

Clack CD, Bakshi VA, Braine L (2016) Smart contract templates: foundations, design landscape and research directions. arXiv:1608.00771. https://doi.org/10.48550/arXiv.1608.00771

Clark J, Marin G, Ardic Alper OP et al (2025) Digital public infrastructure and development: a World Bank Group approach. Digital Transformation White Paper, vol 1. World Bank, Washington, DC

Daraghmi EY, Jayousi S, Daraghmi YA et al (2024) Smart contracts for managing the agricultural supply chain: a practical case study. IEEE Access 12:125462–125479. https://doi.org/10.1109/ACCESS.2024.3439412

GoI (2023) Agri stack: digital infrastructure by the states and the central government. Ministry of Agriculture and Farmers' Welfare, Government of India. https://agristack.gov.in/#/

Kamilaris A, Fonts A, Prenafeta-Boldú FX (2019) The rise of blockchain technology in agriculture and food supply chains. Trends Food Sci Technol 91:640–652. https://doi.org/10.1016/j.tifs.2019.07.034

Krithika LB (2022) Survey on the applications of blockchain in agriculture. Agriculture 12(9):1333. https://doi.org/10.3390/agriculture12091333

Kshetri N (2021) Blockchain and sustainable supply chain management in developing countries. Int J Inf Manag 60:102376. https://doi.org/10.1016/j.ijinfomgt.2021.102376

Kumarathunga M, Calheiros RN, Ginige A (2022) Smart agricultural futures market: blockchain technology as a trust enabler between smallholder farmers and buyers. Sustainability 14(5):2916. https://doi.org/10.3390/su14052916

Leduc G, Kubler S, Georges J (2021) Innovative blockchain-based farming marketplace and smart contract performance evaluation. J Clean Prod 306:127055. https://doi.org/10.1016/j.jclepro.2021.127055

Lin Q, Wang H, Pei X et al (2019) Food safety traceability system based on blockchain and EPCIS. IEEE Access 7:20698–20707. https://doi.org/10.1109/ACCESS.2019.2897792

Mao D, Hao Z, Wang F et al (2018) Innovative blockchain-based approach for sustainable and credible environment in food trade: a case study in Shandong Province, China. Sustainability 10(9):3149. https://doi.org/10.3390/su10093149

Mohapatra S, Sainath B, Anirudh KC et al (2025) Application of blockchain technology in the agri-food system: a systematic bibliometric visualization analysis and policy imperatives. J Agribus Dev Emerg Econ 15(2):288–310. https://doi.org/10.1108/JADEE-10-2022-0237

Mulinge W, Bett C, Makokha S et al (2021) Report on digital platforms in Kenya: case study report—Twiga Foods. Forum Agric Res Afr 8(4):1–81

NDDB (2023) National digital livestock mission blueprint. National Dairy Development Board, Ministry of Fisheries, Animal Husbandry and Dairying, Government of India. https://megahvt.gov.in/notification/National%20Digital%20Livestock%20Mission.pdf

Noor NM, Razali NAM, Malizan NA et al (2022) Decentralized access control using blockchain technology for application in smart farming. Int J Adv Comput Sci Appl 13(9):1–10. https://doi.org/10.14569/IJACSA.2022.0130993

Paliwal V, Chandra S, Sharma S (2020) Blockchain technology for sustainable supply chain management: a systematic literature review and a classification framework. Sustainability 12(18):7638. https://doi.org/10.3390/su12187638

Puthenveettil NR, Sappati PK (2024) A review of smart contract adoption in agriculture and food industry. Comput Electron Agric 223:109061. https://doi.org/10.1016/j.compag.2024.109061

Rijanto A (2021) Business financing and blockchain technology adoption in agroindustry. J Sci Technol Policy Manag 12(2):215–235. https://doi.org/10.1108/JSTPM-03-2020-0065

Sander F, Semeijn J, Mahr D (2018) The acceptance of blockchain technology in meat traceability and transparency. Br Food J 120(9):2066–2079. https://doi.org/10.1108/BFJ-07-2017-0365

Saurabh K, Upadhyay P, Rani N (2023) A study on blockchain-based marketplace governance platform adoption: a multi-industry perspective. Digit Policy Regul Gov 25(6):653–692. https://doi.org/10.1108/DPRG-04-2023-0053

Tripoli M, Schmidhuber J (2020) Emerging opportunities for the application of blockchain in the agri-food industry. Food and Agriculture Organization of the United Nations (FAO). https://openknowledge.fao.org/handle/20.500.14283/ca9934en

Yadav VS, Singh AR, Raut RD et al (2023) Blockchain drivers to achieve sustainable food security in the Indian context. Ann Oper Res 327:211–249. https://doi.org/10.1007/s10479-021-04308-5

Smart Sensing and AI for Monitoring Rancidity in PUFA-Rich Agricultural Products

12

Anupma Sharma, Ritesh Kumar, Rishemjit Kaur, Sudeshna Bagchi, Saurav Kumar, and Amol P. Bhondekar

Abstract

Matrices containing polyunsaturated fatty acids (PUFA), including nuts, edible oils, seeds, and fish oils, are highly susceptible to oxidative rancidity, which negatively impacts quality, safety, and consumer acceptance. This chapter examines smart farming strategies for monitoring rancidity in two representative PUFA-rich food systems: cashew nuts and edible oils. In both case studies, the detection of oxidative degradation products responsible for rancidity utilizes advanced volatile organic compound (VOC) analysis through gas chromatography (GC) and electronic nose (e-nose) technologies. The application of machine learning (ML) models to VOC data has identified key markers in cashews, such as propanal and hexanal, that indicate quality grades. For edible oils, real-time rancidity monitoring on storage containers employs chemoreceptor-based colorimetric sensors, supported by novel algorithms that track changes in color indexes associated with rancidity. This chapter further explores the development and application of chemometric sensor platforms, integrating multi-sensor data fusion and AI-based pattern recognition to enable accurate, automated quality assessment in agriculture. These sensor systems may be integrated with cloud-based Internet of Things (IoT) platforms and blockchain-enabled traceability, facilitating a convergence of sensing, computation, and data transparency essential to smart agricultural ecosystems. Critical considerations such as sensor calibration, interoperability, cost constraints, and field deployment are addressed, along with recommendations for scalable implementation. Such integrated smart agricultural systems have the potential to transform quality monitoring of PUFA-rich foods by linking advanced sensor technologies to practical applications,

A. Sharma (✉) · R. Kumar · R. Kaur · S. Bagchi · S. Kumar · A. P. Bhondekar
CSIR-Central Scientific Instruments Organization, Chandigarh, Punjab, India

H. N. Oo et al. (eds.), *Transforming Agriculture with AI, IoT, and Blockchain*,
https://doi.org/10.1007/978-981-92-1491-4_12

thereby enhancing supply chain reliability, reducing post-harvest losses, and increasing consumer confidence.

Keywords

PUFA · Nuts · AI · Chemometric · Oil · Hexanal

12.1 Introduction

12.1.1 Nutritional and Economic Significance of PUFAs

Polyunsaturated fatty acids (PUFAs), particularly linoleic acid (C18:2, ω-6) and alpha-linolenic acid (C18:3, ω-3), constitute a critical nutrient group with well-documented roles in cardiovascular protection, modulation of inflammation, cognitive development, and immune regulation (Mititelu et al. 2024; Saini and Keum 2018). These PUFAs serve as precursors to eicosanoids and docosanoids, bioactive metabolites that regulate metabolic and immunologic processes at both cellular and systemic levels (Kumar et al. 2019; Dennis and Norris 2015; Calder 2006). Owing to these established health benefits, major health organizations worldwide recommend including PUFA-rich foods in a balanced diet (Molendi-Coste et al. 2011; Lands 2014).

The economic significance of commodities with a high PUFA content is considerable. The global edible oil market is projected to reach USD 223 billion in 2024 and USD 322 billion by 2032, with a compound annual growth rate (CAGR) of 5.4% (OECD-FAO 2023). Major sources of PUFA include vegetable oils (soybean, sunflower, rapeseed, sesame, groundnut, palm oils), tree nuts (cashews, almonds, walnuts), oilseeds, and marine lipids, which collectively account for nearly one-third of global agricultural trade (Calder 2006; OECD-FAO 2023; Dorni et al. 2018). Among PUFA-rich commodities, tree nuts—particularly cashews—represent a significant segment. Ivory Coast and India together produce just under 46% of the global cashew kernel supply, estimated at approximately 3.9 million tonnes annually (Food and Agriculture Organization of United Nations 2025; Belitz et al. 2008). During the 2024–25 fiscal year, cashew exports from India alone totaled USD 404 million. In sub-Saharan Africa, exports of cashews and groundnuts constitute 8%–15% of agricultural export earnings, thereby supporting rural livelihoods and contributing to food security (Food and Agriculture Organization of United Nations 2025).

12.1.2 Oxidative Degradation as a Large Postharvest Bottleneck

Fatty foods with high PUFA content are nutritionally and commercially significant; however, they are biochemically vulnerable to oxidative degradation during harvesting, storage, transportation, and retail distribution (Porter et al. 1995). This susceptibility arises primarily from their molecular structure, specifically the presence of multiple methylene-interrupted double bonds, which are highly prone to radical attack.

Lipid oxidation proceeds through three distinct mechanistic steps (Zielinski and Pratt 2017; Sabarez and Noomhorm 1993):

- *Initiation:* Reactive oxygen species (ROS), heat, or light abstract bis-allylic hydrogens, generating lipid radicals (L•).
- *Propagation:* Lipid radicals react with oxygen to form lipid peroxyl radicals (LOO•), which subsequently abstract hydrogens from adjacent fatty acids.
- *Termination:* Interactions between radicals result in the formation of stable, non-radical products.

The primary oxidation products, hydroperoxides, subsequently decompose into volatile aldehydes such as hexanal, propanal, 2-pentenal, nonanal, and 2,4-decadienal, which impart rancid or cardboard-like off-flavors (Coleman et al. 1994). Peroxide values (PV) exceeding 10 meq O_2/kg in nuts and edible oils are typically associated with significant flavor loss and consumer rejection (Coleman et al. 1994; Franco and Janzantti 2005; Mexis and Kontominas 2009a; Gharby et al. 2025).

Oxidative processes leading to rancidity result in substantial post-harvest losses. Under inadequate storage conditions, the shelf life of oils may decrease by 30%–50%, while nut spoilage can exceed 10%–15% in environments with elevated temperature and humidity, which promote oxidation. Globally, oxidative quality defects in oilseed-based products account for up to 8% of rejected shipments (Tsao et al. 2021; Maszewska et al. 2018; Schneider et al. 2021).

12.1.3 Shortcomings of the Conventional Quality Control Methods

Conventional oxidation assessment methods, such as peroxide value (PV), anisidine value (AV), and thiobarbituric acid reactive substances (TBARS), remain regulatory standards (Gharby et al. 2025; Tsao et al. 2021). However, these methods present several limitations within modern supply chains:

- *Laboratory dependence and sample destructiveness*: Standard assays require laboratory infrastructure, chemical reagents, and destructive sample preparation. A single PV analysis may require 2–3 hours.
- *Temporal gaps:* Batch testing provides only periodic quality measurements, resulting in extensive periods of unmonitored storage and transportation.
- *Cost barriers:* The high expense of laboratory testing restricts routine quality assessments, particularly for smallholders and processors in low-income settings.
- *Absence of predictive capability:* Classical assays do measure current levels of oxidation, but do not predict future damage or shelf life.

These limitations highlight the need for continuous, real time, and predictive quality monitoring tools.

12.1.4 Development of Data-Driven Smart Food Quality Systems

The agri-food sector is experiencing a substantial transformation in food quality testing, moving from conventional laboratory-based methods to intelligent, sensor-driven quality control enabled by the Internet of Things (IoT), machine learning (ML), and artificial intelligence (AI) (Min et al. 2020; Ayaz et al. 2019).

Agriculture 4.0, also known as Smart Farming, uses IoT, cloud analytics, robotics, and automation to enable real-time monitoring, decision support, and automated management throughout agricultural supply chains (Adidrana and Surantha 2019; Mahanti et al. 2024). ML models trained on historical oxidation sensor data can detect subtle patterns in volatile markers and deliver predictive quality analytics that surpass human interpretation (Gu et al. 2025; Anwar et al. 2023; Wang et al. 2021).

Recent advancements include smart sensing systems such as electronic noses (e-noses), graphene aldehyde sensors, FTIR/NIR chemometric platforms, and integrated IoT networks that continuously monitor PUFA oxidation reactions in oils and nuts (Gu et al. 2025; Salgado et al. 2021; Yildiz et al. 2001). These technologies minimize losses, improve traceability, and facilitate early detection of issues within global supply chains.

12.2 Literature Review

12.2.1 Smart Sensing Technologies in Food Quality Monitoring

Sensing platforms are foundational components of digital agriculture and precision post-harvest management in contemporary agri-food systems. These platforms measure oxygen, carbon dioxide, relative humidity, pH, and volatile organic compounds (VOCs) that are indicative of oxidation and microbial spoilage (Adidrana and Surantha 2019; Mahanti et al. 2024). The integration of low-power electronics, miniaturized transducers, and wireless connectivity enables real-time data transmission to predictive analytics platforms (Gu et al. 2025).

Commodities rich in PUFAs are susceptible to oxidative rancidity, which highlights the importance of smart sensors for early detection of deterioration. Early detection facilitates compliance with international export standards. The transition from reactive testing to proactive digital quality assurance has been expedited by the adoption of ML and IoT systems (Anwar et al. 2023; Wang et al. 2021).

12.2.2 Transduction Mechanisms and Sensor Types

Sensors employed in agri-food quality monitoring include optical, electrochemical, piezoelectric, biosensors, and gas sensors. Each sensor type converts a specific chemical or physical change into an electrical signal. The underlying operational principles of these sensors are summarized in Table 12.1.

Optical sensors quantify changes in absorbance, fluorescence, or reflectance associated with oxidation processes, including the formation of conjugated dienes and trienes (Salgado et al. 2021). Electrochemical sensors identify redox-active aldehydes and peroxides (Yildiz et al. 2001). Piezoelectric quartz crystal microbalance (QCM) sensors determine mass changes resulting from VOC adsorption (Lu 2014). Gas sensors detect changes in resistance in metal-oxide-semiconductor (MOS) films upon exposure to aldehydes (Giaretta et al. 2024). Biosensors detect hydroperoxides via enzyme- or antibody-based selectivity (Sharma et al. 2023). Colorimetric sensors exhibit a visible color change upon reaction with lipid oxidation products (Di Pietrantonio et al. 2015).

Metal oxide nanowires, carbon nanotubes, and graphene are nanomaterials that enhance sensor sensitivity and response time. These materials are incorporated into cost-effective, flexible substrates, facilitating their application in food packaging, edible oil quality assessment, and nut storage monitoring (Ponzoni et al. 2012; Bruce-Tagoe et al. 2025).

Table 12.1 Sensor types and their applications in food quality measurement applications

Sensor type	Principle	Target analyte	Example application	Reference
Optical (NIR/UV–Vis/FTIR)	Light absorbance change	Conjugated dienes	Edible oil/nuts oxidation	Lu (2014), Giaretta et al. (2024)
Electrochemical	Redox reaction current	Aldehydes, Peroxides	Nut /fish meat oxidation sensors	Sharma et al. (2023), Di Pietrantonio et al. (2015)
Surface acoustic wave (SAW)	Mass loading frequency	Octenol and carvone	Food contamination	Ponzoni et al. (2012)
Gas (MOS)	Resistance change	VOC	Soft drinks/olive oil	Bruce-Tagoe et al. (2025)
Biosensor	Catalytic reaction signal	Hydroperoxides	Foodborne pathogen detection	Janzen et al. (2006)
Colorimetric	Chromogenic response	Aldehydes/VOC	Smart packaging/milk	Ziyaina et al. (2019), Ye et al. (2021)

12.2.3 Electronic Nose (E-Nose) Applications in Food Quality

Electronic noses (e-noses) employ multisensory arrays to detect VOCs associated with oxidation, while ML algorithms such as linear discriminant analysis (LDA), support vector machines (SVM), random forest, convolutional neural networks (CNN), and long short-term memory (LSTM) networks are used to interpret sensor signals (Li et al. 2024; Wei et al. 2018). The adulteration of peony seed oil with less expensive edible oils has been investigated using an array of metal oxide semiconductor (MOS) sensors in combination with principal component analysis (PCA) and LDA, achieving a differentiation accuracy of 96.72% between peony seed oil and four adulterants. Tang and Yu (2020) developed a prediction model for total volatile basic nitrogen (TVB-N) in chicken meat, attaining a prediction accuracy of 93.3%.

Hybrid arrays combining MOS and QCM sensors demonstrate improved selectivity by integrating electrical resistance and frequency shift data (Julian et al. 2020; Kaur et al. 2012). The dynamic social impact theory-based optimizer (SITO) and moving window time slicing (MWTS) methods have been applied to tea classification tasks, further enhancing electronic nose performance (Tang and Wang 2024). Table 12.2 describes some of the recent e-nose systems used for food quality monitoring.

12.2.4 IoT-Enabled Sensing Networks

IoT technologies integrate distributed sensors, wireless communication, and cloud computing to enable comprehensive monitoring of food quality (Pramanik et al. 2024; Tzounis et al. 2017; Zhang et al. 2017). Most IoT architectures consist of the following layers:

- *Sensing layer*: Smart sensors, such as electronic noses, colorimetric tags, and humidity and temperature sensors, measure real-time parameters.

Table 12.2 Recent e-nose systems for food quality monitoring

Type of system	Technology	Commodity	Year	ML method	Reference
E-Nose (MOS + DL)	CNN	Rice	2024	CNN	(Hou et al. 2025)
E-Nose (MOS array)	Feature optimization	Wheat	2025	Gaussian Process Regression (GPR)	(Prasad 2025)
E-Nose (MOS array)	MOS	Oyster mushrooms	2025	SVM/ANN	(Andre et al. 2022)
E-Nose (nanofibrous)	Amine detection	Fish	2025	PCA/LDA	(Jia et al. 2024)
E-nose (MOS) + IoT	MOS + IoT	Meat	2024	SVM	(Vasanthi et al. 2024)

- *Communication layer*: Wireless standards, including Wi-Fi, Bluetooth Low Energy (BLE), and LoRaWAN, facilitate data transmission. Zigbee mesh networks transmit signals to designated gateways.
- *Edge computing layer*: Microcontrollers, such as Arduino, ESP32, and Raspberry Pi, perform local preprocessing and ML inferences.
- *Cloud layer*: Centralized servers aggregate and analyze sensor data to generate dashboards and predictive alerts.

IoT-connected sensor platforms are currently used to monitor edible oils, grains, and perishable animal products. When integrated with blockchain ledgers, each batch receives a digital identity that documents all quality parameters throughout the chain of custody. Recent reports on these applications are summarized in Table 12.3.

IoT-enabled ecosystems mark a significant shift from static batch testing to dynamic, predictive quality management across the value chain.

12.2.5 Sustainability and Socioeconomic Implications

Smart sensing and IoT–AI ecosystems contribute to sustainability, inclusion, and affordability, particularly in emerging economies facing infrastructure and cost constraints. Smart sensors enable nondestructive analysis, reducing chemical use and energy consumption compared with traditional rancidity tests. Low-power IoT modules utilize energy-harvesting methods such as solar and triboelectric technologies, thereby reducing their carbon footprint. The adoption of biodegradable sensor

Table 12.3 IoT-enabled systems in food/agriculture applications

Commodity	Sensor network	Communication method	Outcome	Reference
Citrus orchards	Soil moisture sensors	IoT cloud-based system	Optimized irrigation, reduced water waste	Alturif et al. (2024)
General crops	Multiple environment sensors (temp, moisture, gas)	Wireless sensor networks	Real-time monitoring, increased yield	Seetha et al. (2023)
Dairy & poultry	Temperature, humidity sensors	Wireless communication	Improved storage conditions, food safety	Khan and Hussain (2019)
Plant health	MOS sensor arrays, gas sensors	IoT sensors integrated	Quality evaluation using ML	Ndunagu et al. (2022)
Rice fields	Soil moisture and water level sensors	LoRa/IoT wireless networks	Water management, sustainable practices	Janaa et al. (2025)
Livestock	Health monitoring sensors	IoT wireless communication	Improved health tracking	Davies and Labuza (1997)

substrates and packaging further reduces electronic waste and supports the development of a circular economy.

Agricultural quality control is increasingly shaped by smart sensing technologies. The integration of sensor science, IoT connectivity, ML, and advanced materials enables real-time, nondestructive monitoring of lipid oxidation and rancidity in foods rich in PUFAs. As the agri-food sector adopts predictive, transparent, and resource-efficient supply chains—incorporating cloud analytics, digital twins, and sustainable design—significant improvements in operational efficiency and product quality are anticipated. To translate these scientific advancements into inclusive and scalable improvements for both producers and consumers, it is essential to address challenges such as calibration stability, affordability, and digital literacy.

Despite these advancements, several gaps remain. Most current systems rely on laboratory settings and are not readily deployable in the field. Few studies have investigated the relationship between chemical markers and sensor responses across diverse PUFA-rich matrices. Persistent challenges include data standardization, maintaining calibration stability, and managing environmental variables such as temperature and humidity. The integration of sensor data into decision-support tools for farmers and traders is still at an early stage. Therefore, it is crucial to present practical case studies that combine analytical chemistry, sensor innovation, and data-driven modeling within real agricultural contexts.

This chapter addresses these gaps by presenting two case studies: (i) an ML-based VOC analysis for detecting rancidity in cashew nuts and (ii) a colorimetric sensor platform for monitoring oxidative degradation in edible oils. These studies provide insights into the design, testing, and development of scalable smart monitoring systems for PUFA-rich agricultural products.

12.3 Case Studies

12.3.1 Case Study I: Application of ML to Find Rancidity in Cashew Nuts

Background

Cashew nuts (*Anacardium occidentale* L.) contain high levels of PUFAs, proteins, vitamins, and minerals, contributing to their significance as both an international trade commodity and a dietary commodity. However, the high degree of unsaturation in cashew lipids renders them particularly susceptible to oxidative rancidity, especially during post-harvest storage and transportation in tropical climates. Even minor oxidative processes can significantly impair the flavor, aroma, and commercial value of cashew products (Mexis and Kontominas 2009b; Su et al. 2004).

Over the past decade, analytical research has focused on profiling VOCs in cashew apples and kernels. Prior studies have examined the effects of irradiation (Ajith et al. 2014), temperature and humidity (Jayalekshmy and Narayanan 1989), and methodological advancements such as gas chromatography (GC) and gas chromatography–mass spectrometry (GC–MS) for VOC detection (Zhang and Suslick 2005). While these methods are accurate, they are time-consuming and not readily adaptable to the high-throughput quality control required in processing environments. Consequently, rapid analytical systems, particularly e-noses, have been adopted to fingerprint holistic VOC profiles and approximate sensory quality in various foods, including fruits, vegetables, nuts, oils, and beverages. However, their application to cashew products remains relatively underexplored.

Traditional methods, such as PV and TBARS, remain industry standards for assessing oxidative stability. However, these techniques require destructive sampling and are unsuitable for real-time monitoring within dynamic supply chains. Recent advances have addressed these limitations through the integration of ML-based data analytics with sensor-based VOC measurements, enabling rapid, non-destructive, and automated assessment of rancidity and flavor acceptability by directly correlating sensor and spectral data (Sharma et al. 2022).

Objectives and Methodology of the Study

This case study presents the development and implementation of an ML-based workflow for detecting rancidity in cashew kernels using electronic nose VOC fingerprints, which are validated by GC–MS. The primary objectives are to (i) identify key VOC oxidation patterns, (ii) correlate VOC and sensor data with both instrumental and sensory freshness indices, and (iii) develop predictive ML models for automated grade prediction.

Collection and Preparation of Samples

A total of 47 distinct cashew kernel grades were systematically sourced from major regional processors and the local market, coded as Bennim, Goa, BSK & Sons, Kalabavi, Kaladhar, and Local Market, with 3 to 15 grades each. All samples underwent random and blind sensory evaluation by an untrained panel using a 9-point hedonic scale to assess flavor, aroma, color, and texture. Multi-attribute cutoffs were applied to classify the samples as TOTAL GOOD, TOTAL BAD, or MIXED.

Chemical and Instrumental Analysis

VOCs, including propanal, hexanal, nonanal, benzaldehyde, and furfural, as well as relevant acids, were obtained from commercial sources. GC headspace analysis (DANI Make) was employed to identify and quantify VOCs in cashew samples. Aqueous solutions (0.01%) of each standard compound (pentane, propanal, IBA, IVA, hexanal, trans-2-pentenal, heptanal, octanal, nonanal, octanoic acid, hexanoic acid, benzaldehyde, and furfural) were prepared in triplicate. These solutions were subsequently analyzed using an e-nose equipped with 18 sensors (P40/1, P40/2, P10/1, P10/2, P30/1, P30/2, PA/2, T40/1, T40/2, T30/1, T70/2, TA/2, LY2/LG, LY2/G, LY2/AA, LY2/gCT1, LY2/gCT, LY2/GH) at an incubation temperature of 50 °C. Individual cashew nuts were placed in 10 ml standard e-nose vials and sealed prior to analysis. Each grade from the respective origin was analyzed in triplicate.

Data Processing and Statistical Analysis
The peak values of the temporal responses from 18 e-nose sensors for cashew kernels and standards were organized into columns and analyzed using PCA. PCA was also applied to the average sensory scores of the cashew samples. Additionally, PCA was performed on responses to VOC standards to evaluate the sensors' affinities for these chemicals. PCA was further conducted on the relative percentages of the VOCs to identify those responsible for rancidity or off-flavors. The relative percentage values for each VOC in each sample were calculated by area normalization of their gas chromatograms. To assess the effects of VOC co-occurrence on cashew quality, new features were created by calculating ratios of propanal to other VOCs, resulting in 10 additional features. These features were derived as follows:

$$f_i = \frac{\%\text{propanal} + 1}{\%\text{propanal} + \%VOC_i + 1}$$

Here, i denotes the ith VOC. The value 1 was added to prevent division-by-zero errors. Feature reduction was applied to both the relative percentage features and the derived feature set to identify the optimal features for classifying GOOD and BAD cashews.

Major Results and Research Implications
Figure 12.1 presents the PCA bi-plot of the average sensory score data. Samples classified as TOTAL BAD, TOTAL GOOD, and MIXED by the sensory panel are situated within the red, green, and blue regions, respectively.

Boundary-forming samples are highlighted. TOTAL BAD samples are defined as those that do not qualify any sensory tests, while TOTAL GOOD samples qualify all

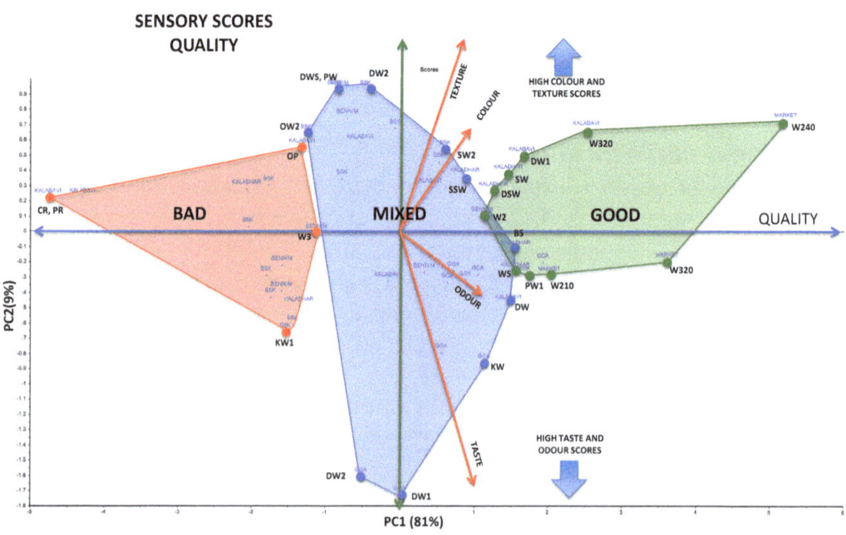

Fig. 12.1 PCA bi-plot for sensory evaluation scores by untrained panel

sensory tests. The total explained variance along the PC1 axis is 81%, attributable to the overall quality of the samples. The PC2 axis accounts for 9% of the explained variance and is associated with appearance and flavor characteristics. Samples with high taste and odor scores are positioned on the negative end of the PC2 axis. Most samples originating from Goa exhibited superior taste and odor attributes compared to other samples, but demonstrated deficiencies in texture and color. TOTAL GOOD samples are primarily from the W320, 210, and 240 grades, although some other grades also met the overall quality criteria. CR and PR samples are positioned at the far left of the PC1 axis, indicating they are the lowest-quality samples in the set. Subsequent axes, including PC3, PC4, and PC5, collectively account for 10% of the variance. Figure 12.2 presents the 3D bi-plot generated using the percentage values of VOCs present in the cashew samples.

Samples marked in blue, red, and green indicate the TOTAL GOOD, TOTAL BAD, and MIXED quality classes, respectively. The red arrows indicate the vectors (loadings) of the peak area values. Samples aligned with these loadings generally exhibit a positive correlation with the corresponding loading. For instance, the TOTAL GOOD samples contain a higher concentration of propanal compared to the TOTAL BAD samples. Conversely, the TOTAL BAD samples display higher concentrations of benzaldehyde, hexanal, and octanal. Notably, some samples with high propanal content were still rated as bad, primarily due to deficiencies in

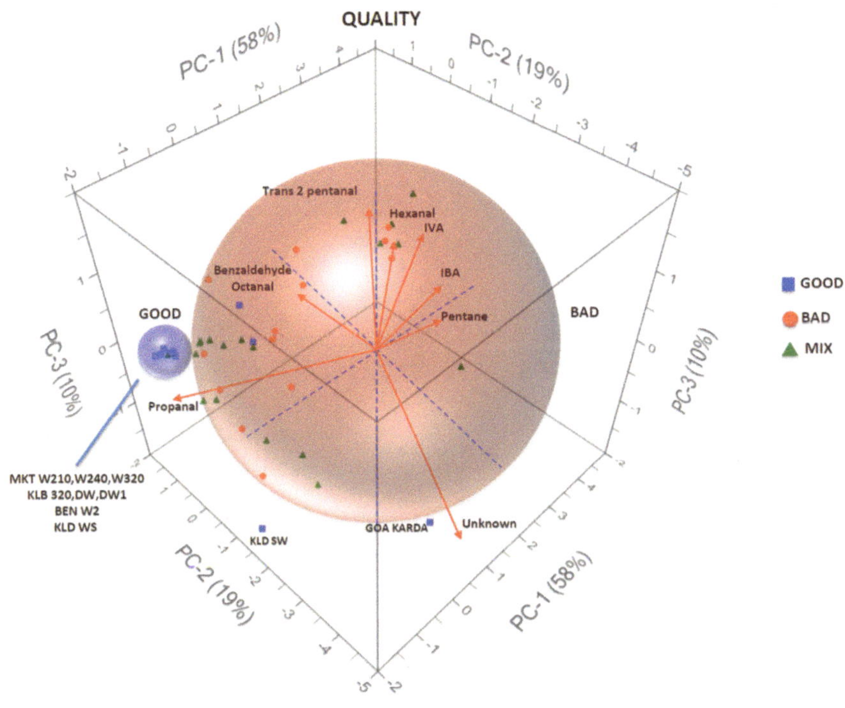

Fig. 12.2 PCA bi-plot using GC peak area values of the cashew samples

appearance despite favorable taste and aroma. To facilitate visualization and interpretation, the data points have been encapsulated in spheres. The GC analysis demonstrates that good and bad quality samples generate distinct VOC signatures, which can be utilized for sample grading and classification.

Additionally, derived features based on relative percentage values were utilized for PCA. Multiple combinations of original and derived features were evaluated, revealing that three features—propanal (%), hexanal (%), and their ratio—provide effective discrimination between GOOD and BAD samples. Figure 12.3 presents the PCA bi-plot generated using these three features.

These findings indicate that two sensors with specificity for propanal and hexanal could be utilized for quality grading of cashew kernels. To characterize the responses of the e-nose sensors to standard compounds, peak sensor response values were used to construct the PCA bi-plot shown in Fig. 12.4. It should be noted that these data do not represent mixtures of standards; sensor behavior may differ in the presence of complex VOC mixtures. The positive direction of principal component 1 (PC1) appears to correspond to molecules arranged by decreasing boiling point, which is associated with decreasing molecular weight from left to right. The odor

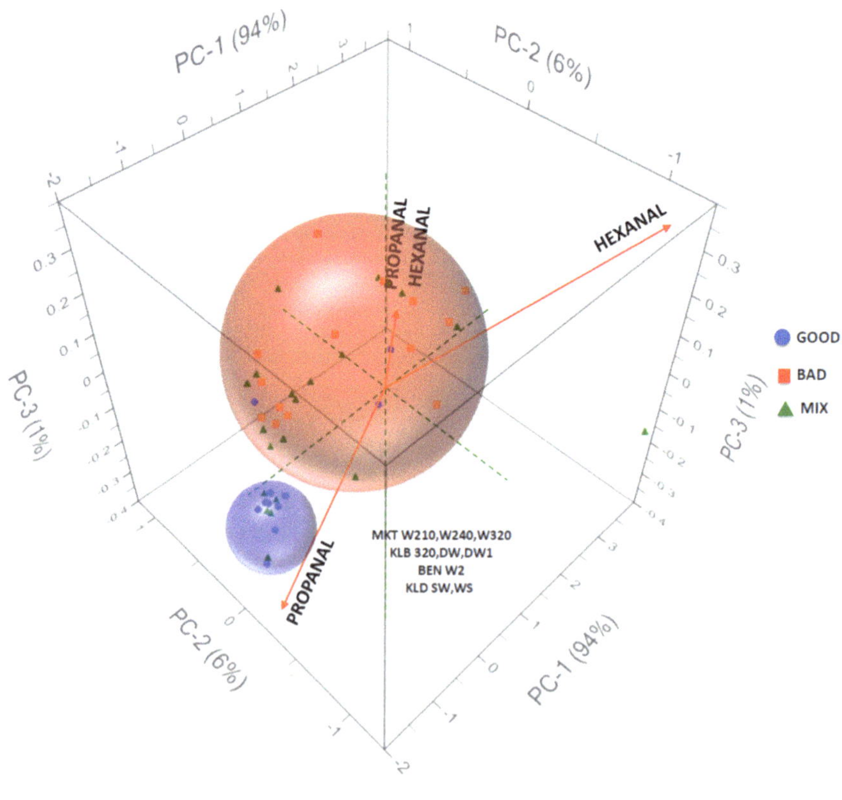

Fig. 12.3 PCA bi-plot using propanal, hexanal, and ratio loadings

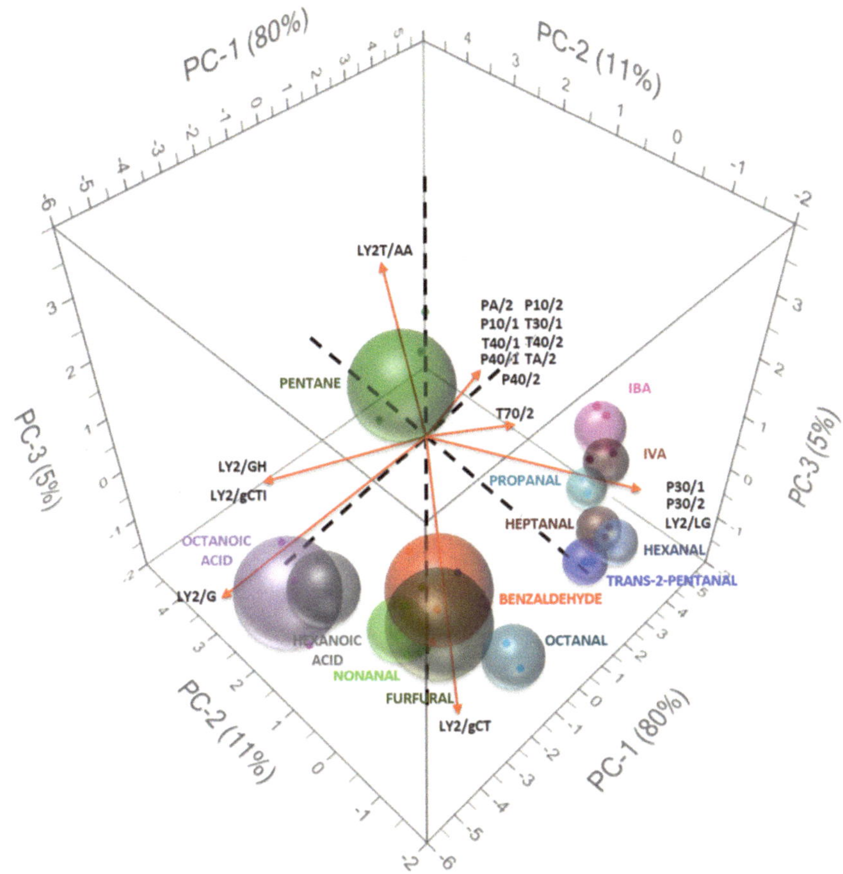

Fig. 12.4 PCA bi-plot of VOC standards using e-nose response

descriptors of the molecules follow a progression from 'rancidity' to 'fruity' to 'malty' aromas. The leftmost molecules, hexanoic and octanoic acids, elicit 'fatty' and 'rancid' odors, followed by nonanal, which also produces a 'fatty' aroma. The descriptors then shift toward the 'almond' and 'nutty' aromas of benzaldehyde and furfural, before returning to the 'aldehydic' and 'fatty' odor of octanal. Subsequent molecules elicit 'fruity' aromas, culminating in hexanal, which has a pungent scent reminiscent of 'freshly cut grass,' followed by isovaleric acid (IVA) and isobutyric acid (IBA), which are characterized by 'husk' and 'malt' odors.

The data presented in Fig. 12.4 provides an overview of sensor behavior and informs the selection of sensors for feature reduction. The LY2/GH, LY2/gCTI, and LY2/G sensors exhibit affinity towards acids, while most of the P and T series sensors respond primarily to pentane. Sensors P30/1, P30/2, and LY2/LG are sensitive to propanal, IVA, IBA, heptanal, and trans-2-pentanal, whereas LY2/gCT responds to nonanal, furfural, benzaldehyde, and octanal. These findings indicate that only four sensors may be sufficient to classify good and bad samples. Accordingly, the responses of four sensors—LY2gCT, P30/1, LY2/LG, and LY2/GH—were analyzed for cashew samples using PCA. Figure 12.5 presents the PCA bi-plot generated from the selected sensor responses.

The samples were categorized according to the taste sensory attribute. Blue and red samples indicate GOOD and BAD taste, respectively. In general, the GOOD samples are concentrated on the negative side of the PC1 axis, while the BAD samples are located on the positive side. A distinct cluster of high-quality samples, primarily sourced from the market, is observed within the green sphere. A similar trend is evident when samples are categorized by odor and overall quality. These findings indicate that the e-nose response can effectively classify GOOD and BAD cashew samples based on their flavor sensory attributes.

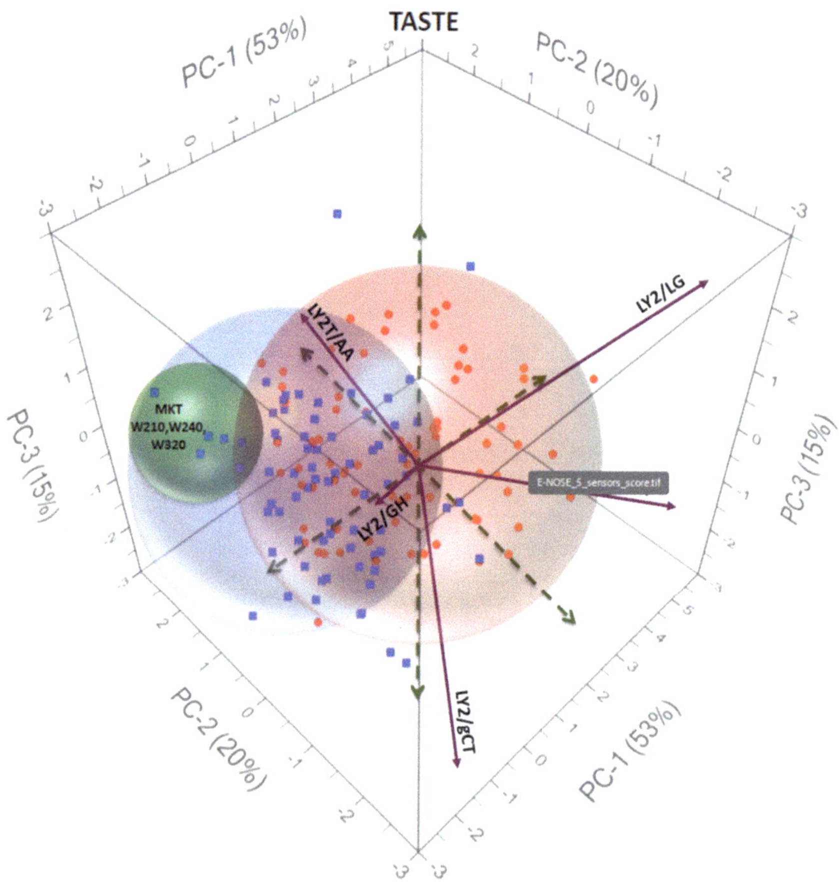

Fig. 12.5 PCA bi-plot of EN response using only five sensors

12.3.2 Case Study II: Colorimetric Smart Sensing for Rancidity in Edible Oils

Background

Edible oils play a critical role in the global food system due to their high content of PUFA and essential nutrients, as well as their widespread use in both direct consumption and processed foods. However, the substantial presence of unsaturated fatty acids, particularly linoleic (C18:2) and linolenic (C18:3) acids, makes these oils highly susceptible to oxidative degradation. The resulting business and health impacts are significant: global losses in the edible oil market from rancidity and oxidative spoilage are estimated at USD 1.8 billion annually, driven primarily by product rejection, recalls, and reduced shelf life (OECD-FAO 2023).

The safety risks and sensory deterioration associated with oil oxidation are well documented. Epidemiological studies have linked outbreaks of gastrointestinal disease, including a notable incident in 2015 in Wyoming and previous cases in India, to the consumption of oxidized or rancid fat-containing foods. Degradation products of oxidized oils, such as peroxides and aldehydes, may not cause acute symptoms but can lead to chronic health effects, including cytotoxicity and gastrointestinal disturbances.

Traditional analytical indices for assessing oil rancidity, such as PV, para-AV, AV, TBARS, and the TOTOX index, provide quantitative data on primary and secondary oxidation products. However, these methods are labor-intensive, time-consuming, and may not correlate with sensory perception. Advanced chromatographic techniques, including GC and high-performance liquid chromatography (HPLC), offer specificity but are unsuitable for rapid, field-based, or large-scale continuous monitoring.

Objectives

To address these limitations, this case study presents the development and validation of a rapid and low-cost platform utilizing colorimetric sensor strips for visual and quantitative determination of rancidity in edible oils. The platform incorporates novel sensing materials, optimized extraction and calibration protocols, and enables real-time, distributed environmental sensing through IoT-enabled digital analysis.

Sensor Chemistry and Theory Justification

Transition-metal tetraphenyl porphyrins (MTPPs) and their derivatives were selected as key sensor materials, with hexanal identified as the primary volatile indicator of secondary lipid oxidation. The specific interaction between the carbonyl oxygen of hexanal and the metal d orbitals of MTPPs was confirmed through theoretical simulations (HOMO–LUMO analysis) and experimental spectroscopic measurements, resulting in observable red and hypochromic shifts in the visible absorption spectrum. The selectivity and sensitivity of this detection approach are attributed to this mechanistic validation.

Optimization of the Analysis and Methodology

For the quantitative determination of hexanal in oil matrices, a systematic headspace extraction approach using GC was employed, with optimization of sample mass, solvent selection, and extraction temperature and time. The optimal protocol utilized a 10 g oil aliquot and a 20-minute extraction at 120 °C, which provided a balance between signal intensity and analyte stability. Fresh soybean oil served as the calibration standard, and matrix matching was performed to ensure the relevance of the analysis of the real sample. Hexanal and other oxidation markers, including PV, AV, para-AV, and total polar compounds (TPCs), were measured in randomly collected market oils and regulatory edible oil samples. A strong and positive Pearson correlation was observed between hexanal concentration and conventional oxidation indices [66].

Colorimetric: Sensing and Digital Analysis

Colorimetric strips were prepared by immobilizing derivatives of MTPP onto an absorbent surface. For analysis, the strips were positioned on the lid of the oil sample container (Fig. 12.6), and sensor images were scanned at fixed intervals under standardized lighting conditions. RGB values were extracted using the OpenCV library in Python and converted into ΔE color difference metrics. The ΔE values were statistically compared with hexanal concentrations and oxidation indices measured by GC, demonstrating their suitability as both visual and quantitative indicators.

Storage and Real-Life Tests

Storage stability was evaluated by monitoring hexanal concentrations and sensor responses in edible oils stored at 25 °C or 45 °C for several months (Fig. 12.6).

After 2 months of storage, hexanal formation increased significantly, as indicated by both sensor color shift and positive Kreis' test results. Nearly 60% of the sampled oil exceeded acceptable PVs at the study endpoint, highlighting the urgent need for practical, deployable detection tools (Table 12.4).

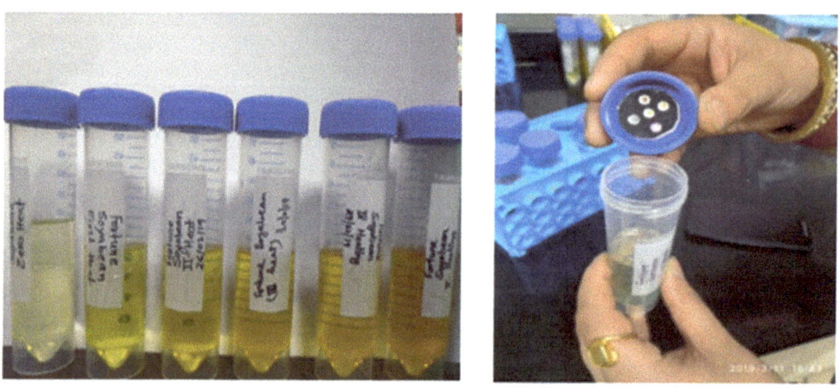

Fig. 12.6 Oil sample containers having colorimetric sensors

Table 12.4 Laboratory test results for the analyzed parameters and sensor response

Kreis test I	Kreis test II	Peroxide value (mEq/kg)	p-Anisidine value	Hexanal (ppm)	Sensor response	
					Before exposure	After exposure
Negative	0.013	1.44	0.36	0.448	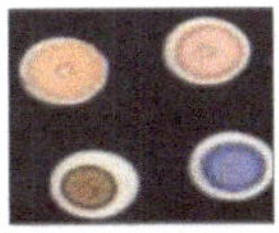	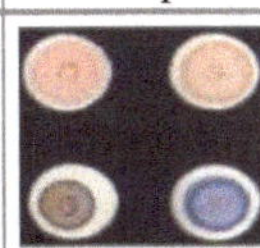
Positive	1.003	12.88	5.46	5.031	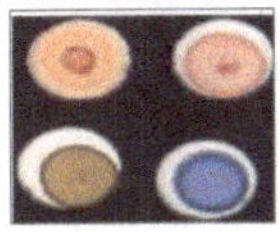	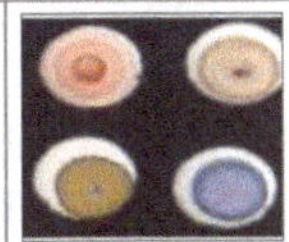
Positive	2.281	22.35	9.94	11.316	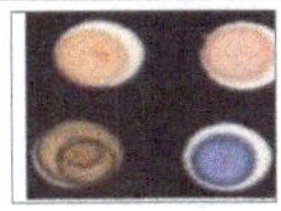	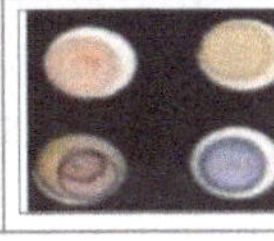

Scientific and Practical Implications

This case study demonstrates the feasibility of nondestructive rancidity monitoring in edible oils through the integration of advanced colorimetric sensor chemistry, digital image processing, and IoT connectivity. This approach is expected to be developed into a commercially accessible platform and offers several significant advantages:

- Hexanal, a well-established secondary oxidation marker, exhibits high selectivity and sensitivity
- Rapid visual and digital quantification enables both on-site and remotely networked quality control
- Compatibility with conventional chemical indices and resilience in complex oil matrices

Adding IoT

A Bluetooth-enabled microcontroller (ESP32) transmits colorimetric response data to a custom Android application that records RGB intensities, time stamps, and temperature. The data are stored in a Firebase cloud database, enabling real-time visualization of oxidation trends. Threshold-based alerting within the system updates the oil quality status.

Connecting to Supply Chain Monitoring

The developed ML model can be deployed on a Raspberry Pi-based microcontroller integrated with the smart jar. This configuration enables on-site classification in under 30 seconds per sample. The results are transmitted via Wi-Fi to a local cloud dashboard, which displays freshness indices and predicted shelf life.

Such IoT integration enables exporters and warehouse personnel to receive real-time alerts as rancidity thresholds approach, thereby reducing product loss and enhancing consumer safety.

Wider Effects

The findings indicate that data-driven modeling can substitute lengthy laboratory tests with real-time predictive diagnostics. From a sustainability perspective, this approach supports the transition to smart agriculture and precision post-harvest management, aligning with global digitalization objectives in the agri-food sector.

Furthermore, the interpretability of ML models enables the identification of key biochemical markers relevant to degradation kinetics. This capability can facilitate sensor optimization and inform material selection for future e-nose designs, such as molecularly imprinted polymers (MIPs) with enhanced aldehyde capture or hybrid MOS sensors capable of operating in higher-humidity environments. Adding AI-powered models to blockchain traceability platforms could further enhance the transparency and security of nut exports, supporting certification standards such as HACCP and ISO 22000.

Problems and Limitations

Although the ML-based rancidity detection framework has demonstrated effectiveness, several challenges remain to be addressed:

- *Environmental cross-sensitivity:* Variations in temperature and humidity can influence MOS sensors. To enhance robustness, signal drift correction algorithms and compensation sensors are required.
- *Data standardization:* Sensor responses vary with device age and calibration status, making it challenging to reproduce results across different laboratories.
- *Scalability:* Integration into large-scale supply chains requires reliable wireless communication and robust data security protocols.
- *Model transferability:* ML models developed for a specific nut type or geographic origin may not generalize to other nut varieties due to compositional differences.

As demonstrated by the aforementioned case studies, oxidative degradation in PUFA-rich foods can now be monitored using an integrated approach that combines chemistry, sensor design, and data analytics. Smart sensing frameworks enable continuous, nondestructive, and predictive rancidity analysis, in contrast to conventional titrimetric or chromatographic methods that provide only discrete measurements. This advancement reflects the broader transition of agri-food systems toward digitization and precision control within the framework of Agriculture 4.0.

12.4 Conclusion and Future Scope

The development of smart sensing for PUFA-enriched agricultural products represents a significant advancement in food quality assurance. Integrating chemical marker identification with AI-based sensor analytics enables industries to detect the onset of rancidity earlier than sensory evaluation allows. The integration of IoT and blockchain technologies ensures that quality information is tracked across the entire supply chain, from farms and processing plants to retailers, thereby safeguarding nutritional value and trade integrity. Evidence presented in this chapter demonstrates that intelligent sensing systems are faster, more sensitive, and more scalable than traditional assays. Predictive quality management is effectively modeled by electronic noses, colorimetric platforms, and spectroscopic probes when these are connected to adaptable machine-learning models. These innovations not only enhance industry competitiveness but also contribute to public health by ensuring the consistent quality of lipid-rich diets.

Future research should focus on developing universal calibration procedures for measuring volatile markers across diverse matrices to achieve cross-product comparability. The implementation of deep transfer learning and edge AI models is essential for creating self-adaptive systems that remain robust to sensor drift and environmental variability. Advancements in smart packaging systems utilizing bioresponsive nanomaterials that visually indicate oxidative changes would enhance consumer awareness and improve supply chain responsiveness. Additionally, integrating sustainability metrics, such as carbon footprint and energy consumption, into IoT-enabled dashboards will facilitate holistic product assessment beyond traditional quality parameters. In summary, the integration of chemistry, engineering, and data science in smart sensing offers a significant opportunity to translate laboratory accuracy into field reliability, ensuring that predictive smart sensing becomes a universal standard for monitoring the quality and shelf life of PUFA-rich agricultural products.

References

Adidrana D, Surantha N (2019) Hydroponic nutrient control system based on internet of things and K-nearest neighbours. In: International conference on computer, control, informatics and its applications (IC3INA). IEEE, pp 166–171

Ajith S, Pramod S, Prabha Kumari CP et al (2014) Effect of storage on cashew nuts. J Food Sci Technol 52:4631–4636. https://doi.org/10.1007/s13197-014-1476-6

Alturif G, Saleh W, El-Bary AA et al (2024) Towards efficient IoT communication for smart agriculture: a deep learning framework. PLoS One 19:e0311601. https://doi.org/10.1371/journal.pone.0311601

Andre RS, Facure MHM, Mercante LA et al (2022) Electronic nose based on nanofibrous mats for meat spoilage monitoring. Sens Actuators B Chem 353:131114. https://doi.org/10.1016/j.snb.2021.131114

Anwar H, Anwar T, Murtaza S (2023) Review on food quality assessment using machine learning and electronic nose system. Biosens Bioelectron X 14:100365. https://doi.org/10.1016/j.biosx.2023.100365

Ayaz M, Ammad-Uddin M, Sharif Z et al (2019) Internet-of-things (IoT)-based smart agriculture: toward making the fields talk. IEEE Access 7:129551–129583. https://doi.org/10.1109/ACCESS.2019.2932609

Belitz HD, Grosch W, Schieberle P (2008) Edible fats and oils. In: Food chemistry. Springer, Berlin, Heidelberg. https://doi.org/10.1007/978-3-540-69934-7_15

Bruce-Tagoe TA, Jeevanandam J, Pan S et al (2025) Aptamer-based biosensors for monitoring food quality and safety. In: Sensor technologies for food quality and safety. RSC, pp 91–114

Calder PC (2006) Polyunsaturated fatty acids and inflammation. Biochem Soc Trans 33(Pt 2):423–427. https://doi.org/10.1042/BST0330423

Coleman WM, White JL, Perfetti TA (1994) Characteristics of heat-treated aqueous extracts of peanuts and cashews. J Agric Food Chem 42:190–194. https://doi.org/10.1021/jf00037a034

Davies K, Labuza T (1997) The Maillard reaction: application to confectionery products. Pennsylvania State University

Dennis EA, Norris PC (2015) Eicosanoid storm in infection and inflammation. Nat Rev Immunol 15:511–523. https://doi.org/10.1038/nri3859

Di Pietrantonio F, Benetti M, Cannatà D et al (2015) A surface acoustic wave bio-electronic nose for detection of volatile odorant molecules. Biosens Bioelectron 67:516–523. https://doi.org/10.1016/j.bios.2014.09.027

Dorni C, Sharma P, Saikia G et al (2018) Fatty acid profile of edible oils and fats consumed in India. Food Chem 238:9–15. https://doi.org/10.1016/j.foodchem.2017.05.072

Food and Agriculture Organization of United Nations (2025) Crops and livestock products. https://worldpopulationreview.com/country-rankings/cashew-production-by-country

Franco MRB, Janzantti NS (2005) Aroma of minor tropical fruits. Flavour Fragr J 20:358–371. https://doi.org/10.1002/ffj.1515

Gharby S, Asbbane A, Nid Ahmed M et al (2025) Vegetable oil oxidation: mechanisms, impacts on quality, and approaches to enhance shelf life. Food Chem X 28:102541. https://doi.org/10.1016/j.fochx.2025.102541

Giaretta J, Oveissi F, Naficy S et al (2024) Thread-based bienzymatic biosensor for linoleic acid detection. ACS Omega 9:43184–43192. https://doi.org/10.1021/acsomega.4c07394

Gu Z, Xu Q, Wang X et al (2025) Food freshness prediction platform utilizing deep learning-based multimodal sensor fusion of volatile organic compounds and moisture distribution. ACS Sens 10:3091–3100. https://doi.org/10.1021/acssensors.5c00254

Hou Y, Xiong L, Luo X et al (2025) Detection of pest infestation in stored grain using electronic nose system. J Food Meas Charact 19:439–452. https://doi.org/10.1007/s11694-024-02980-2

Janaa R, Dixit S, Sharma M et al (2025) An explainable AI based approach for monitoring animal health. arXiv:2508.10210. https://doi.org/10.48550/arXiv.2508.10210

Janzen MC, Ponder JB, Bailey DP et al (2006) Colorimetric sensor arrays for volatile organic compounds. Anal Chem 78:3591–3600. https://doi.org/10.1021/ac052111s

Jayalekshmy A, Narayanan CS (1989) Chapter 33: Flavor constituents of roasted cashew nuts. In: Thermal generation of aromas. ACS, pp 355–365. https://doi.org/10.1021/bk-1989-0409.ch033

Jia W, Lv H, Liu Y et al (2024) Automated detection of stale beef from electronic nose data. Food Sci Nutr 12:9856–9865. https://doi.org/10.1002/fsn3.3910

Julian T, Hidayat SN, Rianjanu A et al (2020) Intelligent mobile electronic nose system comprising a hybrid polymer-functionalized quartz crystal microbalance sensor array. ACS Omega 5:29492–29503. https://doi.org/10.1021/acsomega.0c04433

Kaur R, Kumar R, Gulati A et al (2012) Enhancing electronic nose performance using feature selection approach. Sens Actuators B Chem 166–167:309–319. https://doi.org/10.1016/j.snb.2012.02.067

Khan S, Hussain MM (2019) IoT enabled plant sensing systems for small and large scale automated horticultural monitoring. In: 2019 IEEE 5th world forum on internet of things (WF-IoT), Limerick, Ireland. IEEE, pp 303–308. https://doi.org/10.1109/WF-IoT.2019.8767309

Kumar NG, Contaifer D, Madurantakam P et al (2019) Dietary bioactive fatty acids as modulators of immune function: implications on human health. Nutrients 11:2974. https://doi.org/10.3390/nu11122974

Lands B (2014) Historical perspectives on the impact of n-3 and n-6 nutrients on health. Prog Lipid Res 55:17–29. https://doi.org/10.1016/j.plipres.2014.04.002

Li Y, Wang Z, Zhao T et al (2024) Electronic nose for detection and discrimination of volatile organic compounds: application, challenges, and perspectives. Trends Anal Chem 180:117958. https://doi.org/10.1016/j.trac.2024.117958

Lu Y (2014) Characterization of rapeseed oil using FTIR-ATR spectroscopy. J Food Sci Eng 4:244–249

Mahanti NK, Shivashankar S, Chhetri KB et al (2024) Enhancing food authentication through e-nose and e-tongue technologies: current trends and future directions. Trends Food Sci Technol 150:104574. https://doi.org/10.1016/j.tifs.2024.104574

Maszewska M, Florowska A, Dłużewska E et al (2018) Oxidative stability of selected edible oils. Molecules 23:1746. https://doi.org/10.3390/molecules23071746

Mexis SF, Kontominas MG (2009a) Effect of gamma irradiation on the physico-chemical and sensory properties of raw shelled peanuts (*Arachis hypogaea* L.) and pistachio nuts (*Pistacia vera* L.). J Sci Food Agric 89:867–875. https://doi.org/10.1002/jsfa.3526

Mexis SF, Kontominas MG (2009b) Effect of gamma irradiation on peanuts and pistachio nuts. J Sci Food Agric 89:867–875. https://doi.org/10.1002/jsfa.3526

Min W, Jiang S, Liu L et al (2020) A survey on food computing. ACM Comput Surv 52:1–36. https://doi.org/10.1145/3329168

Mititelu M, Lupuliasa D, Neacşu SM et al (2024) Polyunsaturated fatty acids and human health: a key to modern nutritional balance in association with polyphenolic compounds from food sources. Foods 14:46. https://doi.org/10.3390/foods14010046

Molendi-Coste O, Legry V, Leclercq IA (2011) Why and how meet n-3 PUFA dietary recommendations? Gastroenterol Res Pract 2011:1–11. https://doi.org/10.1155/2011/364040

Ndunagu JN, Ukhurebor KE, Akaaza M et al (2022) Development of wireless sensor network and IoT-based smart irrigation system. Appl Environ Soil Sci 2022:1–13. https://doi.org/10.1155/2022/7678570

OECD-FAO (2023) Agricultural outlook 2023–2032. OECD-FAO

Ponzoni A, Comini E, Concina I et al (2012) Nanostructured metal oxide gas sensors: a survey of applications. Sensors 12:17023–17045. https://doi.org/10.3390/s121217023

Porter NA, Caldwell SE, Mills KA (1995) Mechanisms of free radical oxidation of unsaturated lipids. Lipids 30:277–290. https://doi.org/10.1007/BF02536034

Pramanik S, Kadam V, Bhatlawande S (2024) Real time food monitoring and quality alert system using IoT and Streamlit. In: Kulkarni AJ, Cheikhrouhou N (eds) Intelligent systems for smart cities. Springer, Singapore, pp 171–183. https://doi.org/10.1007/978-981-99-6984-5_11

Prasad A (2025) Real-time food quality assessment of oyster mushrooms using an electronic nose. Food Sci Nutr 11:1–9. https://doi.org/10.24966/FSN-1076/100211

Sabarez HT, Noomhorm A (1993) Performance testing of an experimental screw conveyor dryer for roasting cashew nuts. Postharvest Biol Technol 2:171–178. https://doi.org/10.1016/0925-5214(93)90045-5

Saini RK, Keum YS (2018) Omega-3 and omega-6 polyunsaturated fatty acids: dietary sources, metabolism, and significance—a review. Life Sci 203:255–267. https://doi.org/10.1016/j.lfs.2018.04.049

Salgado PR, Di Giorgio L, Musso YS et al (2021) Recent developments in smart food packaging focused on biobased and biodegradable polymers. Front Sustain Food Syst 5:630393. https://doi.org/10.3389/fsufs.2021.630393

Schneider P, Rochell V, Plat K et al (2021) Circular approaches in small-scale food production. Circ Econ Sustain 1:1231–1255. https://doi.org/10.1007/s43615-021-00129-7

Seetha J, Senthil Kumar S, Nruthya K et al (2023) Enhancement of agriculture productivity and automation of agriculture using IoT. In: 2023 International conference on inventive computation technologies (ICICT), Lalitpur, Nepal. IEEE, pp 1411–1417. https://doi.org/10.1109/ICICT57646.2023.10134441

Sharma A, Bhardwaj A, Khanduja G et al (2022) Determination of hexanal and its correlation with oxidative rancidity in edible oils. Food Anal Methods 15:2652–2663. https://doi.org/10.1007/s12161-022-02320-4

Sharma P, Thakur D, Kumar D (2023) Novel enzymatic biosensor utilizing a MoS2/MoO3 nanohybrid for the electrochemical detection of xanthine in fish meat. ACS Omega 8:31962–31971. https://doi.org/10.1021/acsomega.3c03776

Su M, Venkatachalam M, Teuber SS et al (2004) Impact of irradiation and thermal processing on nut proteins. J Sci Food Agric 84:1119–1125. https://doi.org/10.1002/jsfa.1748

Tang X, Wang N (2024) Electronic nose combined with deep learning to identify rice quality. Sens Actuators A Phys 379:115930. https://doi.org/10.1016/j.sna.2024.115930

Tang X, Yu Z (2020) Rapid evaluation of chicken meat freshness using gas sensor array. Int J Food Prop 23:297–305. https://doi.org/10.1080/10942912.2020.1716797

Tsao CH, Chang CW, Ho YC et al (2021) Application of OXITEST for prediction of shelf-lives of selected cold-pressed oils. Front Nutr 8:763524. https://doi.org/10.3389/fnut.2021.763524

Tzounis A, Katsoulas N, Bartzanas T et al (2017) Internet of things in agriculture: recent advances and future challenges. Biosyst Eng 164:31–48. https://doi.org/10.1016/j.biosystemseng.2017.09.007

Vasanthi R, Ashika K, Archana D (2024) IoT-based food quality monitoring system using machine learning. In: International conference on networking and communications (ICNWC). IEEE, pp 1–7

Wang C, Sun Y, Zhou Y et al (2021) Dynamic monitoring oxidation process of nut oils through Raman technology combined with PLSR and RF-PLSR model. LWT 146:111290. https://doi.org/10.1016/j.lwt.2021.111290

Wei X, Shao X, Wei Y et al (2018) Rapid detection of adulterated peony seed oil by electronic nose. J Food Sci Technol 55:2152–2159. https://doi.org/10.1007/s13197-018-3132-z

Ye Z, Liu Y, Li Q (2021) Recent progress in smart electronic nose technologies enabled with machine learning methods. Sensors 21:7620. https://doi.org/10.3390/s21227620

Yildiz G, Wehling RL, Cuppett SL (2001) Method for determining oxidation of vegetable oils by near-infrared spectroscopy. J Am Oil Chem Soc 78:495–502. https://doi.org/10.1007/s11746-001-0292-1

Zhang C, Suslick KS (2005) A colorimetric sensor array for organics in water. J Am Chem Soc 127:11548–11549. https://doi.org/10.1021/ja052606z

Zhang X, Zhang J, Li L et al (2017) Monitoring citrus soil moisture and nutrients using an IoT-based system. Sensors 17:447. https://doi.org/10.3390/s17030447

Zielinski ZAM, Pratt DA (2017) Lipid peroxidation: kinetics, mechanisms, and products. J Org Chem 82:2817–2825. https://doi.org/10.1021/acs.joc.7b00152

Ziyaina M, Rasco B, Coffey T et al (2019) Colorimetric detection of volatile organic compounds for shelf-life monitoring of milk. Food Control 100:220–226. https://doi.org/10.1016/j.foodcont.2019.01.018

Socioeconomic Impact of Smart Agriculture

13

Debi Kalyan Jayasingh and Abhijeet Satapathy

Abstract

Smart agriculture, supported by deep learning (DL), Internet of Things (IoT), and blockchain technologies, is transforming agrarian systems by enabling real-time data analytics, precision resource management, and transparent value chains. These advancements increase farm productivity, improve market access, and provide greater income stability for smallholders. This chapter analyzes the socioeconomic impacts of smart agriculture, including the inclusion of marginalized groups, shifts in employment toward technology-enabled roles, rural economic diversification, risk mitigation through predictive tools and parametric insurance, and enhanced value chain integration. These effects are shown through Indian case studies such as Tamil Nadu's IoT irrigation (50% water savings, 30% yield gains), Maharashtra's PoCRA (15–20% productivity increase), and Odisha's Ama Krushi (90% higher harvests in rain-deficit areas). However, challenges like digital divides, affordability constraints, algorithmic biases, data privacy concerns, labor displacement due to automation, and increased technological dependency may undermine equitable outcomes without targeted interventions. To fully realize the transformative potential of smart agriculture, policy priorities should include developing inclusive digital infrastructure, farmer-centric data governance, public–private partnerships for capacity building, ethical Artificial Intelligence (AI) frameworks, and integrating indigenous knowledge to promote resilient and equitable agricultural futures.

D. K. Jayasingh (✉)
Odisha University of Agriculture and Technology, Bhubaneswar, Odisha, India

A. Satapathy
Birsa Agricultural University, Ranchi, Jharkhand, India

H. N. Oo et al. (eds.), *Transforming Agriculture with AI, IoT, and Blockchain*,
https://doi.org/10.1007/978-981-92-1491-4_13

Keywords

Smart agriculture · IoT · Blockchain agriculture · Socioeconomic impact · Digital divide · Precision farming · Rural livelihoods

13.1 Introduction

Smart agriculture exemplifies a paradigm shift in the conceptualization and regulation of agrarian systems, driven by the integration of advanced digital technologies such as deep learning (DL), the Internet of Things (IoT), and blockchain. These tools enable real-time data collection, decentralized record-keeping, and predictive analytics, thereby enhancing decision-making processes and resource utilization while promoting transparency throughout the value chain. Socioeconomic transformations are evident as digital devices become more prevalent in rural areas, offering prospects for increased farm output and improved market access at reduced costs. For instance, IoT-based sensors can optimize irrigation and fertilization schedules, Artificial Intelligence (AI) can predict weather patterns and pest outbreaks, and farmers can implement preventive strategies accordingly. Blockchain applications further ensure transaction traceability and trust, which is particularly beneficial for smallholders seeking equitable market access.

However, the implementation of smart agriculture raises critical concerns regarding equity, access, and sustainability. The digital divide, characterized by disparities in literacy, affordability, and connectivity, continues to disadvantage marginalized groups unless addressed through inclusive deployment strategies and policy development. Furthermore, the adoption of new systems and the development of relevant competencies are essential to facilitate the transition to data-driven agriculture and to ensure the equitable distribution of benefits.

13.2 Inclusion of Smallholders, Women, and Marginalized Communities

Smart agriculture holds significant transformative potential to enhance access to agronomic innovation through democratization, particularly benefiting smallholders, women, and marginalized groups. However, these populations frequently encounter systemic barriers such as insecure land tenure, insufficient extension services, and limited digital literacy, challenges that are often exacerbated by unequal technological diffusion.

13.2.1 IoT and Mobile Platforms

IoT and mobile platforms are key drivers in addressing these gaps. Increasing customization of low-cost sensors, mobile-based weather notifications, and voice-based advisory services targets farmers with limited resources. Initiatives such as

Digital Green and e-Sagu in India demonstrate how vernacular content and community-based video distribution can expand the reach and relevance of extension services. When integrated with mobile networks, IoT-enabled devices provide remote farmers with real-time information on soil conditions, pest infestations, and irrigation schedules, which were previously inaccessible or delayed (Prashanth et al. 2025).

13.2.2 Blockchain Technology

Blockchain technology has significant potential to strengthen land tenure protection and improve transparency in benefit transfers. In many regions, women and other marginalized groups lack formal land titles, which restricts their access to credit and subsidies. Blockchain-based land registries, as implemented in states such as Andhra Pradesh, provide decentralized, tamper-proof records that verify rights to possession and tenancy (Kshetri 2021). Additionally, blockchain-enabled smart contracts can automate and monitor the disbursement of subsidies and input vouchers, thereby reducing leakages and corruption.

13.2.3 Deep Learning (DL) Tools

DL applications are increasingly tailored to address language and literacy barriers. AI-powered chatbots and image-based diagnostic tools deliver recommendations in users' native languages and utilize audio-visual formats accessible to nonliterate individuals. For example, convolutional neural networks (CNNs) trained on local crop images enable farmers to identify diseases using smartphone images, while natural language processing (NLP) models translate complex agronomic information into culturally relevant formats (Prashanth et al. 2025).

However, technological advancements do not automatically ensure inclusion. Without deliberate policy and design interventions, smart agriculture may reinforce existing inequities. Technology design should incorporate gender sensitivity, participatory development models, and community-based digital literacy programs to ensure the equitable distribution of the benefits of digital transformation. Engagement of civil society and public–private partnerships is essential for scaling inclusive innovations and integrating them into local institutional frameworks.

13.3 Employment Generation and Skill Development

The implementation of DL, IoT, and blockchain technologies in agriculture is transforming production processes while simultaneously creating new employment opportunities and shaping emerging skill requirements. As farming becomes increasingly data-driven, rural economies are experiencing a transition from traditional labor-intensive activities to technology-based services, presenting both opportunities and challenges for employment generation and human capital development.

13.3.1 Novel Employment Avenues

Emerging employment opportunities in agri-technical services and agri-technical agriculture span the entire agricultural value chain. Positions such as drone operator, sensor technician, data annotator, AI model trainer, blockchain auditor, and digital extension facilitator are increasingly in demand. Agribusinesses and startups are recruiting young professionals in rural areas to manage IoT devices for irrigation systems, implement AI on crop sensors, and facilitate digital transactions for online markets and farmers. The rise of platform-based agriculture and e-commerce has also created roles such as online marketing agents, remote advisory staff, and logistics coordinators.

13.3.2 Youth Upskilling

Upskilling rural youth is essential to capitalize on emerging opportunities. Training in sensors, drone navigation, and AI-based analytics is increasingly offered at agricultural universities, Krishi Vigyan Kendras (KVKs), and dedicated agri-tech centers. For example, initiatives such as the Digital Agriculture Fellowship and Smart Farming Bootcamps equip young professionals with competencies in Python-based data analysis, GIS mapping, and blockchain ledger management. These programmes integrate practical fieldwork with computer simulations, thereby ensuring both contextual relevance and practical proficiency.

13.3.3 Institutional Setup

Digital extension and vocational training are evolving to meet the requirements of smart agriculture through various institutional models. A critical factor in enhancing digital literacy and technical education is the active participation of both the public and industry stakeholders. Government initiatives such as Skill India and PM-Kisan have introduced modules on IoT, AI, and blockchain applications within the agricultural sector. Non-governmental organizations (NGOs) and cooperatives are piloting community-based training models, in which local champions are trained as digital facilitators to support their peers in adopting new technologies. These approaches emphasize inclusivity, ensuring that women and marginalized groups are not excluded during the digital transition.

Despite these advancements, significant challenges remain. High technology costs, limited internet connectivity, and inadequate access to training facilities continue to impede widespread adoption. Addressing these barriers requires targeted investments in rural digital infrastructure, revisions to the agricultural education curriculum, and the promotion of agri-tech entrepreneurship.

13.4 Effects on Rural Economies and Agricultural Livelihoods

DL, IoT, and blockchain technologies are driving significant changes in the economies of rural regions. These technologies not only enhance productivity but also transform the organization of agricultural livelihoods and the functioning of rural markets. Their collective impact is evident in increased income stability, improved market access, and the emergence of new economic models that promote diversification and resilience.

13.4.1 Output Enhancements

A primary method for quantitatively assessing rapid outcomes in smart agriculture is to measure increases in production per hectare using predictive analytics and precision farming. IoT sensors deliver real-time data on soil moisture, nutrient concentrations, and crop health, enabling informed decisions regarding irrigation, fertilization, and pest control. Predictive models facilitate yield forecasting, disease outbreak detection, and optimized planting schedules. Enhanced accuracy reduces input costs, mitigates environmental degradation, and increases per-acre returns, which is particularly beneficial for smallholder farmers operating on marginal lands.

13.4.2 Market Integrations

Blockchain technology is transforming consumer engagement in rural producer-value chain relationships by enabling market integration through traceability and smart contracts. These solutions provide end-to-end transparency in the flow of agricultural products, allowing consumers to trace products to their origin. Enhanced traceability increases consumer trust and facilitates access to high-value markets for organic and sustainably grown products. Smart contracts, which are self-executing agreements programmed on the blockchain, automate transactions, ensure adherence to quality standards, and minimize reliance on intermediaries. Collectively, these mechanisms improve price realization for farmers while reducing transaction costs and payment delays.

13.4.3 Income Diversification

A significant transformation in rural economies is the diversification of income sources facilitated by digital cooperatives and platform-based farming. Digital platforms enable farmers to share resources, utilize shared equipment, and collectively market their produce. These cooperatives typically operate through mobile applications that support input procurement, logistics coordination, and collective bargaining. Additionally, platform models such as contract-farming portals and farm-to-fork e-commerce systems allow farmers to engage in value-added activities, including processing, packaging, and direct sales. These diversification strategies reduce dependence on monocropping and enhance resilience to market and climate uncertainties. Collectively, these innovations foster a more dynamic and inclusive rural economy, provided that adequate infrastructure, digital literacy, and supportive policy frameworks are in place. Without these conditions, the risk of exacerbating rural inequality remains a pressing concern.

13.5 Income Enhancement and Risk Reduction

The integration of DL, IoT, and blockchain technologies within smart agriculture is transforming the economic framework of farming by enhancing income stability and mitigating production and market risks. The combination of resource efficiency, decentralized financial tools, and predictive capabilities provided by these technologies strengthens the financial resilience of farmers, particularly smallholders who operate under volatile climatic and market conditions.

13.5.1 Prediction and Forecasting

Predictive yield estimation and market forecasting using AI represent highly impactful applications for improving agricultural incomes. DL architectures trained on historical weather, soil, and harvest data enable highly precise yield predictions, allowing farmers to develop more efficient plans for harvesting, storage, and sales. AI-driven market analytics also facilitate the prediction of price fluctuations, demand surges, and supply-chain disruptions. Machine learning (ML) platforms such as IBM's Watson Decision Platform for Agriculture and AgNext in India provide accessible, localized insights, empowering farmers to determine optimal market entry times and negotiate more favorable prices (Prashanth et al. 2025).

13.5.2 Input Optimization

IoT devices, such as soil moisture sensors, automated drip irrigators, and weather-controlled irrigation systems, facilitate the operation of smart irrigation systems and optimize input costs. These technologies reduce water and fertilizer waste, lower energy consumption, and enhance crop health. When integrated with AI algorithms, they generate site-specific recommendations to further optimize input costs while maximizing output. The adoption of precision agriculture, particularly through reduced consumption of water and fertilizers, can result in significant cost savings for small- and medium-sized farmers.

13.5.3 Risk Coverage

Financial vulnerability in agriculture can be mitigated through innovative approaches such as blockchain-based crop insurance and risk pooling. Conventional insurance plans are often characterized by delayed payments, lack of transparency, and limited market penetration. Blockchain platforms address these challenges by deploying smart contracts that automate claims processing and provide tamper-resistant data records. For instance, weather-linked parametric insurance models can trigger immediate payouts when predefined thresholds are exceeded. Furthermore, blockchain technology facilitates community-based risk pooling, allowing farmers to contribute resources to decentralized funds and receive support in the event of crop failures or market shocks (Jouini and Sethom 2023; Iyer et al. 2021; Lambey and Alamsyah 2025).

The implementation of these innovative solutions is expected to enhance the safety and profitability of the agrarian ecosystem by reducing uncertainties and enabling more effective planning. This, in turn, empowers farmers to make evidence-based decisions. However, the success of such initiatives depends on the availability of reliable data, robust digital infrastructure, and the institutionalization of scalable, inclusive financial technologies.

13.6 Labor Dynamics and Employment Shifts

The adoption of smart agriculture technologies, particularly automation, AI, and the IoT, results in significant changes to rural labor dynamics. While these innovations are anticipated to enhance efficiency and productivity, they also tend to disrupt traditional employment structures and necessitate the resettlement and reorganization of skills, work processes, and support systems.

13.6.1 Influence of Automation

A prominent consequence of digital transformation in agriculture is the automation of key processes and its impact on rural labor demand. Tasks such as sowing, weeding, irrigation, and harvesting are increasingly automated through technologies, including autonomous tractors, robotic weeders, and drone-based spraying. This trend reduces the demand for manual labor, particularly for low-skilled, repetitive tasks that have traditionally employed large numbers of rural workers, including women and seasonal migrants (Apicella 2025). While automation can alleviate physical burdens and enhance workplace safety, it also poses risks of job loss and unemployment in rural areas unless accompanied by substantial investments in reskilling programs.

13.6.2 Tech-Based Agricultural Services

The emergence of technologically mediated agricultural services and gig jobs is generating new, specialized, and digitally oriented employment opportunities. These roles include drone pilots, data collectors, remote sensing analysts, digital extension agents, and blockchain verifiers. Most of these services operate within decentralized, platform-based ecosystems, where individuals offer on-demand services to farmers or agribusinesses. The rural gig economy exemplifies both flexibility and inherent precarity, as farmers increasingly rely on freelance soil testers, pest scouts, and irrigation technicians enabled by mobile applications.

13.6.3 Skillset Importance

The skills necessary for digital agriculture adoption are critical. Smart agriculture represents a significant shift that demands a workforce equipped with technical, analytical, and digital communication skills. Training programs should incorporate AI-based diagnostics, sensor calibration models, blockchain literacy, and data visualization. These competencies should be delivered through short-term certifications and practical workshops offered by vocational institutions, agricultural universities, and Krishi Vigyan Kendras (KVKs). Local and inclusive capacity-building strategies are essential to address persistent disparities in access to these opportunities.

Overall, Smart Agriculture constitutes not only a technological transformation but also a socioeconomic reorganization of rural labor. Policy interventions to ensure a fair and sustainable transition should include inclusive skill development, social protection for displaced workers, and the creation of dignified, technology-enabled employment opportunities in rural areas.

13.7 Market Access and Value Chain Integration

The advancement of agricultural market access and integration into value chains increasingly relies on smart agriculture technologies, including IoT, blockchain, and AI. These technologies enhance transparency, efficiency, and inclusivity, particularly benefiting smallholders and marginalized producers.

13.7.1 IoT-Enabled Traceability and Quality Assurance

The adoption of IoT technologies, such as smart sensors, RFID tags, and GPS trackers, is increasing for tracking the condition and movement of farm produce within the supply chain. These technologies enable real-time monitoring of temperature, humidity, and handling practices, which is critical for perishable products such as dairy, fruits, and vegetables. The resulting data quality supports compliance with safety inspections and facilitates access to high-quality domestic and export markets. Furthermore, IoT-driven traceability enhances consumer confidence and supports branding initiatives for organic and sustainably produced goods.

13.7.2 Blockchain for Direct-to-Consumer Models and Fair Pricing

Blockchain technology facilitates peer-to-peer transactions by circumventing intermediaries. Smart contracts allow farmers to transact directly with consumers, retailers, or processors, resulting in prompt pricing and payments. These contracts automatically enforce delivery terms and quality standards, thereby reducing conflicts and transaction delays. Projects such as Digital Farm Link in India demonstrate that blockchain can enable trust-based marketplaces, ensuring greater benefits for farmers while providing consumers with access to information on provenance and fair trade (Wagh et al. 2025).

13.7.3 AI-Driven Logistics and Demand Aggregation Platforms

AI streamlines logistics and market coordination by analyzing historical sales data, weather patterns, and consumer behavior. AI-based applications forecast demand, recommend optimal harvesting times, and support the aggregation of farm outputs from multiple producers. These functions facilitate large-scale sales, minimize post-harvest losses, and enhance bargaining power. Farmer-Producer Organizations (FPOs) and cooperatives are increasingly adopting these platforms to bridge rural supply and urban demand, resulting in higher incomes and more stable demand.

Collectively, these technologies contribute to the development of a resilient and equitable agri-food landscape. They facilitate smallholder access to high-value markets, reduce dependence on exploitative intermediaries, and foster trust among value chain participants. However, the proliferation of such innovations necessitates capital investment in capacity building, digital infrastructure, and supportive regulatory frameworks to ensure inclusive growth.

13.8 Social Equity and Digital Divide Challenges

While smart agriculture offers significant potential to enhance productivity and livelihoods, its benefits are not distributed equitably. The rapid integration of digital technologies such as the IoT, AI, and blockchain may exacerbate existing socioeconomic disparities if challenges related to access, affordability, and digital literacy remain unaddressed. This section provides a critical analysis of the structural constraints and ethical considerations that shape the inclusiveness of smart agriculture.

13.8.1 Barriers to Adoption: Affordability, Connectivity, and Digital Literacy

The adoption of smart agriculture tools is constrained by the high cost of digital devices, limited internet penetration, and low levels of digital literacy, particularly among smallholder farmers, women, and marginalized groups. Technologies such as IoT sensors, drones, and AI-powered advisory platforms require significant upfront investments that are often out of reach for resource-poor farmers. In many rural areas, especially in low- and middle-income countries, internet connectivity is unreliable and prohibitively expensive, limiting access to cloud-based tools and real-time data. Additionally, digital literacy is unevenly distributed across gender, caste, and age groups. Older women and farmers frequently face additional barriers due to limited formal education or sociocultural norms that discourage the use of technology. Without targeted interventions, these disparities risk creating a digital elite who benefit disproportionately from smart agriculture, while others remain excluded.

13.8.2 Risks of Exclusion and Techno-Dependency in Underserved Regions

Uncritical reliance on digital technologies in agriculture may exacerbate existing inequalities. Farmers lacking digital identification, smartphones, or formal land ownership risk exclusion from blockchain-based markets, AI-driven advisory services, and digital insurance enrolment. This dynamic creates a two-tier system, where digitally connected farmers gain competitive advantages while others are marginalized from emerging value chains. Furthermore, overreliance on proprietary platforms and algorithmic decision-making can foster technological dependence, reducing farmers to passive recipients of digital recommendations rather than active contributors of knowledge. Such trends threaten traditional ecological knowledge and may undermine local autonomy in agricultural decision-making. The limited explainability of automated recommendations and the opacity of AI model operations further complicate issues of trust and accountability.

13.8.3 Policy Frameworks for Inclusive Digital Infrastructure and Ethical AI Deployment

Addressing these challenges requires robust policy frameworks that prioritize inclusive digital infrastructure and ethical technology governance. Governments should invest in rural broadband, subsidize access to smart devices, and support community-based digital literacy programs. Public–private partnerships can play a significant role in scaling affordable and contextually relevant solutions, such as voice-based advisory services in local languages or offline-compatible applications. The implementation of artificial intelligence and blockchain in agriculture must adhere to principles of transparency, accountability, and fairness. This includes ensuring algorithmic explicability, participatory data governance, and robust consent mechanisms. Ultimately, the socioeconomic impacts of smart agriculture will depend not only on technological innovation but also on the inclusivity of design, implementation, and regulation. Equity must remain a core value in the digital transformation of agriculture.

13.9 Institutional and Policy Implications

The socioeconomic impacts of smart agriculture are influenced not only by technological innovation but also by the institutional and policy frameworks that guide its implementation. The integration of digital tools such as the IoT, AI, and blockchain into agricultural systems raises critical concerns about data ownership, infrastructure equity, and capacity development. Addressing these challenges necessitates collaborative governance structures and broad multi-stakeholder partnerships.

13.9.1 Governance Models for Data Ownership and Digital Commons

Sensors, drones, and digital platforms are increasingly employed to generate agricultural data, ranging from soil health metrics to comprehensive records of agricultural product transactions. However, the control and ownership of this data remain contested. Frequently, agri-technology companies are privately owned and maintain exclusive control over data generated by farmers, raising concerns regarding exploitation, surveillance, and monopolistic practices (Sustainability Directory 2025).

Emerging governance models propose treating agricultural information as a digital common, managed collectively in accordance with principles of transparency, informed consent, and the common good. Federated data exchanges, such as the proposed Agri. DPI (Digital Public Infrastructure in Agriculture) in India aims to grant farmers ownership of their data and autonomy in sharing it with service providers under equitable conditions (WEF 2024). These models prioritize open standards, interoperability, and community-based governance to ensure equal access and prevent data colonization.

13.9.2 Public–Private Partnerships for Infrastructure and Capacity Building

Scaling smart agriculture necessitates substantial investment in digital infrastructure, including broadband connectivity, cloud services, sensor networks, and human capital development. Public-Private Partnerships (PPPs) have emerged as strategic mechanisms for resource mobilization, risk distribution, and fostering innovation. Governments typically provide enabling policy environments and foundational infrastructure, while private-sector partners contribute technical expertise, platforms, and funding (Patra and Babu 2023).

Successful PPPs in Telangana and Maharashtra illustrate that collaborative models can facilitate agri-data exchanges, establish farmer training centers, and deliver digital extension services. These partnerships also enable the localization of technology, ensuring adaptation to regional languages, cropping systems, and sociocultural contexts. Furthermore, robust accountability structures are essential for governing PPP operations, safeguarding farmers' rights, and regulating commercial interests.

13.9.3 Policy Recommendations for Inclusive Digital Agriculture

To ensure that smart agriculture fosters inclusive rural development, policy frameworks should:

- Establish farmer-based data governance systems that include consent procedures, grievance compensation mechanisms, and profit-sharing schemes.
- Prioritize digital infrastructure investments in rural areas, with a particular focus on underserved regions and marginalized communities.
- Support the development of interoperable and open platforms to prevent vendor lock-in and encourage innovation.
- Integrate digital literacy and ethical principles into agricultural education to prepare future farmers and extension agents for responsible technology use.
- Promote multilevel stakeholder involvement, including farmer organizations, civil society groups, academic institutions, and local governments.

These policies and institutional interventions are essential for realizing the digital potential of agriculture and achieving equitable socioeconomic outcomes.

13.10 Challenges and Ethical Considerations

The agricultural sector is undergoing a significant transformation due to the integration of IoT, AI, and blockchain technologies. However, these advancements also generate complex ethical and sociocultural dilemmas. If unaddressed, such challenges may undermine trust, equity, and sustainability within rural economies.

13.10.1 Data Privacy, Algorithmic Bias, and Techno-Dependency

The widespread adoption of sensors, drones, and AI platforms generates substantial volumes of farmer-specific data, such as soil health, crop yields, and transaction records. Data privacy becomes a critical concern when farmers lack control over how their information is collected, stored, and commercialized. In the absence of effective regulation, such data may be exploited by external companies for commercial gain, thereby undermining farmers' autonomy and exacerbating power imbalances.

Algorithmic bias represents a significant challenge. AI models trained on incomplete or biased datasets may generate recommendations that favor specific crops, regions, or large-scale operations, thereby disadvantaging smallholder farmers. Such discrimination can reinforce existing inequalities in access to credit, insurance, and markets (Dara et al. 2022). For instance, predictive analytics may be less accurate in diverse small-scale farming systems than in monoculture operations.

Techno-dependency is a growing concern, as farmers may become overly reliant on proprietary platforms and automated advisory systems. While these technologies can enhance efficiency, they may simultaneously reduce resilience when services become inaccessible due to financial constraints or network disruptions. Overdependence on digital recommendations can marginalize local ecological knowledge and diminish local decision-making capacity.

13.10.2 Sociocultural Resistance and Trust in Digital Systems

Beyond technical challenges, sociocultural factors significantly influence technology adoption. Farmers frequently exhibit resistance to digital systems due to limited confidence in their reliability, concerns about losing autonomy, and apprehension regarding the intentions of external technology promoters. In many rural communities, agriculture is integral to cultural identity, and rapid digitalisation may be perceived as a threat to both heritage and local traditions.

Trust in online systems is further diminished by opaque algorithms and AI-generated recommendations that lack transparency. Farmers are often hesitant to adopt recommendations they cannot comprehend or validate, particularly when these suggestions contradict established local knowledge. Building trust requires participatory design, transparent communication, and culturally sensitive approaches that integrate digital tools with existing agricultural practices rather than replacing them (Dara et al. 2022).

13.10.3 Toward Ethical and Inclusive Smart Agriculture

Several strategies are necessary to address these challenges:

- Implement cultivator-centric information management systems that incorporate open grievance mechanisms and consent protocols.
- Promote algorithmic fairness by acquiring diverse datasets, establishing audit mechanisms, and developing transparent artificial intelligence systems.
- Develop digital literacy programs that empower farmers to become informed users rather than passive recipients.

- Facilitate sociocultural integration to ensure that online resources effectively complement traditional agricultural practices.
- Establish ethical frameworks for agricultural innovation that prioritize responsibility, inclusivity, and sustainability.

Integrating principles of morality and fairness into smart agriculture is essential for building stakeholder confidence and ensuring that technological advancements benefit all participants.

13.11 Case Studies

13.11.1 Case Study 1: Tamil Nadu—IoT-Based Smart Irrigation

The semi-arid and coastal regions of Tamil Nadu, particularly the Coimbatore and Thanjavur districts, have emerged as key sites for testing smart irrigation technologies within Climate-Smart Agriculture (CSA) models. These regions experience recurrent challenges, including unpredictable rainfall, groundwater depletion, and saline water intrusion, necessitating precise management of water and input resources. To address these issues, farmers in these districts adopted IoT-based drip irrigation systems equipped with integrated soil moisture sensors, automated valves, and weather advisory systems. These technologies enable real-time monitoring of field conditions and facilitate precise irrigation scheduling tailored to crop requirements. Tamil Nadu Agricultural University (TNAU), in collaboration with local KVKs, provided training and supported the implementation of these technologies.

Pilot interventions demonstrated a 50% reduction in water usage and a 30% increase in crop yields (Sankari and Sarayu 2024). Farmers also reported decreased electricity consumption due to optimized pump operation and reduced fertilizer runoff, resulting in both economic and environmental benefits. In addition to hardware, mobile-based pest alert systems and weather advisories delivered timely information in Tamil through platforms such as Uzhavan and mKisan. These recommendations helped mitigate pest outbreaks and informed decisions regarding sowing, irrigation, and harvesting, thereby increasing productivity and reducing input costs. The interventions produced socioeconomic benefits, including greater income stability, reduced production risk, and enhanced resilience to climate variability. The success of these pilot projects has facilitated scaling through the Tamil Nadu Precision Farming Project and the Climate Resilient Agriculture Mission, demonstrating the viability of region-specific digital agriculture models.

13.11.2 Case Study 2: Maharashtra—Climate Resilient Agriculture Through PoCRA

The agrarian sector in Maharashtra, particularly in drought-prone districts such as Yavatmal, Jalna, and Osmanabad, has been severely affected by climate variability, groundwater depletion, and recurrent crop failures. In response, the Government of Maharashtra, in collaboration with the World Bank, launched the Project on Climate Resilient Agriculture (PoCRA), a large-scale initiative to enhance the adaptive capacity of small and marginal farmers through integrated climate-smart interventions. PoCRA introduced a suite of institutional and digital innovations, including digital soil-health cards, weather-forecasting applications, mobile-based agro-advisories, and grievance-redressal mechanisms.

These tools were designed to support informed decision-making in crop planning, irrigation scheduling, and pest management. The project also facilitated community-based water budgeting, the adoption of micro-irrigation systems, and crop diversification tailored to local agro-ecological conditions. According to the official PoCRA Impact Assessment Report (Government of Maharashtra 2025), the program benefited over 700,000 farmers across 4000 villages. Key outcomes included a 15–20% increase in crop productivity, reduced input costs, and enhanced resilience to drought and variable rainfall. Farm communities reported improved access to timely advisory services and greater trust in digital extension platforms, particularly when advisory content was delivered in the local Marathi dialect by FPOs and Krishi Mitras.

PoCRA notably emphasized institutional convergence by integrating the efforts of the departments of agriculture, water resources, and rural development. This comprehensive approach ensured both the implementation and institutionalization of climate-smart technologies within local governance frameworks. The success of the project has established it as a model for other states seeking to mainstream climate resilience in agriculture. This case illustrates that well-coordinated, state-led digital transformation, combined with participatory planning and local capacity building, can yield measurable socioeconomic benefits in vulnerable agricultural regions.

13.11.3 Case Study 3: Odisha—Digital Advisory Via Ama Krushi Platform

Odisha is an agriculture-based state with a significant proportion of small and marginal farmers, and it has been a leader in adopting digital technologies to enhance agricultural extension. The Ama Krushi portal, a digital advisory platform launched in 2018 by the Department of Agriculture and Empowerment of Farmers, Government of Odisha, in collaboration with Precision Development (PxD), represents a major advancement in this area. Ama Krushi provides farmers with access through Interactive Voice Response (IVR) calls and SMS, offering personalized, timely, and localized agro-advisory services in Odia and the Sambalpuri dialect. The platform covers more than 29 value chains, including paddy, pulses, vegetables, and horticultural crops.

By 2025, Ama Krushi had reached over 3.2 million farmers across all 30 districts of Odisha, demonstrating a high degree of inclusivity, particularly among female farmers and Scheduled Tribe groups (Precision Development 2025).

Recommendations are tailored based on agroclimatic regions, crop life cycles, and seasonal risks. Topics addressed include seed selection, pest and disease control, irrigation scheduling, and market prices. A randomized assessment conducted by PxD and J-PAL South Asia found that farmers receiving Ama Krushi advisories achieved a 90% higher harvest in rain-deficit regions and experienced 21% less crop loss, especially during kharif seasons affected by unpredictable monsoons. The success of the platform is attributed to its human-centered design, local language interface, and integration with state extension systems. Ama Krushi complements traditional extension services by facilitating last-mile coverage, particularly in remote tribal blocks where physical barriers limit access. Additionally, the platform enables a feedback loop, allowing farmers to pose questions and receive expert responses.

This case exemplifies how digital public infrastructure can promote more equitable access to agricultural knowledge and improve socioeconomic outcomes for vulnerable agricultural communities when designed with contextual sensitivity and supported by institutions.

13.12 Conclusion

13.12.1 Synthesizing Socioeconomic Outcomes

The case studies and thematic analyses presented in this chapter demonstrate the potential of smart agriculture to increase productivity, enhance resource efficiency, and improve resilience to climate change. Across diverse agroclimatic regions, such as the semi-arid plains of Tamil Nadu and the rural hinterlands of Odisha, digital technologies enable farmers to make more informed decisions, reduce input costs, and access new markets. Programs utilizing drone-monitored crops, mobile advisories, and IoT irrigation systems have reported measurable improvements in yield, water-use efficiency, and income stability.

However, these advantages are not distributed equally. Access to infrastructure, digital literacy, institutional support, and cultural acceptance significantly influence socioeconomic outcomes. Marginalized groups, particularly smallholders, women, and tribal communities, require targeted interventions to ensure equitable participation in the digital transformation of agriculture. The widespread adoption of platforms such as Ama Krushi and PoCRA illustrates that inclusive design, accessible language interfaces, and community-based extension models are critical for the sustainable scaling of smart agriculture.

13.12.2 Future Research Directions

Future research should prioritize the following areas to enhance understanding and implementation:

- Conduct long-term impact analyses of smart agriculture interventions across diverse agro-ecological zones and socioeconomic strata.
- Examine algorithmic fairness and transparency within AI-driven advisory and credit scoring systems.
- Investigate data sovereignty and digital commons by identifying institutional frameworks that empower farmers as data owners.
- Conduct gender analysis within the digital agricultural sector, identifying barriers and facilitators to women's participation.
- Facilitate the integration of digital platforms with indigenous knowledge systems.
- Integration of indigenous knowledge systems with digital platforms to preserve ecological wisdom and enhance contextual relevance.

13.12.3 Policy Recommendations

Key policy priorities include the following:

- Development of people-centered digital systems, including device subsidies, rural broadband expansion, and offline-compatible tools.
- Implementation of compliance-based AI standards that ensure consent, explainability, and effective grievance redressal mechanisms.
- Promotion of public–private joint ventures to support localized innovation, capacity building, and delivery of farmer-centric services.
- Strengthening digital cooperatives and farmer-producer organizations to facilitate collective negotiation and federated data governance.
- Encouragement of multi-sectoral convergence by harmonizing agricultural, educational, and rural development policies to foster resilient agricultural futures.

Therefore, smart agriculture should be understood primarily as a socioeconomic transformation rather than solely a technological advancement. Its long-term impact will depend on the extent to which it incorporates inclusive design, ethical governance, and resilient responses to the diverse and evolving needs of farming communities.

References

Apicella A (2025) Automation in agriculture: occupational trends, worker outcomes, and labor market implications. Int J Acad Res Bus Soc Sci 15:398–418. https://doi.org/10.6007/ijarbss/v15-i8/26128

Dara R, Hazrati Fard SM, Kaur J (2022) Recommendations for ethical and responsible use of artificial intelligence in digital agriculture. Front Artif Intell 5:884192. https://doi.org/10.3389/frai.2022.884192

Government of Maharashtra (2025) Project on climate resilient agriculture impact assessment report. In collaboration with World Bank, Sambodhi Research and Communications Pvt Ltd, TERI and NABARD Consultancy Services (P) Ltd. https://mahapocra.gov.in/assets/docs/mne/PoCRA%20Impact%20Assessment%20Report%20%28ETR%29.pdf

Iyer V, Shah K, Rane S et al (2021) Decentralised peer-to-peer crop insurance. In: Proceedings of 3rd ACM international symposium blockchain secure critical infrastructure (BSCI '21). ACM, pp 3–12. https://doi.org/10.1145/3457337.3457837

Jouini O, Sethom K (2023) AgriBIoT: a blockchain-based IoT architecture for crop insurance. In: Barolli L (ed) Advanced information networking and applications, Lecture notes in networks and systems, vol 655. Springer, Cham. https://doi.org/10.1007/978-3-031-28694-0_32

Kshetri N (2021) Blockchain as a tool to facilitate property rights protection in the global south: lessons from India's Andhra Pradesh state. Third World Q 43:371–392. https://doi.org/10.1080/01436597.2021.2013116

Lambey FR, Alamsyah A (2025) Parametric crop insurance with FarmFund pool: a blockchain-based approach for climate resilience in Indonesia agriculture. In: International conference on data science and its applications (ICoDSA), Jakarta. IEEE, pp 932–937. https://doi.org/10.1109/ICoDSA67155.2025.11157645

Patra NK, Babu SC (2023) Institutional and policy process for climate-smart agriculture: evidence from Nagaland state, India. J Water Clim Change 14:1–16. https://doi.org/10.2166/wcc.2022.024

Prashanth JS, Krishna GB, Prasad AVK et al (2025) Smart farming revolution: a cutting-edge review of deep learning and IoT innovations in agriculture. Oper Res Forum 6:35. https://doi.org/10.1007/s43069-025-00434-z

Precision Development (2025) Ama Krushi—Scaling advisory services to millions of farmers in Odisha, India. https://precisiondev.org/project/ama-krushi/

Sankari M, Sarayu PS, Kamatchi KS et al (2024) Smart drip irrigation using IoT enabled water management in agriculture. In: 4th International conference on ubiquitous computing intelligent information system (ICUIS), Gobichettipalayam, pp 1774–1780. https://doi.org/10.1109/ICUIS64676.2024.10866965

Sustainability Directory (2025) Automated agriculture and rural employment shifts: scenario. https://prism.sustainability-directory.com/scenario/automated-agriculture-and-rural-employment-shifts/

Wagh C, Bavane H, Hirade M et al (2025) Digital farm link: a trust-centric marketplace integrating AI, blockchain, IoT, and federated learning for transparent farmer-to-consumer trade. Int J Adv Res Sci Commun Technol 5:172–196. https://doi.org/10.48175/ijarsct-29826

WEF (2024) Advancing digital public infrastructure for the agriculture sector—briefing paper. World Economic Forum, pp 1–12. https://dataforpublicgood.org.in/wp-content/uploads/2025/02/Advancing_Digital_Public_Infrastructure_Agriculture_Sector_2024.pdf

Transformative Egg Industry Revolution: A Comprehensive Framework for AI, IoT, Machine Learning, and Blockchain Integration

14

Saurav Kumar, Binay Kumar Singh, Vikash K. Chaudhary, Akshita Sharma, Anupma Sharma, and Nandkishore Thombare

Abstract

Agriculture has advanced beyond the transition from traditional methods, such as bullocks and low-yield crops, to tractors and high-yield varieties. It now represents a sophisticated integration of art and science, utilizing advanced technologies such as Machine Learning (ML), Artificial Intelligence (AI), the Internet of Things (IoT), and blockchain to improve the production quality and reduce losses through predictive analytics. These technologies are increasingly impacting various agricultural sectors, including the poultry industry. While the poultry sector faces unique challenges compared to conventional agriculture, adopting these technologies has not been particularly difficult. Modern innovations, from predictive algorithms to blockchain applications, are transforming the egg industry to meet future demands. This chapter reviews the global state of the egg industry, highlighting major producers such as the United States, China, and India. It examines the economic factors that determine profit boundaries and

S. Kumar (✉) · A. Sharma
CSIR-Central Scientific Instruments Organization, Chandigarh, Punjab, India

Academy of Scientific and Innovative Research, Ghaziabad, Uttar Pradesh, India

B. K. Singh
Indian Institute of Wheat and Barley Research, Karnal, Haryana, India

V. K. Chaudhary
Divyansh Agro, Ranchi, Jharkhand, India

A. Sharma
CSIR-Central Scientific Instruments Organization, Chandigarh, Punjab, India

N. Thombare
ICAR-National Institute of Secondary Agriculture, Ranchi, Jharkhand, India

H. N. Oo et al. (eds.), *Transforming Agriculture with AI, IoT, and Blockchain*,
https://doi.org/10.1007/978-981-92-1491-4_14

margins for producers. The current level of technological integration has reshaped production and supply chains in the egg industry. This chapter provides a comprehensive analysis of the role of the egg industry within agriculture, exploring how digital transformation is reshaping production, distribution, and consumption patterns in key markets. Additionally, it discusses profit margin ranges as a critical factor for technological adoption and identifies markets with the greatest potential for future technological development.

Keywords

Egg industry · Poultry farming · AI integration · IoT applications · Machine learning · Blockchain agriculture

14.1 Introduction

Agriculture has been fundamental to human civilization, with ongoing innovations developed to address increasing food demands. Within this context, animal husbandry has emerged as a significant branch of agriculture, encompassing sectors such as poultry, dairy, and aquaculture (Neethirajan 2023). Poultry, in particular, has evolved into a structured global industry over the past century. As nutritional demands rise alongside population growth, the poultry industry, especially egg production, faces considerable pressure to improve efficiency, quality, and sustainability through technological innovation (Bist et al. 2024; Taleb et al. 2025). Over the past four decades, egg production has demonstrated a marked upward trend compared to major cereals, oilseeds, pulses, and fruits. This growth necessitates advanced solutions to manage increasingly complex operations. The United States, China, and India have made notable progress in adopting modern poultry farming techniques, including automated feeding systems, environmental controls, and advanced monitoring tools to optimize egg production and animal welfare (Fan et al. 2025). China currently leads global egg production, followed by India and the United States. These countries serve as important case studies for analyzing the structure of the egg industry, profit margins, operational capacity, technological advancements, and farm-level adaptation (Essien and Neethirajan 2025). Annual egg production exceeds 560 billion in China, 120 billion in India, and 100 billion in the United States, highlighting the global scale and economic significance of the sector. This scale highlights the urgent need for advanced technological integration to sustain growth and meet rising consumer demand. Production costs remain the primary driver of the industry, influencing the adoption of new technologies amid fluctuating global commodity prices, disease outbreaks, and labor shortages (Cruz et al. 2024). The price of one dozen eggs is highest in the United States at approximately \$7 USD and lowest in India at around \$1.46 USD. India also reports the highest profit margins, ranging from 42% to 47%, compared with the narrower margins observed in the United States and China. These differences in profitability are directly linked to varying rates of technological adoption. To address earnings disparities and meet the growing global demand for food, farms and businesses must adopt modern tools to increase output, reduce costs, and improve performance across regions (Kostenko 2021). The United States currently leads in integrating advanced technologies such as artificial intelligence (AI), the Internet of Things (IoT), and blockchain into its poultry sector. However, the full potential of these solutions for tasks like egg grading, sorting, and farm management has yet to be realized (Akkoyun et al. 2022; Wu et al. 2025; Yang et al. 2023). Leading technologies in poultry farm management include sensor systems for real-time environmental monitoring, computer vision for individual bird and egg assessment, and machine learning (ML) algorithms for predictive analytics (Wu et al. 2025; Yang et al. 2023; Naeem et al. 2025). These integrations are particularly important given that feed accounts for 70%–80% of total production costs, making diet optimization a key factor in profitability (Haruna et al. 2025). The use of AI and IoT enables precise feed management, reducing waste and ensuring optimal nutrient delivery (Essien and Neethirajan 2025). Additionally, these technologies facilitate comprehensive monitoring of animal health and behavior, replacing traditional, labor-intensive inspection methods with continuous, data-driven insights into flock well-being and productivity (Taleb et al. 2025; Essien and Neethirajan 2025). The poultry industry faces persistent challenges related to egg shelf life, gender determination, fertility assessment, quality indexing, and the detection of anomalies such as cracked eggs, all of which significantly affect economic viability and consumer satisfaction. Addressing these challenges requires innovative technological interventions to refine operational processes, ensuring sustained economic output and consumer confidence in egg products (Essien and Neethirajan 2025; Shi et al. 2022; Wang et al. 2025). Recent advancements in AI and the IoT provide promising solutions to these complex issues by enabling real-time data acquisition, advanced analytics, and automated decision-making, thereby transforming the egg industry from production to market (Taleb et al. 2025; Panagi et al. 2025). This chapter explores the transformative potential of AI and IoT, examining their diverse applications in improving efficiency, sustainability, and quality throughout the stages of egg production, from environmental control and animal monitoring to advanced supply chain management.

14.2 Review of Literature

Agriculture comprises diverse sectors, including the cultivation of staple food crops such as rice and wheat, cash crops such as cotton, livestock, fisheries, and forestry. Each of these sectors plays a vital role in ensuring food security, driving economic growth, and sustaining rural livelihoods. Within this multifaceted agricultural system, the poultry and egg industries represent a distinct yet essential component (Rauw et al. 1998). While traditional agriculture emphasizes land-based crop production, egg production is a specialized branch of animal husbandry focused on raising laying hens for eggs and broilers for meat. This sector is significant both commercially and nutritionally, as eggs provide an affordable, widely available, and high-quality source of protein, essential amino acids, vitamins such as B_{12} and D, and trace minerals for populations worldwide (Akkoyun et al. 2022; Myers and Ruxton 2023).

Despite its importance, the poultry industry faces several challenges, including price volatility, disease outbreaks, susceptibility to adverse weather, and distribution inefficiencies from farm to market. Recently, technological advancements, particularly in AI and ML, have begun to transform farm operations. Predictive technologies facilitate disease management and reduce the incidence of non-fertile eggs in hatcheries. Although identifying chick gender remains a contentious issue, it is essential for hatchery efficiency. Optimizing farm operations by regulating feed rates, temperature, humidity, and other key parameters enhances overall yield (Panagi et al. 2025; Rowe et al. 2019). In addition to predictive and quality assessment methods, transportation logistics significantly influence product loss. Blockchain-enabled supply chains are emerging as effective tools for ensuring traceability, authenticity, and transparency in egg distribution, which is increasingly important to meet the expectations of consumers and regulatory bodies (Kamilaris et al. 2019). However, blockchain technology is resource-intensive, requiring substantial power and data to operate each node. For small and medium-sized egg production units, it is necessary to develop neo-blockchain techniques that minimize resource demands. Both offline and online systems can be effectively utilized in this context. The classification of agriculture and the positioning of the egg industry within the poultry sector are illustrated in Fig. 14.1.

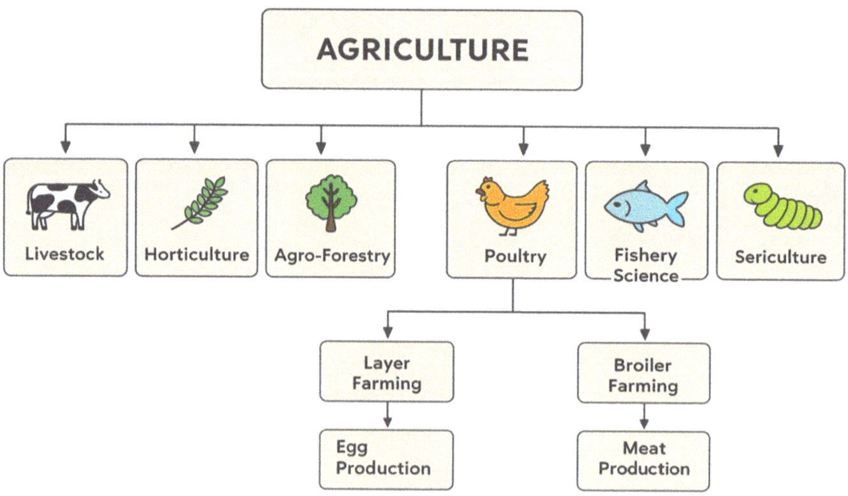

Fig. 14.1 Schematic overview of agricultural classification, highlighting the placement of the egg industry within the poultry segment

14.2.1 Production and Consumption Overview of Eggs

Eggs are hard-shelled reproductive structures laid by birds. While their primary biological function is reproduction, most eggs from domesticated poultry, particularly chickens, are not fertilized and are produced specifically for human consumption. Although eggs from other birds, such as ducks, geese, turkeys, guinea fowl, pigeons, pheasants, and quail, are consumed in various cultures, chicken eggs account for the majority of global production.

Bird eggs have served as a significant food source since prehistoric times. The domestication of chickens began in Southeast Asia and subsequently spread to ancient civilizations such as Sumer, Egypt, and Greece. Historically, eggs were used in rituals, preserved for storage, and incorporated into various meals. The dried egg industry emerged in the nineteenth century and expanded significantly during World War II. Through selective breeding, modern hens now lay over 300 eggs annually.

Advancements in poultry breeding and increased egg production have established the egg industry as a reliable source of both economic activity and nutrition. Global egg production has exhibited a consistent upward trajectory. This trend is especially evident when compared to the growth of other nutritional commodities, as illustrated in Table 14.1. Over the five-decade period from 1970 to 2021, the production shares of various agricultural commodities have shifted significantly. Egg production, as a percentage of total agricultural output, increased from 0.5% in 1970–71 to 1.2% in 2020–21, representing a 58.34% rise. This growth highlights the growing recognition of eggs as an essential nutritional commodity, particularly in populations with protein deficiencies. In contrast, the shares of staples such as cereals, pulses, and oilseeds have declined, while the increase in eggs, fruits, and vegetables reflects a shift from quantity-driven to quality-driven dietary patterns, influenced by urbanization, rising incomes, and greater health awareness.

The expansion of the global egg industry is primarily attributed to substantial growth in Asia, rising incomes, and the development of commercial layer operations. According to FAO and Our World in Data, global egg production increased from tens of millions of tonnes in the 1980s to approximately 97 million tonnes in 2023, with hen eggs comprising the majority of output. Per-capita consumption has also risen, particularly in Asia, although significant regional variation persists. For example, some European Union countries, including the Netherlands, exhibit high consumption levels, while consumption in parts of Africa remains comparatively low. The global per-person supply has increased alongside production and improved

Table 14.1 Share of different commodities (%) in the agriculture sector

Agri commodity/group	1970–1971	1995–1996	2020–2021	Change in percentage
Cereals	37.1	25.3	16.7	54% reduction
Pulses	4.7	4.2	3.9	17% reduction
Oilseeds	8.2	8.2	5.2	36% reduction
Fruits and vegetables	12.4	14.6	19.4	56% increase
Eggs	0.5	0.9	1.2	58.34% increase

distribution systems. Since the 1980s, China has emerged as the leading egg producer, with the most significant growth in production occurring between 1991 and 2000. Prior to 1985, the United States was the primary producer, but China surpassed the United States after 1985 and has maintained the highest production levels to date. In India, egg production continues to rise, supported by government initiatives and private investment in modern layer units. Currently, China is the largest producer, followed by India and the United States. Factors contributing to the egg industry's transformation into a large-scale global sector include demographic trends, technological advancements in poultry farming, investments in market infrastructure, and evolving consumer preferences.

Over the past two decades, global egg production, consumption, and prices have all increased. The primary contributors to the egg industry are China, India, and the United States. China accounts for approximately 40% of global egg production, and its domestic consumption has also risen by about 40% during this period. Egg prices have generally remained stable, with only minor fluctuations resulting from changes in feed costs and demand cycles.

In India, egg production has grown at an average annual rate of 7% over the past 20 years, while per capita consumption has increased by 5%. Egg prices in India fluctuated from approximately 360 INR per 100 eggs to 630 INR per 100 eggs by the end of 2024. The Indian egg market, valued at USD 7 billion in 2023, is influenced by urbanization and the expansion of the middle class. In the United States, egg production has remained steady for several years, but prices have fluctuated due to avian influenza outbreaks and increased demand during pandemics. Recently, India began exporting eggs to the United States in response to rising retail prices, leading to increased imports from India.

14.2.2 Egg Consumption and Future Trends

Global demand for eggs continues to rise, driven by population growth and the perception of eggs as an affordable protein source. Over the past two decades, worldwide egg consumption has increased by approximately 40%, with per capita consumption reaching about 10 kg per year by 2019. China leads in per capita egg consumption at approximately 18 kg, while Hong Kong and Macao exceed 20 kg. In the United States, projections for 2025 indicate a per capita consumption of 270.7 eggs, a figure influenced by outbreaks such as Highly Pathogenic Avian Influenza. Approximately 70% of eggs are consumed fresh, while 30% are processed; however, demand for processed products, including egg powder and liquid mixes, has grown, particularly in Asia and Latin America, due to limited availability of fresh eggs. In the 1980s, egg consumption in India was very low. The expansion of commercial poultry farming and rising incomes since the 1990s have led to a rapid increase in consumption, with per capita egg consumption reaching approximately 101–103 eggs per person per year by 2022–2024. China experienced a substantial rise in egg consumption from 1985 to 2020, increasing from a minimal per capita supply in the mid-twentieth century to over 20 kg per person per year

Table 14.2 Comparison of per capita egg consumption and future growth potential in key global markets

Country	Current per capita consumption	Equivalent in (kg/year)	Growth potential
China	~400–440 eggs/person/year (2020–21)	~20 kg	Already saturated; consumption peaked; marginal future growth expected
United States	~270–290 eggs/person/year (2020–2023)	~17.5–19 kg	Mature and stable market; modest future growth possible via health-focused or processed egg products
India	~101–103 eggs/person/year (2022–2024)	~6.3–6.5 kg	High potential for growth due to a growing middle class, nutritional programs (mid-day meals), and rising awareness. The current per capita income is far below that of global leaders

(approximately 400–440 eggs) by 2020–2021, positioning China among the highest egg-consuming nations globally. In contrast, egg consumption in the United States has remained consistently high and stable compared to developing countries, averaging 230–260 eggs per person per year during the 1980s–2000s and increasing slightly to around 270–290 eggs per person per year in recent years, with the 2023 USDA estimate at 287.4 eggs per person. A comparative analysis of per capita egg consumption and growth potential across major global markets is presented in Table 14.2.

Initial technological interventions focused on disease detection and prevention in poultry, with AI-driven image analysis and IoT sensors significantly enhancing early detection, thereby reducing the risk of widespread outbreaks and economic losses (Natsir et al. 2025). Various methods have been developed to study disease onset, particularly the early detection of avian influenza and other infectious diseases, through continuous monitoring of bird behavior, vocalizations, and physiological parameters (Kalita et al. 2025). Although eggs may appear non-living externally, they are living entities and, like other perishable food items, undergo decay over time. This perishability necessitates robust monitoring systems to maintain optimal conditions throughout the egg's lifecycle, from farm to consumer, thereby extending shelf life and ensuring food safety. Following the successful modeling of disease prediction in eggs, attention has shifted to identifying shelf-life parameters and estimating quality from the consumer perspective to inform pricing. The detection of gas release from eggs, for example, serves as a critical indicator of freshness and potential spoilage, prompting the development of advanced IoT sensors capable of real-time gas analysis (Neethirajan 2022). Differentiating between fertile and non-fertile eggs remains a significant challenge for the hatchery industry, as misidentification results in substantial economic losses and inefficiencies in chick production. In addition to these issues, further challenges include precise environmental control within poultry houses, accurate yield prediction, and the optimization of cold-chain logistics for egg distribution (Astill et al. 2020).

The challenges in the poultry industry can be broadly classified into three categories: real-time adaptability and model generalization, cost-effectiveness and scalability, and the integration of multimodal systems (Tina et al. 2025). Implementing these technologies across diverse operational scales, from small independent farms to large commercial enterprises, presents significant hurdles that require innovative solutions. In countries such as the United States and China, large farms predominate, whereas in India, the poultry industry consists mainly of small and medium-sized farms, necessitating adaptable and cost-effective AI and IoT solutions (Ojo et al. 2022). Farm size and egg production volume are key determinants for selecting appropriate technological solutions (Dones et al. 2024). Implementation and cost-effectiveness are critical for widespread adoption, especially in regions with varying economic conditions and technological infrastructures (Senoo et al. 2024). Technological solutions for small farms typically require a longer period to achieve return on investment, while large-scale operations can absorb initial capital costs more readily due to economies of scale. Return on investment (ROI) is a decisive factor in the feasibility and success of adopting advanced AI and IoT systems across diverse poultry farming contexts (Finistrosa et al. 2025). Production costs are primarily composed of feed, labor, energy, and veterinary care, all of which can be optimized through AI and IoT to enhance economic viability. In India, labor costs are relatively low, making technological integration for labor replacement less compelling; instead, efforts focus on improving feed efficiency and disease prevention to maximize output and quality. Precision feeding systems, which dynamically adjust dietary components based on real-time data, are already improving feed efficiency by 1.5%–5% annually (Moss et al. 2021). Disease detection accuracy and specificity for common avian pathogens now range from 90% to 95%, significantly reducing mortality rates and antibiotic usage (Mohyuddin et al. 2024). Additionally, AI and IoT contribute to animal welfare by continuously monitoring environmental parameters such as temperature, humidity, and ammonia levels, thereby ensuring optimal living conditions and healthier flocks (Salihu et al. 2024). For egg quality assessment, including fertility, numerous studies have focused on noninvasive imaging techniques and machine learning algorithms to accurately distinguish fertile from infertile eggs at early stages of incubation, thereby enhancing hatchery efficiency (Akkoyun et al. 2022; Çelik and Tekin 2024). Shelf-life determination and the identification of physical anomalies, such as cracked eggs, remain persistent challenges in quality control. Advanced AI-powered vision systems are being developed to detect micro-cracks and structural integrity issues that are often missed by human inspection (Wu et al. 2025). These systems employ sophisticated algorithms to analyze subtle visual cues, thereby improving the accuracy and speed of defect identification, which is essential for maintaining product quality and reducing waste in the egg industry.

14.2.3 AI Tools and Techniques Used in Agriculture

IoT enhances agricultural productivity by collecting real-time data on crop health, soil conditions, and livestock using cameras, drones, and sensors. AI utilizes these data to support informed decision-making, including resource optimization, yield prediction, and pest detection, thereby increasing production and efficiency (Mansoor et al. 2025; Rajak et al. 2023). Blockchain technology serves as a trust mechanism that maintains transparent and secure records throughout the supply chain, from farmers to consumers. It also improves food safety and facilitates automated processes such as payments (Mansoor et al. 2025).

14.2.3.1 AI Tools Used in Poultry Farms

In poultry farming, both commercial and core ML approaches are utilized as AI technologies. Key commercial tools include computer vision systems for monitoring behavior and bird health, smart climate controllers for maintaining optimal environmental conditions, and AI applications for disease detection, feeding management, growth prediction, health assessment, and automated hatchery operations such as egg grading and collection (Natsir et al. 2025; Merenda et al. 2024). Core techniques include RNNs for time-series analysis, CNNs for image processing, anomaly detection for behavioral pattern analysis, expert systems for decision support, and reinforcement learning for strategy optimization. Customization of these technologies is possible based on the specific requirements of broilers and layers (Salomon 2025).

14.2.3.2 AI Tools for the Detection of Egg Quality

In the egg industry, AI tools and techniques are employed for quality control, efficiency improvements, and reducing manual labor. Key applications include automated egg grading, counting, defect detection, and hatchery optimization, which encompasses temperature regulation and identification of fertilized and non-fertilized eggs (Akkoyun et al. 2022; Ojo et al. 2022; Çelik and Tekin 2024). The AI-based methods used for egg defect detection are illustrated in Fig. 14.2.

AI is a powerful tool for enhancing various processes in the egg industry. It supports the entire egg production chain, including egg grading, breeding, disease detection, and data analysis, more efficiently.

14.2.4 Applications of AI and IoT in Optimizing the Poultry Industry

- Computer vision technologies are employed for quality control and grading by detecting and analyzing the size, shape, shell defects, and color of eggs. Crack detection systems identify cracks on eggshells and facilitate sorting of eggs based on whether they are cracked or intact (Wu et al. 2025; Yang et al. 2023).

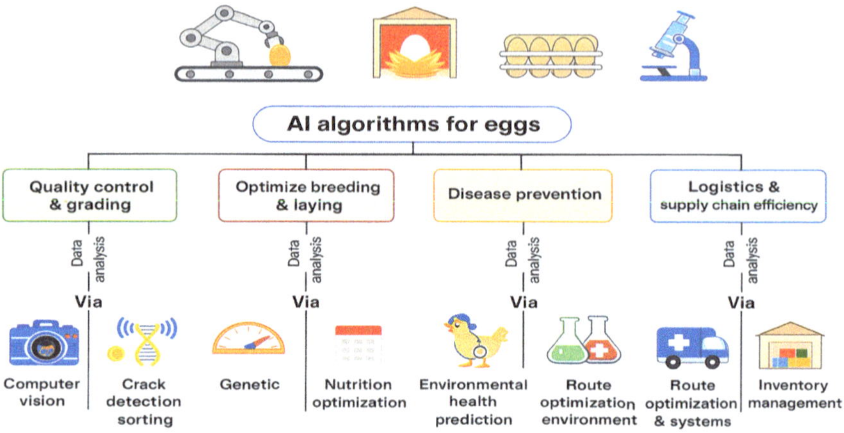

Fig. 14.2 Overview of AI-based algorithms and methods used for detecting defects in eggs

- Genetic analysis is utilized to select birds capable of producing higher quality and greater quantities of eggs. For nutrition optimization, AI recommends appropriate feed to enhance egg production and improve hen health.
- AI predicts and prevents health issues in poultry farms through environmental health monitoring, which tracks temperature, ammonia, and humidity levels. AI also optimizes farm design and flock management to improve environmental conditions. Additionally, health monitoring systems detect infections or stress at early stages (Natsir et al. 2025; Ahmed et al. 2024; Cakic et al. 2023).
- Artificial intelligence enhances logistics and supply chain efficiency by optimizing transportation routes and systems for faster and safer egg delivery. AI-driven inventory management reduces storage losses and maintains balanced stock levels.

AI has increased global egg production and efficiency, while reducing disease outbreaks and stabilizing price fluctuations (Natsir et al. 2025; Cakic et al. 2023). Major egg-producing countries such as China, India, and the United States have widely adopted AI-driven techniques. The United States and China utilize more advanced and costly AI systems, including sensors, automated grading systems, and fully automated smart farms. India is also beginning to implement AI-driven tools in the egg industry (Yang et al. 2023). Table 14.3 provides a comparison among these three countries.

India exhibits a higher growth rate in the egg industry. However, India has not adopted AI tools and techniques as extensively as the United States and China. Increased investment in AI technologies could benefit India's egg industry, given the rapidly expanding poultry sector, rising consumer demand, and the ongoing shift toward data-driven farming. The country's large market size, substantial domestic consumption, and export potential position it as a competitive player globally. The Indian government should promote the adoption of AI tools across both large-scale and small-scale farms. Although many AI tools are costly, small farms

Table 14.3 Comparative analysis of AI adoption in the egg industry: India, China, and the United States

Aspect	India	China	The United States
AI adoption stage	Emerging and rapidly growing with a focus on automation for farm efficiency and sustainability. Emphasis on integrating AI in broiler and layer production with sensor data, robotics, and real-time monitoring	More advanced usage of AI robotics, image processing, and automation in commercial layer and hatchery operations. Implementation of autonomous egg collection robots and advanced sorting systems	Mature AI systems with cutting-edge optical inspection, hyperspectral imaging for embryo detection, and highly integrated IoT-enabled smart farms. Strong emphasis on AI for quality control and large-scale operation optimization
Technologies	Robotics for egg collection, environmental monitoring using AI sensors, AI data analytics for feed and health management	DL-based image recognition for egg sorting, autonomous mobile robots for egg collection	AI-based optical inspection with NVIDIA platforms for real-time imaging, ML for internal egg analysis, automated egg counting systems
Primary focus	Increasing productivity, automating manual tasks, improving welfare and sustainability	High-throughput, precision sorting, and hatchery efficiency	Comprehensive quality assurance, defect and embryo detection, automated data-driven operations
Challenges	Integration with smallholder farms, high initial investment	Mobility within poultry houses, egg breakage prevention in robotic systems	The cost of advanced technology, scalability for small producers

can utilize low-cost sensors and smartphones equipped with applications that function as cameras and AI-based health monitoring assistants. Government support, collaborative initiatives, affordable and inclusive technology models, and pressures related to labor and sustainability can further facilitate AI adoption among small and marginal farmers. These factors collectively create strong investment incentives for the egg industry in India. The applications of AI and ML in egg quality and hatchability assessment are summarized in Table 14.4.

Various techniques are employed to assess egg quality and detect defects (Fig. 14.3). Photoplethysmography and convolutional neural networks have been utilized to identify fertility defects. The AlexNet, VGG16, and VGG19 CNN models are applied to analyze images for cracks, internal defects, and contamination. Hyperspectral imaging is used to assess the physical properties and freshness of eggs. Feature extraction yields information about key egg properties, while ensemble learning integrates multiple models to improve data accuracy. Real-time measurements and random forest algorithms can predict egg weight based on size and shell characteristics. Collectively, these methods address defects such as freshness, fertility, shell cracks, and contamination.

Table 14.4 AI and ML applications in egg quality and hatchability assessment

Application	Sensors/input	ML technique(s)	Reported accuracy/ performance	Reference
Eggshell defect detection (cracks, dirt, blood)	Input data is RGB images of the egg's surface	CNN architectures: AlexNet, VGG16, VGG19, SqueezeNet, GoogleNet, Inceptionv3, ResNet18, Xception	The highest accuracy is approximately 96.25% (VGG19) for classifying among dirty, bloody, cracked, and intact	Türkoğlu et al. (2021)
Non-destructive detection of egg internal quality (freshness, scattered yolk, cracks)	Hyperspectral imaging (Vis–NIR), morphology & crack detection + egg freshness via spectral data	Multivariate analysis Image processing (feature wavelength selection, etc.)	Reported good classification of freshness, scattered yolk, and cracks	Yao et al. (2022)
Eggshell crack detection	Image patches from egg surface images, some with micro-cracks	CNN, compared with SVM, uses traditional image features (HOG, LBP)	Accuracy is approximately 95.38% for classifying full-egg images; precision is 98.28%, and recall is ~92.41%	Botta et al. (2022)
Freshness detection via hyperspectral scattering Ensemble learning	Scattering hyperspectral imaging; different incident angles; spectral preprocessing, etc.	Ensemble learning, feature wavelength extraction, and weak classifiers stacking	Accuracy is approximately 96.25% (0° incident source) for freshness classification; the stacking approach raised some models to 100% under certain settings	Dai et al. (2020)
Egg grading, defect detection, weight estimation	Real images of eggs, features extracted for defect detection & size/ weight measurement.	RTM Det (a CNN object-detector) Random Forest for weight regression	Classification accuracy is around 94.8%, regression (weight) R^2 is 0.96 for weight prediction.	Yang et al. (2023)

(continued)

Table 14.4 (continued)

Application	Sensors/input	ML technique(s)	Reported accuracy/ performance	Reference
Fertility detection of hatching eggs	Heartbeat signal via photoplethysmography (PPG) + images of the heartbeat waveform	CNN models (E-CNN, SR-CNN, including SE-Res modules)	Very high reported accuracy, up to 99.50% and 99.62%, for classifying fertile vs. dead eggs	Geng et al. (2019)
Predicting hatchability (batch/ environmental and animal factors)	Historic data: breeder strain, breeder age, egg storage, egg weight loss, etc.	Linear Regression, Random Forest, Gradient Boosting Machine	Performance is measured by R^2 and RMSE (values depend on model and dataset)	Bouba et al. (2021)

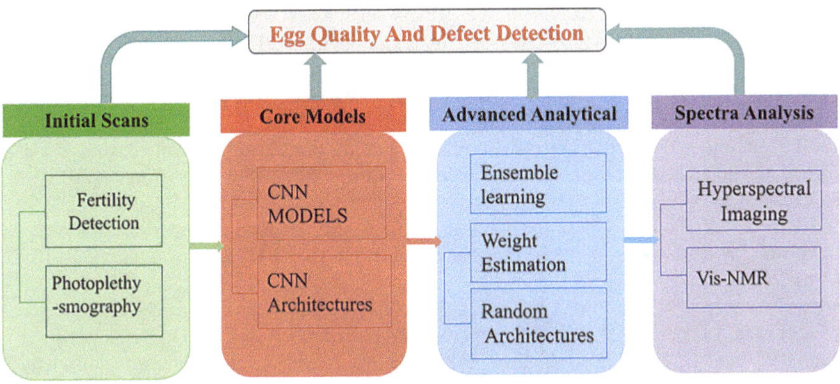

Fig. 14.3 Schematic representation of egg quality assessment and defect detection using AI and imaging techniques

14.3 Conclusion and Future Scope

The integration of AI, ML, IoT, and blockchain technologies is transforming the egg industry by enhancing productivity, transparency, and decision-making throughout the supply chain. Eggs represent a critical and affordable source of high-quality protein and essential nutrients for the expanding global population, necessitating the development of efficient and sustainable production systems. The adoption of these advanced technologies enables real-time monitoring of avian health, early disease detection, automated environmental control, and precision management through sensor-based systems. These innovations reduce losses and improve operational efficiency. Leading producers, including the United States and China, have used such technologies to achieve significant gains in production and quality. India is also progressively adopting these innovations, although high initial investment costs, particularly for smallholder farmers, remain a constraint. However, higher profit margins in the Indian poultry sector, largely due to lower labor costs, present a strategic opportunity for increased technological investment without substantially raising production expenses. Advances in computer vision, imaging, and sensor technologies have enabled automated egg grading, defect detection, and production forecasting, thereby strengthening quality assurance mechanisms. Despite these benefits, scalability remains a challenge in developing economies, where adaptation to small and medium-scale farming systems is essential. Emerging approaches, such as shared-farm models and cost-effective neo-blockchain frameworks, may provide viable pathways to broader adoption.

References

Ahmed MM, Hassanien EE, Hassanien AE (2024) A smart IoT-based monitoring system in poultry farms using chicken behavioral analysis. Internet Things 25:101010. https://doi.org/10.1016/J.IOT.2023.101010

Akkoyun F, Ozcelik A, Arpaci I et al (2022) A multi-flow production line for sorting of eggs using image processing. Sensors 23:117. https://doi.org/10.3390/S23010117

Astill J, Dara RA, Fraser EDG et al (2020) Smart poultry management: smart sensors, big data, and the internet of things. Comput Electron Agric 170:105291. https://doi.org/10.1016/J.COMPAG.2020.105291

Bist RB, Bist K, Poudel S et al (2024) Sustainable poultry farming practices: a critical review of current strategies and future prospects. Poult Sci 103:104295. https://doi.org/10.1016/j.psj.2024.104295

Botta B, Gattam SSR, Datta AK (2022) Eggshell crack detection using deep convolutional neural networks. J Food Eng 315:110798. https://doi.org/10.1016/J.JFOODENG.2021.110798

Bouba I, Visser B, Kemp B et al (2021) Predicting hatchability of layer breeders and identifying effects of animal-related and environmental factors. Poult Sci 100:101394. https://doi.org/10.1016/J.PSJ.2021.101394

Cakic S, Popovic T, Krco S et al (2023) Developing edge AI computer vision for smart poultry farms using deep learning and HPC. Sensors 23:3002. https://doi.org/10.3390/S23063002

Çelik A, Tekin E (2024) Classification of hatchery eggs using a machine learning algorithm based on image processing methods: a comparative study. Braz J Poult Sci 26:eRBCA-2023-1882. https://doi.org/10.1590/1806-9061-2023-1882

Cruz E, Hidalgo-Rodriguez M, Acosta-Reyes AM et al (2024) AI-based monitoring for enhanced poultry flock management. Agriculture 14:2187. https://doi.org/10.3390/AGRICULTURE14122187

Dai D, Jiang T, Lu W et al (2020) Nondestructive detection for egg freshness based on hyperspectral scattering image combined with ensemble learning. Sensors 20:5484. https://doi.org/10.3390/S20195484

Dones AN, Nadela JA, Sisbreño CL et al (2024) CluckTrack: Manok Bisaya monitor using computer vision YOLO V8 and IoT GSM alert for sustainable farming. In: IEEE international conference on Computing (ICOCO). IEEE, pp 337–343. https://doi.org/10.1109/ICOCO62848.2024.10928195

Essien D, Neethirajan S (2025) Multimodal AI systems for enhanced laying hen welfare assessment and productivity optimization. Smart Agric Technol 12:101564. https://doi.org/10.1016/J.ATECH.2025.101564

Fan W, Peng H, Yang D (2025) Review: the application and challenges of advanced detection technologies in poultry farming. Poult Sci 104:105870. https://doi.org/10.1016/J.PSJ.2025.105870

Finistrosa R, Mañoso C, de Madrid ÁP et al (2025) Low-cost IoT and LoRaWAN-based system for laying hen identification in family poultry farms. Appl Sci 15:4856. https://doi.org/10.3390/APP15094856

Geng L, Hu Y, Xiao Z et al (2019) Fertility detection of hatching eggs based on a convolutional neural network. Appl Sci 9:1408. https://doi.org/10.3390/APP9071408

Haruna AW, Salam M, Sulaiman AA et al (2025) Examining the effect of farmer's characteristics and production costs on the income of the layer chicken farm. Front Sustain Food Syst 9:1555709. https://doi.org/10.3389/FSUFS.2025.1555709

Kalita AJ, Subba M, Adil S et al (2025) Application of artificial intelligence and machine learning in poultry disease detection and diagnosis: a review. Lett Anim Biol 5:01–06. https://doi.org/10.62310/liab.v5i1.155

Kamilaris A, Fonts A, Prenafeta-Boldú FX (2019) The rise of blockchain technology in agriculture and food supply chains. Trends Food Sci Technol 91:640–652. https://doi.org/10.1016/J.TIFS.2019.07.034

Kostenko O (2021) Placement and concentration in agriculture: latest trends (case study by Russian egg poultry). E3S Web Conf 247:01022. https://doi.org/10.1051/e3sconf/202124701022

Mansoor S, Iqbal S, Popescu SM et al (2025) Integration of smart sensors and IoT in precision agriculture: trends, challenges and future prospectives. Front Plant Sci 16:1587869. https://doi.org/10.3389/FPLS.2025.1587869

Merenda VR, Bodempudi VUC, Pairis-Garcia MD et al (2024) Development and validation of machine-learning models for monitoring individual behaviors in group-housed broiler chickens. Poult Sci 103:104374. https://doi.org/10.1016/J.PSJ.2024.104374

Mohyuddin G, Khan MA, Haseeb A et al (2024) Evaluation of machine learning approaches for precision farming in smart agriculture system: a comprehensive review. IEEE Access 12:60155–60184. https://doi.org/10.1109/ACCESS.2024.3390581

Moss AF, Parkinson G, Crowley TM et al (2021) Alternatives to formulate laying hen diets beyond the traditional least-cost model. J Appl Poult Res 30:100137. https://doi.org/10.1016/J.JAPR.2020.100137

Myers M, Ruxton CHS (2023) Eggs: healthy or risky? A review of evidence from high quality studies on hen's eggs. Nutrients 15:2657. https://doi.org/10.3390/NU15122657

Naeem M, Jia Z, Wang J et al (2025) Advancements in machine learning applications in poultry farming: a literature review. J Appl Poult Res 34:100602. https://doi.org/10.1016/J.JAPR.2025.100602

Natsir MH, Mahmudy WF, Tono M et al (2025) Advancements in artificial intelligence and machine learning for poultry farming: applications, challenges, and future prospects. Smart Agric Technol 12:101307. https://doi.org/10.1016/J.ATECH.2025.101307

Neethirajan S (2022) Automated tracking systems for the assessment of farmed poultry. Animals 12:232. https://doi.org/10.3390/ANI12030232

Neethirajan S (2023) Artificial intelligence and sensor innovations: enhancing livestock welfare with a human-centric approach. Hum Centric Intell Syst 4:77–92. https://doi.org/10.1007/S44230-023-00050-2

Ojo RO, Ajayi AO, Owolabi HA et al (2022) Internet of things and machine learning techniques in poultry health and welfare management: a systematic literature review. Comput Electron Agric 200:107266. https://doi.org/10.1016/J.COMPAG.2022.107266

Panagi P, Karatsiolis S, Mosphilis K et al (2025) Poultry farm intelligence: an integrated multi-sensor AI platform for enhanced welfare and productivity. arXiv:2510.15757. https://arxiv.org/pdf/2510.15757

Rajak P, Ganguly A, Adhikary S et al (2023) Internet of things and smart sensors in agriculture: scopes and challenges. J Agric Food Res 14:100776. https://doi.org/10.1016/J.JAFR.2023.100776

Rauw WM, Kanis E, Noordhuizen-Stassen EN et al (1998) Undesirable side effects of selection for high production efficiency in farm animals: a review. Livest Prod Sci 56:15–33. https://doi.org/10.1016/S0301-6226(98)00147-X

Rowe E, Dawkins MS, Gebhardt-Henrich SG (2019) A systematic review of precision livestock farming in the poultry sector: is technology focussed on improving bird welfare? Animals 9:614. https://doi.org/10.3390/ANI9090614

Salihu S, Jauro SS, Salihu MN (2024) Enhancing poultry welfare and production through advanced chicken detection, tracking and counting: a deep learning approach. In: MakeLearn 2024: artificial intelligence for human-technologies-economy sustainable development. ToKnowPress

Salomon AR (2025) Machine learning in commercial poultry farming: comparing standard and oriented bounding boxes in YOLOv11 for broiler detection. agriRxiv. https://doi.org/10.31220/AGRIRXIV.2025.00318

Senoo EEK, Anggraini L, Kumi JA et al (2024) IoT solutions with artificial intelligence technologies for precision agriculture: definitions, applications, challenges, and opportunities. Electronics 13:1894. https://doi.org/10.3390/ELECTRONICS13101894

Shi C, Wang Y, Zhang C et al (2022) Nondestructive detection of microcracks in poultry eggs based on the electrical characteristics model. Agriculture 12:1137. https://doi.org/10.3390/AGRICULTURE12081137

Taleb HM, Mahrose K, Abdel-Halim AA et al (2025) Using artificial intelligence to improve poultry productivity: a review. Ann Anim Sci 25:23–33. https://doi.org/10.2478/AOAS-2024-0039

Tina FW, Afsarimanesh N, Nag A et al (2025) Integrating AIoT technologies in aquaculture: a systematic review. Future Internet 17:199. https://doi.org/10.3390/FI17050199

Türkoğlu M, Sinir E et al (2021) Detection of eggshell defects using convolutional neural networks. Turk J Agric Food Sci Technol 9:559–567. https://doi.org/10.24925/TURJAF.V9I3.559-567.4046

Wang Q, Yang Z, Liu C et al (2025) Research progress on non-destructive testing technology and equipment for poultry eggshell quality. Foods 14:2223. https://doi.org/10.3390/FOODS14132223

Wu Z, Zhang H, Fang C (2025) Research on machine vision online monitoring system for egg production and quality in cage environment. Poult Sci 104:104552. https://doi.org/10.1016/J.PSJ.2024.104552

Yang X, Bist RB, Subedi S et al (2023) A computer vision-based automatic system for egg grading and defect detection. Animals 13:2354. https://doi.org/10.3390/ANI13142354

Yao K, Sun J, Chen C et al (2022) Non-destructive detection of egg qualities based on hyperspectral imaging. J Food Eng 325:111024. https://doi.org/10.1016/J.JFOODENG.2022.111024

Policy Frameworks for Capacity Building and Education in Agriculture

15

Astha Gupta, Pardeep Singh Tiwana, and Gagandeep Kaur

Abstract

Policy frameworks, capacity building, and education are essential for improving the livelihood of the farmers and achieving sustainable agriculture. Key challenges, including food security, resource management, market access, and climate change, must be addressed through effective policy frameworks, which provide structured approaches for stakeholders and governments. An agricultural policy framework can promote equitable access to land and modern technology, while also ensuring environmental protection and social integration. Capacity building enables farmers and agricultural organizations to acquire knowledge, resources, practical skills, and tools necessary for modern farming. It also supports yield improvement through targeted training programs. Education facilitates the adoption of advanced technologies, enhances problem-solving and critical thinking skills, and deepens understanding of fundamental agricultural concepts. Integrating agricultural education into both formal and informal learning systems strengthens knowledge dissemination and supports the adoption of sustainable, climate-friendly technologies. Collectively, policy frameworks, capacity building, and education establish a robust foundation for growth, resilience, and economic development. Strengthening these areas equips farmers to address global challenges such as market fluctuations, climate change, and overpopulation, thereby supporting long-term food security and rural development. This chapter examines the foundations and evolution of agriculture, as well as its policy frameworks. It also presents government initiatives and institutional

A. Gupta (✉)
Chitkara University Institute of Engineering and Technology, Chitkara University, Rajpura, Punjab, India

P. S. Tiwana · G. Kaur
CGC, College of Engineering, Mohali, Punjab, India

H. N. Oo et al. (eds.), *Transforming Agriculture with AI, IoT, and Blockchain*,
https://doi.org/10.1007/978-981-92-1491-4_15

mechanisms in Indian agriculture, with a focus on technology-driven and digitally transformed agricultural education. Additionally, the roles of community participation and public–private partnerships (PPP) in capacity building are discussed, along with associated challenges and future directions.

Keywords

Agricultural education · Policy frameworks · Capacity building · Digital transformation · Public–Private Partnerships (PPP)

15.1 Foundation of Agriculture

For centuries, agriculture has been regarded as a fundamental component of social and economic systems, serving as the primary livelihood for many individuals. In India, agriculture is often described as the backbone of rural regions, with millions of farmers dependent on farming for their livelihoods. The sector contributes to poverty reduction, promotes social cohesion, and ensures food and nutritional security. Despite its significance, agriculture in India faces persistent challenges, including inadequate infrastructure, declining soil quality, fragmented landholdings, and inefficient water usage. Farmers also encounter difficulties in accessing capital markets and have limited financial resources. Although agriculture and related activities contribute substantially to India's economy, the sector's share of Gross Domestic Product (GDP) has declined by 16–18%, even as approximately 40–46% of the population remains reliant on agriculture for their livelihoods (Rajpurohit et al. 2023). This decline highlights the urgent need for agrarian transformation in India. Furthermore, climate variability exacerbates farming challenges, with unpredictable temperature fluctuations, irregular rainfall, floods, and heat waves. As these interconnected issues intensify, it is essential to move beyond traditional farming practices. Establishing platforms for knowledge sharing, improved governance models, and robust capacity-building technologies is necessary to enhance agricultural resilience and profitability. Consequently, the development of policies aimed at increasing productivity, human capital, innovation, and strengthening agricultural organizations is imperative (Arora 2013).

A significant transformation is underway in Indian agriculture through agricultural education and capacity building. Agricultural education, delivered through formal channels such as universities and colleges, nonformal avenues like training programs and short courses, and informal community-based initiatives, enables farmers and rural youth to acquire systematic, analytical knowledge and essential farming skills. Additionally, these educational opportunities foster entrepreneurial skills, supporting the establishment of new agricultural ventures. Capacity building further enhances the ability of farmers and organizations to plan, adopt, and utilize modern technologies effectively. When agricultural education and capacity building are integrated, farmers are better equipped to articulate their needs, challenges, limitations, goals, and objectives to the government (Ministry of Agriculture and

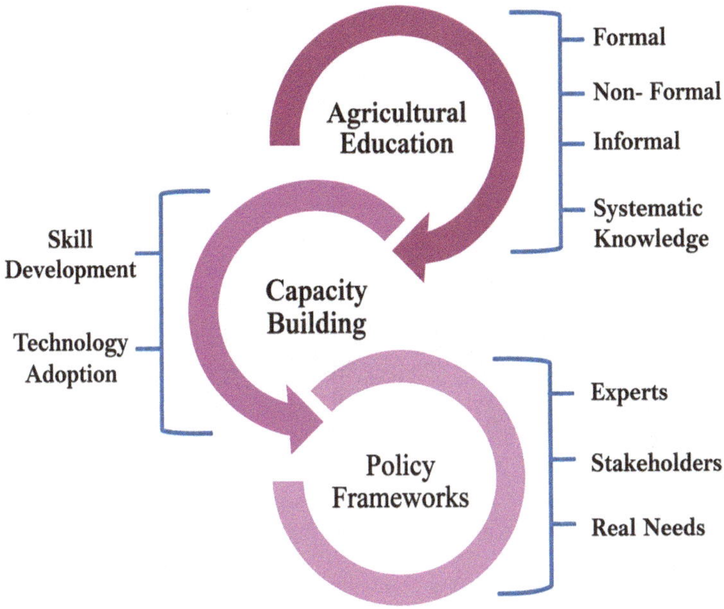

Fig. 15.1 Interconnection of agricultural education, capacity building, and policy frameworks

Farmers Welfare 2024). This, in turn, assists policymakers in formulating policies that are practical, evidence-based, and relevant to the agricultural sector. Figure 15.1 illustrates the interconnections among agricultural education, capacity building, and the policy framework.

15.2 Agriculture Policy Framework and Its Evolution

In recent years, India has developed a robust agricultural education system, particularly following the Green Revolution, with the establishment of various research centers, universities, and training agencies. These initiatives have enhanced farmers' understanding of agricultural practices and improved their skills, resulting in increased crop yields. The primary policy frameworks that have supported and strengthened Indian agriculture include the National Policy for Farmers (2007), the Doubling Farmers' Income Strategy (2018), and the National Education Policy (2020). Collectively, these policies highlight the importance of integrating education and innovation to foster entrepreneurship in the agricultural sector. Each policy contributes uniquely to the field and is linked, both directly and indirectly, to the United Nations Sustainable Development Goals (SDGs) (Swaminathan 2017).

The National Policy for Farmers (2007), which is directly related to SDG 2 (zero hunger), emphasizes improving the welfare of farmers by providing access to advanced technologies, ensuring fair pricing, and promoting environmentally sustainable farming practices. These measures aim to increase farmers' incomes and

Table 15.1 Contribution of Indian policy frameworks to the SDGs through direct and indirect linkages

Policy	Contribution	Direct link	Indirect link
National Policy for Farmers (2007)	Improving farmers welfare	SDG 2 (zero hunger)	SDG 1 (no poverty) SDG 12 (responsible consumption and production) SDG 13 (climate action)
Doubling Farmers' Income Strategy (2018)	Improving livelihood of rural regions	SDG 8 (decent work and economic growth)	SDG 1 (no poverty) SDG 2 (zero hunger) SDG 13 (climate action) SDG 17 (partnerships for the goals)
National Education Policy (2020)	Understand whole system from production level to marketing	SDG 4 (quality education)	SDG 5 (gender equality) SDG 8 (decent work and economic growth) SDG 9 (industry, innovation and infrastructure)

enhance their quality of life. The Doubling Farmers' Income Strategy (2018), aligned with SDG 8 (decent work and economic growth), seeks to improve rural livelihoods by expanding market opportunities, reducing production costs, and fostering an efficient, technology-driven agricultural system. The National Education Policy (2020), which aligns with SDG 4 (quality education), advocates the adoption of digital tools, smart farming techniques, and online learning platforms to facilitate a comprehensive understanding of the agricultural value chain, from production to marketing. Table 15.1 summarizes the main contributions of these Indian policies and their direct and indirect connections to all SDGs.

In addition to these policies, the post-independence evolution of India's agriculture sector reflects the broader development initiatives of the nation. The transformation phases of agricultural development are illustrated in Fig. 15.2 and outlined below.

15.2.1 Green Revolution and Post-independence Era (1950s–1980s)

Following independence, India experienced acute food scarcity, low agricultural output, and a predominantly traditional farming system. During this period, the primary objective was to increase food production and ensure food security for the rapidly growing population. To address these challenges, the government initiated programs such as the Intensive Agricultural District Programme (1960) and the Community Development Programme (1952) (Swaminathan 2017). These initiatives represented the first significant efforts to enhance rural livelihoods, develop agricultural services, and promote evidence-based farming techniques. In the mid-1960s, during the Green Revolution era, expanded irrigation systems were

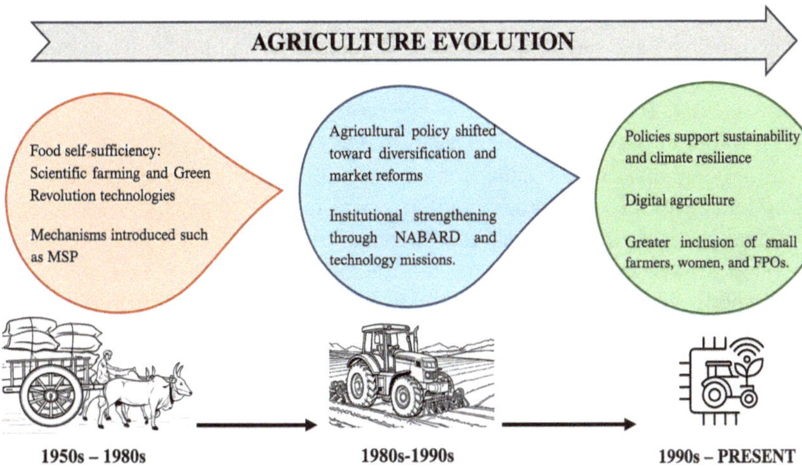

Fig. 15.2 Year-wise evolution of agriculture

introduced, and High-Yielding Varieties (HYVs) of wheat and rice were adopted. Additionally, collaboration with national scientific institutes and international research partners facilitated the development and widespread availability of fertilizers and agricultural chemicals, further boosting productivity. As a result, India transitioned from a food-deficit to a food-surplus nation within two decades. In 1965, the government established the Agricultural Prices Commission, which introduced the Minimum Support Price (MSP) to regulate the agricultural sector and assure farmers of adequate returns. The MSP also encouraged the adoption of modern technologies by reducing concerns about price volatility (Vaidyanathan 2010). However, these advancements were accompanied by certain drawbacks. States such as Punjab, Uttar Pradesh, and Haryana benefited disproportionately, resulting in regional inequalities. Moreover, excessive use of groundwater, synthetic fertilizers, and monocropping practices contributed to soil degradation and water scarcity. These environmental challenges highlight the need for more sustainable, stable, and globally informed agricultural policies in the future.

15.2.2 Modernization of Agriculture and Agrobiodiversity (1980s–1990s)

During this period, agricultural practices shifted from cultivating a limited range of food grains to a more diverse array, reflecting a market-oriented approach. This transition was influenced by the Green Revolution, which initiated environmental degradation. To increase farmers' incomes and reduce the use of land and water resources, the government promoted agricultural expansion in sectors such as oilseeds, dairy, horticulture, fisheries, and pulses. In 1982, the establishment of the National Bank for Agriculture and Rural Development (NABARD) provided financial support for rural infrastructure. The Technology Mission on Oilseeds, launched in 1986 and later expanded to include cotton and horticulture, aimed to increase yields through improved seeds, advanced farming techniques, and research initiatives (FAO 2020). Simultaneously, the economic liberalization of 1991 heightened the importance of international trade, markets, and private investment. Policies increasingly emphasized the enhancement of rural infrastructure, including irrigation systems, roads, cold storage, and improved market connectivity for farmers. Additionally, services promoting the adoption of sustainable practices and new technologies were introduced, laying the foundation for modern agricultural systems. Overall, this era was characterized by a focus on market-driven, diversified, and financially supported agriculture.

15.2.3 Sustainable and Comprehensive Agriculture (1990s–Present)

By the early 2000s, the limitations of previous agricultural growth models had become apparent, as productivity and farmer livelihoods were adversely affected by soil degradation, groundwater depletion, high input costs, and climate change. Consequently, agricultural policy shifted toward environmentally sustainable and accessible farming practices. The government introduced policies promoting smart farming, balanced resource use, and support for smallholder farmers. The emphasis on sustainable activities marked a significant transition during this period. The National Mission for Sustainable Agriculture (NMSA) advanced initiatives such as organic farming, integrated farming, water conservation, and soil health management. Additionally, the Soil Health Card Scheme (2015) enabled farmers to assess soil nutrient levels and reduce fertilizer usage. This era also witnessed a substantial adoption of digital and technology-driven agricultural techniques (ICAR 2022). Digital platforms, weather-based crop alerts, and real-time decision-making tools played critical roles. Schemes such as PM-KISAN (2019) and e-NAM (2016) introduced direct income support and a national digital marketplace, respectively, to help farmers secure better prices. Overall, from the 2000s to the present, agriculture has been redefined as a socially inclusive, economically stable, and environmentally sustainable sector rather than solely a production activity.

15.3 Institutional Mechanisms and Government Initiatives in Indian Agriculture

The agricultural development in India is strengthened by a robust institutional framework and a diverse array of government initiatives. These mechanisms are critical for enhancing productivity, increasing farmers' incomes, ensuring food security, and promoting sustainable agricultural practices. Central and state government agencies, research institutions, financial organizations, and extension agencies constitute the primary pillars of agricultural development in India (Tiwari 2008).

Despite accounting for only 10% of national Gross Value Added (GVA), the agriculture sector employs 35% of the workforce, indicating significant untapped potential for productivity improvement. The sector is at a key point, as rapid technological advancements such as AI-driven crop planning and drone-based monitoring contrast sharply with a workforce in which 35% lack basic literacy, and the majority have not received formal technical training. Furthermore, women farmers, who represent 41% of agricultural workers, experience pronounced marginalization (Bathla and Hussain 2022). Addressing these challenges requires urgent, targeted skill-development interventions, particularly for women and in specific geographical clusters, to bridge the gap between technological progress and human capacity. Such efforts are essential to transform agriculture from a subsistence activity into a driver of inclusive economic growth. Some of the key institutions dedicated to agricultural development in India include the following:

1. *Ministry of Agriculture and Farmers Welfare (MoA&FW)*: It serves as the central authority responsible for policy formulation, planning, and implementation of agricultural development programs in India. This ministry operates through several departments, including:
   - Department of Agriculture and Farmers' Welfare (DAFW).
   - Department of Agricultural Research and Education (DARE).
   - Department of Animal Husbandry and Dairying.
2. *Indian Council of Agricultural Research (ICAR)*: It serves as the apex institution responsible for coordinating agricultural education and research in India. Its primary functions are as follows:
   - Facilitating agricultural research and fostering innovation.
   - Overseeing the administration of agricultural universities and research institutes.
   - Developing improved crop varieties and advanced agricultural technologies.
   - Supporting the development of human resources within the agricultural sector.
3. *State Agricultural Universities (SAUs)*: These universities function as the principal institutions for agricultural education, research, and extension at the state level. These universities enhance local agricultural systems by delivering region-specific research outcomes and tailored training programs.
4. *Krishi Vigyan Kendras (KVKs)*: These institutions serve as grassroots-level extension centers that facilitate the direct transfer of agricultural technologies to farmers. Their services include the following:
   - Skill-based training programs for farmers.
   - Demonstrations of contemporary farming techniques.
   - Dissemination of information regarding government agricultural schemes.

5. *NABARD and Rural Financial Institutions*: NABARD plays a central role in agricultural financing. It provides support for the following areas:
    - Rural credit systems.
    - Farmer-Producer Organizations (FPOs).
    - Infrastructure development projects.
    - Self-help groups (SHGs).
6. Agricultural marketing institutions such as the *Food Corporation of India (FCI)*, *Agricultural Produce Market Committees (APMCs)*, and *eNAM* platforms play a central role in regulating agricultural marketing, procurement, and price stabilization.

15.3.1 Major Government Initiatives in Indian Agriculture

The government is strategically developing a Digital Public Infrastructure (DPI), referred to as Agri-Stack, to modernize governance and service delivery within the agricultural sector. This initiative aims to establish an interoperable ecosystem that delivers personalized and timely information to farmers, thereby enhancing efficiency and transparency in government schemes (Darjee 2023), as illustrated in Fig. 15.3.

- *Pradhan Mantri Kisan Samman Nidhi (PM-KISAN)*: Provides direct income support to small and marginal farmers by transferring financial assistance directly into their bank accounts.
- *Pradhan Mantri Fasal Bima Yojana (PMFBY)*: Crop insurance scheme that protects farmers against losses due to natural calamities, pests, and diseases.

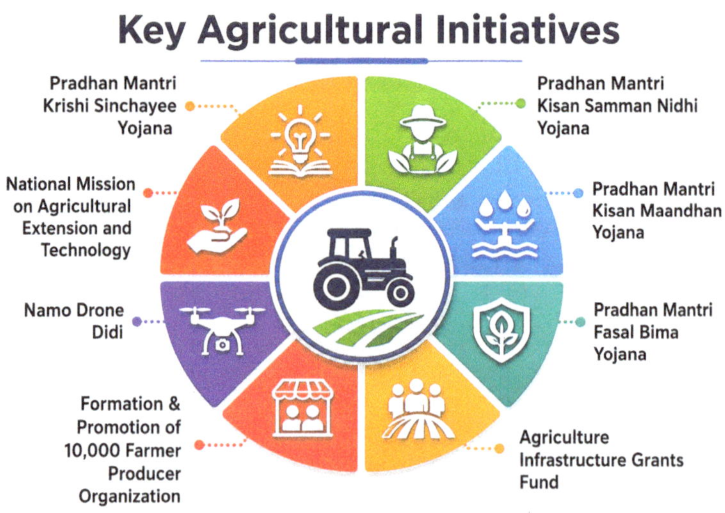

Fig. 15.3 Government initiatives in Indian agriculture

- *Soil Health Card Scheme*: Aims to provide farmers with information on soil fertility status and suitable nutrient management practices.
- *Pradhan Mantri Krishi Sinchai Yojana (PMKSY)*: Focused on expanding irrigation coverage and improving water-use efficiency through micro-irrigation systems like drip and sprinkler.
- *National Mission on Sustainable Agriculture (NMSA)*: Promotes climate-resilient farming practices, ecosystem conservation, and sustainable agricultural techniques.
- *e-NAM (Electronic National Agriculture Market)*: A digital platform to integrate APMC markets, enabling better price discovery and transparent trading.
- *National Food Security Mission (NFSM)*: Aims to increase production of rice, wheat, pulses, and coarse grains to ensure food security and nutritional improvement.
- *National Rural Livelihood Mission (NRLM)*: Supports skill development, self-employment, and livelihood diversification among rural communities, especially women farmers.

The Digital Agriculture Mission aims to create digital identities for 110 million farmers over 3 years and to implement a nationwide Digital Crop Survey (DCS) across all districts. This initiative marks a transition from traditional manual methods to data-driven decision-making. The adoption of drones (UAVs) for precision farming, particularly in spraying and crop health monitoring, represents a significant advancement by addressing labor shortages and enhancing input use efficiency. The liberalized Drone Rules 2021 and associated financial assistance policies demonstrate the government's commitment to making this technology accessible to both individual farmers and FPOs. Drone usage enables pesticide application up to five times faster than manual methods and substantially reduces chemical exposure for farm workers. The government provides a 40–50% subsidy for drone purchases by FPOs and specific farmer categories. The Agri-Tech sector in India has experienced rapid growth, expanding beyond e-commerce to encompass advanced solutions such as artificial intelligence (AI), Internet of Things (IoT)-based precision farming (Agri-IOT), quality management, and supply chain technologies. These start-ups address knowledge and market gaps, thereby making farming more profitable and predictable for smallholders. As of December 2023, nearly 2800 Agri-Tech start-ups were recognized by Startup India, collectively raising approximately ₹6600 crore in funding over the past 4 years, which highlights significant private sector investment and innovation. There is a notable shift toward crop diversification away from water-intensive staples such as rice and wheat, with a focus on promoting Nutri-cereals, referred to as "Shree Anna" (Millets). This transition is essential for enhancing climate change resilience, improving nutrition, and mitigating groundwater depletion. The designation of 2023 as the International Year of Millets aligns with an estimated total millet (Shree Anna) production of 174.08 LMT in 2023–2024, reflecting policy-driven momentum for increased cultivation (Raina et al. 2022).

Rural-to-urban migration has resulted in labor shortages, making farm mechanization essential for sustaining productivity and reducing cultivation costs, particularly for small and marginal farmers. Policy initiatives supporting Custom Hiring Centers (CHCs) are crucial for improving access to expensive agricultural machinery. Through the Sub-Mission on Agricultural Mechanization (SMAM), the government has provided financial assistance, leading to the establishment of thousands of CHCs and the training of 57,139 farmers in mechanization between 2021 and 2025. FPOs are transforming the agricultural market by organizing small and marginal farmers, thereby increasing their collective bargaining power and facilitating improved access to inputs, technology, and formal credit. The government is actively supporting the creation of 10,000 new FPOs nationwide, which is expected to strengthen market linkages and enable farmers to achieve better price realization for their produce. The horticulture sector, which includes fruits, vegetables, and spices, continues to demonstrate strong growth, driven by rising domestic and export demand, diversification strategies, and improved postharvest management. This sector is vital for increasing farm income and promoting value addition. Recent estimates indicate horticulture production reached 352.23 million tonnes in 2023–2024 (Second Advance Estimates), with fruit production projected at 112.63 million tonnes, confirming horticulture as a high-value growth segment (Pathak 2023).

15.3.2 Agricultural Extension Reforms and Policies

Agricultural extension encompasses the systems, institutions, and processes that provide farmers with timely information, training, new technologies, and advisory services. Over time, agriculture has been significantly transformed by climate change, market globalization, technological advancements, and evolving socioeconomic frameworks. In response to these challenges, reforms in agricultural extension and related policies have played a crucial role in strengthening the linkages between research institutions and farming communities (Bathla and Hussain 2022). Traditional extension systems primarily employed a top-down approach focused on technology transfer, often excluding farmer participation. In contrast, contemporary agricultural practices emphasize:

- Advisory services that are farmer-based and responsive to demand.
- Integration of digital tools and ICTs.
- Market-oriented knowledge dissemination.
- Adoption of climate-resilient agricultural practices.

Agricultural extension reforms aim to improve the effectiveness and accessibility of extension workers by strengthening institutional frameworks and enhancing technical and communication skills. These reforms promote partnerships between public and private sectors to deliver diversified, demand-driven agricultural services, ensuring that farmers receive timely and relevant support. Additionally, reforms focus on building farmers' capacity through training programs, field demonstrations, and farm field schools, thereby increasing their knowledge, skills, and decision-making abilities. Emphasis is also placed on mainstreaming information

Table 15.2 Utilization of agricultural extension policies

Year	Utilization of policies (%)	Major focus areas
2018	45	Traditional extension methods, limited ICT use
2019	52	Increased farmer training and awareness programs
2020	58	Digital extension initiatives, mobile advisory services
2021	63	FPOs support e-extension
2022	70	Climate-smart agriculture promotion
2023	78	Strengthened PPP models and AI-based advisory systems
2024	85	Integrated extension services using ICT and farmer-centric reforms

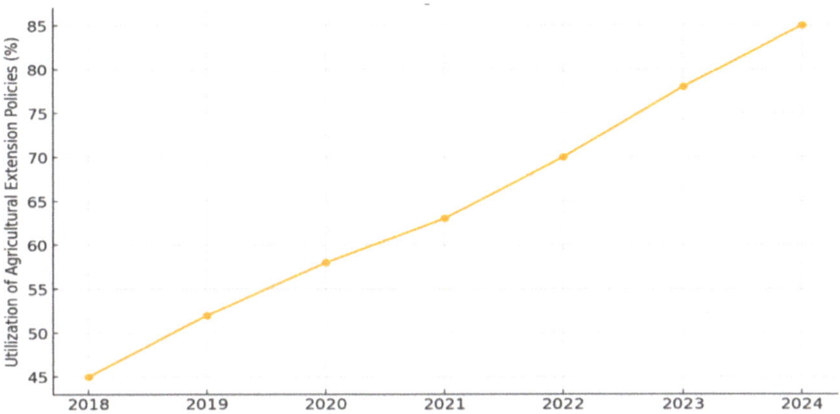

Fig. 15.4 Adoption of agricultural extension policies across the years

and communication technology (ICT)-enabled extension, utilizing digital tools such as mobile applications, AI-based advisory services, and digital kiosks to provide real-time information and solutions. Furthermore, extension reforms seek to strengthen linkages among agricultural research, education, and extension systems to facilitate the rapid dissemination of innovations. These initiatives also promote inclusivity by empowering women, marginal and small farmers, and rural youth to participate actively in agribusiness development and expansion. Table 15.2 presents the year-wise utilization of agricultural extension policies. Figure 15.4 illustrates the adoption of agricultural extension policies over the years.

15.3.2.1 Major Agricultural Extension Reforms in India

Indian agricultural extension reforms have been implemented to decentralize the extension system and make it more farmer-oriented, technology-driven, and inclusive. These reforms aim to improve the efficiency of knowledge transfer, strengthen linkages between research institutions and farmers, and promote sustainable agricultural development. Over time, the government has shifted from a solely government-dominated model to a pluralistic system that includes public agencies, private sector entities, nongovernmental organizations (NGOs), and agricultural organizations. A key change in these reforms is the emphasis on decentralized planning and execution, allowing districts and states to develop extension programs tailored to local agro-climatic and socioeconomic conditions. Another significant transformation is the promotion of ICT-based extension services, which utilize digital tools to efficiently reach large numbers of farmers (Priyadarshini and Abhilash 2020). The reforms also prioritize capacity building, farmer participation, and market linkages. Collectively, these measures have improved technology dissemination, increased farmer engagement, and strengthened the agricultural innovation system in India. Table 15.3 presents the key agricultural extension reforms and the years in which they were introduced.

Table 15.3 Key agricultural extension reforms

Reform/initiative	Year launched	Objectives	Features	Effects on farming
Kisan Call Centers (KCC)	2004	Providing direct advisory services	Toll-free numbers, expert consultation in regional languages	Quick and easy access to expert agricultural advice
Agricultural Technology Management Agency (ATMA)	2005	Decentralization and integration of extension services	District-level planning, Farmer Interest Groups (FIGs), participatory approach	Increased farmer involvement and localized extension planning
FPOs Promotion Scheme	2013	Strengthening collective farming and market linkages	Formation and support of farmer groups and cooperatives	Improved income, bargaining power, and access to markets
National Agricultural Extension Policy Framework	2014	Strengthening extension delivery system	Pluralistic, demand-driven extension, involvement of private sector	Improved service delivery and diversified extension approaches
National Mission on Agricultural Extension & Technology (NMAET)	2014	Enhancing technology dissemination	Focus on mechanization, extension, and training	Strengthened technology outreach and adoption
Digital India—Agricultural Initiatives	2015	Digital transformation of agricultural services	e-NAM, mobile apps, online portals, AI advisory tools	Enhanced access to market and technical information
Farmer FIRST Programme	2016	Farmer-focused research-extension linkage	Farmer-centric technology development and demonstrations	Better adoption of suitable technologies

Recent agricultural extension reforms in India have shifted the traditional extension system toward a decentralized, participatory, and technology-driven framework. Initiatives such as the ATMA, NMAET, KCC, and digital agriculture platforms have expanded farmers' access to information and innovation. These reforms have strengthened linkages among research institutions, extension agencies, and farmers, while also fostering the involvement of private sector entities, farmer organizations, and NGOs. The integration of ICT, a focus on capacity building, and support for farmer collectives have collectively improved the efficiency and reach of agricultural extension services in India.

15.3.2.2 Digital Transformation and Technology-Enabled Agricultural Education

Digital transformation in agricultural education refers to the integration of advanced digital technologies into teaching, learning, training, and knowledge dissemination within the agricultural sector. This transformation aims to enhance the quality, accessibility, and efficiency of educational processes by using ICT, AI, IoT, data analytics, and online educational platforms, as illustrated in Fig. 15.5. Technology-driven education is essential for preparing competent professionals, informed farmers, and future agripreneurs in response to rapidly evolving agricultural practices and climate challenges. Traditionally, agricultural education relied on classroom instruction, physical demonstrations, and field visits. While these methods remain valuable, digital transformation has significantly expanded opportunities through virtual classes, training modules, interactive simulations, and remote-sensing-based learning. Massive Open Online Courses (MOOCs), virtual laboratories, agricultural

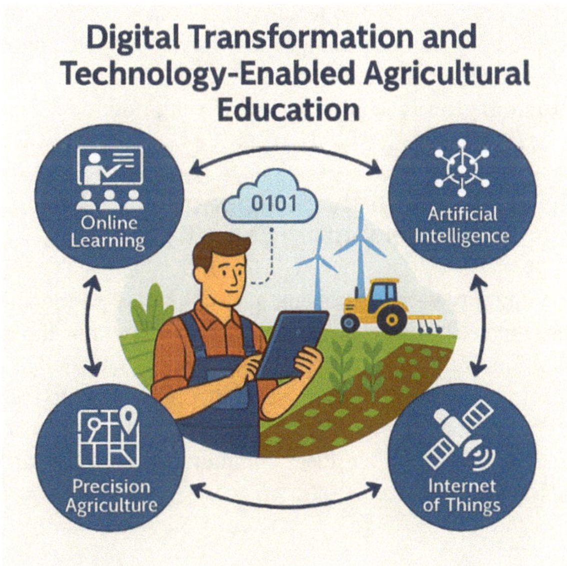

Fig. 15.5 Digital ecosystem supporting modern agricultural education

mobile applications, and digital libraries provide students and farmers with real-time access to current knowledge, research, and best practices from any location.

Smart technologies, including AI-based crop advisory systems, precision farming equipment, and IoT-powered sensors, have been integrated into agricultural curricula. These tools enable learners to monitor crops, assess soil health, forecast weather, and manage pests using real-time, technology-driven methods (Gupta 2024). Practical experience in modern agricultural management encourages the use of digital tools such as Geographic Information Systems (GIS), drone mapping, and satellite imaging, making the learning process more experiential and applied. Technologically enabled agricultural education significantly expands outreach to rural and remote regions. Digital platforms eliminate geographic barriers by allowing farmers, extension workers, and rural youth to access training programs, webinars, and expert consultations via mobile devices and Internet connectivity. E-extension services, virtual farmer field schools, and virtual knowledge hubs support continuous learning and skill development, particularly among small and marginal farmers (Tiwari 2008). Furthermore, digital transformation enhances inclusive education by increasing engagement among women and youth, offering flexible learning opportunities, and localizing digital materials in regional languages. It also fosters innovation, entrepreneurship, and a start-up culture within the agricultural sector by introducing learners to agri-technology solutions, digital markets, and data-driven decision-making. Nevertheless, challenges such as digital illiteracy, limited Internet connectivity in rural areas, inadequate infrastructure, and the high cost of modern technology continue to hinder the full potential of technology-enabled agricultural education. Addressing these challenges requires government policies and institutional initiatives focused on improving rural online infrastructure, building capacity, and promoting affordable Agri-tech solutions (Saleth 2022). Digital transformation and technology-based processes are shaping the future of agriculture by making education more accessible, practical, and innovation-oriented. These advancements contribute to developing a technologically skilled agricultural workforce capable of adopting modern scientific practices, increasing productivity, promoting sustainability, and supporting rural development.

15.4 Community Participation and Public–Private Partnerships in Capacity Building

Community participation and public–private partnerships (PPP) play a critical role in enhancing capacity-building programs within the agricultural sector. Capacity building refers to the improvement of knowledge, skills, competencies, and resources among farmers and rural communities to increase agricultural productivity, sustainability, and livelihoods. These programs are more effective, inclusive, and sustainable when local communities actively participate, and multiple stakeholders collaborate.

15.4.1 Community Participation Role

Community participation involves the active engagement of farmers, SHGs, FPOs, cooperatives, youth groups, and local institutions in the planning, implementation, and evaluation of capacity-building programs. Through this approach, farmers become coproducers of knowledge and solutions rather than passive recipients of training. Participatory methods ensure that training programs align with local needs, socioeconomic conditions, crop requirements, and resource availability. Tools such as Participatory Rural Appraisal (PRA), Farmer Field Schools (FFS), village knowledge centers, and community-led demonstration farms facilitate experiential learning and collaborative problem-solving. Community participation also enhances social capital, fosters peer learning, and increases trust between farmers and extension agencies (Bahinipati et al. 2024). Special emphasis is placed on including women farmers, marginal farmers, and rural youth in the capacity-building process to promote equity and social inclusion.

15.4.2 Role of Public–Private Partnerships (PPPs)

PPPs involve collaboration among government agencies, private companies, NGOs, research institutions, and educational institutions to enhance agricultural capacity building. The public sector contributes through policy support, funding, infrastructure, and organizational structures. The private sector provides technological innovation, market expertise, investment, and efficient service delivery mechanisms. NGOs and civil society organizations facilitate community mobilization and support implementation at the grassroots level (Gupta et al. 2023). Table 15.4 outlines several benefits of community participation and PPPs in capacity building. PPP models contribute significantly to capacity building:

- Modern farming technologies development.
- Agribusiness/entrepreneurship training.
- Diffusion of digital agricultural equipment.
- Market integration and value chain integration.
- Fiscal education and credit availability.

15.5 Challenges and Policy Recommendations

Despite notable advancements in agricultural extension, digital education, and capacity-building initiatives, several persistent challenges hinder the effective implementation and reach of these interventions. Addressing these obstacles and establishing strategic future directions is essential to promote agricultural development and ensure inclusive, sustainable growth (Kumar et al. 2024).

A major challenge is the lack of adequate infrastructure in rural areas, particularly regarding Internet connectivity, electricity supply, and access to digital devices.

Table 15.4 Benefits of community participation and public–private partnerships in capacity building

Aspect	Community participation	PPPs
Decision-making	Involves local people in planning and implementation, ensuring programs match real needs	Enables shared decision-making between government and private sector for efficient program design
Training effectiveness	Improves relevance of training based on local conditions and traditional knowledge	Brings modern technologies, expert trainers, and industry knowledge into capacity-building programs
Technology adoption	Encourages peer learning and collective adoption of new techniques	Accelerates adoption of advanced tools like AI, IoT, and precision farming practices
Resource mobilization	Utilizes local human and natural resources effectively	Attracts financial investment, infrastructure, and technological resources
Inclusion and equity	Supports participation of women, marginal farmers, and rural youth	Helps design inclusive programs with corporate social responsibility (CSR) initiatives
Innovation promotion	Encourages locally innovative solutions and community-based farming models	Facilitates research-based innovations and commercialization of Agri-technologies
Sustainability	Strengthens community ownership, ensuring long-term sustainability	Ensures continuity through financial and institutional support mechanisms
Market linkages	Strengthens direct farmer-to-consumer and collective marketing initiatives	Develops strong market connections, value chains, and export opportunities
Capacity enhancement	Builds local leadership, communication skills, and technical knowledge	Enhances professional skills, entrepreneurship, and agribusiness management
Risk management	Promotes collective risk-sharing and resilience strategies	Introduces financial tools such as crop insurance, access to credit, and risk mitigation services

This limitation restricts the delivery of extension and educational services via technology, especially for small and marginal farmers (Kaur et al. 2025a). Low levels of digital literacy among farmers and extension officials further impede the effective use of mobile applications and online platforms, particularly among older and less-educated farmers, thereby reducing the efficacy of online initiatives. Additionally, there is a shortage of trained extension workers and professionals capable of delivering modern, technology-driven services. Extension services are often understaffed and overburdened, resulting in limited outreach to farmers (Kaur et al. 2025b). Furthermore, inadequate communication among stakeholders, including governmental institutions, NGOs, and local entities, hampers the implementation of integrated capacity-building programs. Fragmented efforts frequently lead to duplication and inefficient use of resources. Financial constraints, limited awareness of government schemes, and sociocultural barriers, such as gender inequality, also restrict the full participation of women and marginalized groups in agricultural development programs.

15.6 Future Directions

To address these obstacles, future initiatives should focus on enhancing rural infrastructure by expanding digital connectivity, improving access to electricity, and providing affordable smart devices to farming communities. There is an urgent need to establish digital literacy and skills development through village-level training programs, farmer field schools, and community learning centers. Training youth and women as digital extension volunteers is recommended. The adoption of emerging technologies such as AI, IoT, remote sensing, and big data analytics in agricultural extension and education systems should be promoted. These technologies can deliver real-time, location-specific solutions to farmers.

Furthermore, increased collaboration among agricultural universities, research institutions, government agencies, private companies, and farmer organizations will contribute to a more coordinated and efficient extension system. Promoting agri-entrepreneurship, start-ups, and innovation hubs in rural areas can generate new employment opportunities for youth and support technology-based farming. To enhance the effectiveness of agricultural extension and capacity-building efforts, the following policy recommendations are proposed:

- The government should establish a unified framework of integrated policies encompassing agricultural extension, digital education, and skill development.
- Financial investment in rural digital infrastructure and extension services should be significantly increased.
- Policies should be implemented to promote PPPs that deliver technology-based training and consultancy services to farmers.
- Targeted policies should be introduced to enhance the participation of women farmers, marginal farmers, and rural youth.
- Continuous monitoring and evaluation systems should be developed to assess the impact of extension programs and facilitate the adoption of best practices.
- Agricultural training programs should mandate the inclusion of climate change education and climate-resilient farming practices.

15.7 Conclusion

The agricultural sector in India is undergoing a significant transformation driven by policy restructuring, institutional empowerment, digital innovation, and capacity-building initiatives. This chapter examines the evolution of the agricultural ecosystem through integrated policy frameworks, extension reforms, digital agricultural education, and community-based strategies. Key institutions such as the MoA&FW, ICAR, SAUs, and KVKs have played a central role in strengthening research-extension-farmer linkages. Initiatives including PM-KISAN, PMFBY, PMKSY, e-NAM, and the Digital Agriculture Mission have enhanced support systems for farmers, increasing their resilience to socioeconomic and climatic challenges. Agricultural extension reforms have shifted toward a decentralized, farmer-centered, and technology-enabled model, replacing the traditional top-down approach. The adoption of ICT, AI-driven advisory systems, Agri-Stack, drone technologies, and digital platforms has improved farmers' access to real-time information, market opportunities, and climate-resilient solutions. Concurrently, community participation and PPP have strengthened grassroots engagement, promoted inclusivity for women and marginal farmers, and supported sustainable capacity building. Nevertheless, challenges such as digital illiteracy, inadequate infrastructure, insufficiently trained extension personnel, and unequal access to technology require urgent attention. Future progress should focus on enhancing rural digital infrastructure, expanding skill development programs, improving inter-institutional collaboration, and fostering agri-entrepreneurship among rural youth and women. A comprehensive strategy encompassing institutional support, policy innovation, technological adoption, community engagement, and private sector collaboration is essential to ensure that Indian agriculture becomes more inclusive, resilient, and future-oriented. These measures will not only increase farm productivity and income but also contribute to achieving national development objectives and SDGs related to poverty reduction, food security, gender equality, and sustainable economic growth.

References

Arora VPS (2013) Agricultural policy in India: retrospect and prospect. Agric Econ Res Rev 26:135–157

Bahinipati CS, Viswanathan P, Singh AK (2024) Do institutions, incentives, and information enhance adoption of climate smart agriculture practices? Empirical evidence from India. Environ Dev 50:100982. https://doi.org/10.1016/j.envdev.2024.100982

Bathla S, Hussain S (2022) Structural reforms and governance issues in Indian agriculture. In: Chand R, Joshi P, Khadka S (eds) Indian agriculture towards 2030. Springer, Singapore. https://doi.org/10.1007/978-981-19-0763-0_9

Darjee D (2023) A review on policy initiatives, institutional mechanism and support for organic farming in India with special reference to North-east states: an exploratory study. Int J Res Anal Rev 10:193–211

FAO (2020) Farmer producer organizations in India: policy and practice. Food and Agriculture Organization of the United Nations

Gupta S (2024) Government initiatives and policies for smart agriculture. In: Agriculture 4.0. CRC Press

Gupta A et al (2023) An automatic speech recognition system in Indian and foreign languages: a state-of-the-art review analysis. Intell Decision Technol 17(3):1–19

ICAR (2022) Annual report 2021–22. Indian Council of Agricultural Research, New Delhi

Kaur G, Tiwana PS, Kommula AV (2025a) Advancements in water quality monitoring: leveraging machine learning and artificial intelligence for environmental management. In: Dubey AK, Srivastav AL, Kumar A et al (eds) Computational automation for water security. Elsevier, pp 11–26

Kaur G, Sardar S, Tiwana PS et al (2025b) Role of blockchain in transforming the water supply chain. In: Blockchain and the water supply chain: opportunities, challenges and innovations. ISTE LTD, London, pp 187–207. https://doi.org/10.1002/9781394417803.ch10

Kumar RK, Senthamizhkumaran VR, Alagendran S et al (2024) Advances in agricultural biotechnology: enhancing productivity and sustainability in India: a review. J Sci Res Rep 30:366–383. https://doi.org/10.9734/jsrr/2024/v30i72154

Ministry of Agriculture and Farmers Welfare (2024) Draft national policy framework on agricultural marketing. Government of India. https://agricoop.nic.in/sites/default/files/Draft%20NPFAM.pdf

Pathak H (2023) Impact, adaptation, and mitigation of climate change in Indian agriculture. Environ Monit Assess 195:52. https://doi.org/10.1007/s10661-022-10537-3

Priyadarshini P, Abhilash PC (2020) Policy recommendations for enabling transition towards sustainable agriculture in India. Land Use Policy 96:104718. https://doi.org/10.1016/j.landusepol.2020.104718

Raina R, Mishra S, Ravindra A et al (2022) Reorienting India's agricultural policy: millets and institutional change for sustainability. J Ecol Soc 34:1–15. https://doi.org/10.54081/JES.028/01

Rajpurohit T, Singh D, Kumar R (2023) Capacity building in agriculture. In: Kritika, Singh D, Sonia (eds) Human development: in perspective of agriculture. GH Publication, Prayagraj

Saleth R (2022) Institutions for agriculture and rural development: a case study of water institutions in India. NABARD Research and Policy Series No.9/2022

Swaminathan MS (2017) 50 Years of green revolution. World Scientific. https://doi.org/10.1142/10279

Tiwari SP (2008) Information and communication technology initiatives for knowledge sharing in agriculture. Indian J Agric Sci 78:737–747

Vaidyanathan A (2010) Agricultural growth in India: the role of technology, incentives, and institutions. Oxford University Press, New Delhi

Case Studies of Transforming Agriculture in Developing and Developed Nations

16

Htet Ne Oo, Heena Wadhwa, Srikanta Kumar Mohapatra, Binay Kumar Singh, and Saurav Kumar

Abstract

Agriculture plays a key role in the economies of both developing and developed countries. It is currently experiencing a significant transformation through the integration of advanced digital technologies. This chapter analyzes the roles of Artificial Intelligence (AI), the Internet of Things (IoT), and blockchain in facilitating precision agriculture and modernizing agri-food systems. It presents a comparative analysis of technological adoption across diverse economic contexts, emphasizing differences in implementation strategies, outcomes, and systemic constraints. In developed countries, the adoption of AI-driven analytics, advanced robotics, and IoT-enabled precision farming systems primarily aims to optimize resource efficiency, increase productivity, and promote environmental sustainability, supported by strong private sector involvement and innovation-focused ecosystems. Conversely, developing countries prioritize accessible and cost-effective technological solutions, such as IoT-based monitoring, mobile-enabled advisory systems, and blockchain-supported traceability, to enhance

H. N. Oo (✉)
Department of Computer Science and Engineering, Chitkara University Institute of Engineering and Technology (CUIET), Chitkara University, Rajpura, Punjab, India
e-mail: htet.neo@chitkara.edu.in

H. Wadhwa · S. K. Mohapatra
Chitkara University Institute of Engineering and Technology, Chitkara University, Rajpura, Punjab, India

B. K. Singh
Indian Institute of Wheat and Barley Research, Karnal, Haryana, India

S. Kumar
CSIR-Central Scientific Instruments Organization, Chandigarh, Punjab, India

Academy of Scientific and Innovative Research, Ghaziabad, Uttar Pradesh, India

H. N. Oo et al. (eds.), *Transforming Agriculture with AI, IoT, and Blockchain*,
https://doi.org/10.1007/978-981-92-1491-4_16

farm-level decision-making, lower transaction costs, and empower smallholder farmers. Policy frameworks also vary considerably, with government-led initiatives being central to digital agriculture adoption in developing regions, while market-driven approaches are more prevalent in developed economies. This analysis demonstrates that although digital agricultural technologies have transformative potential worldwide, the extent of their adoption and impact is influenced by socioeconomic conditions, infrastructure availability, and institutional support systems. Therefore, context-specific strategies are required to achieve inclusive and sustainable agricultural development.

Keywords

Precision agriculture · Sustainable farming · Artificial Intelligence (AI) · Blockchain · Internet of Things (IoT)

16.1 Technology Adoption Patterns in High- and Low-Income Countries

Technology adoption in agriculture results from a complex interplay of economic, structural, and institutional factors. National income levels, farm structure, resource endowment, market integration, policy support, and institutional capacity are particularly influential (Manono et al. 2025). In high-income countries, adoption levels are dominated by capital-intensive innovations such as precision agriculture, automation, and digitally integrated farm management systems (Finger 2023). These technologies are typically deployed in highly commercialized farming systems with large landholdings, strong access to capital, advanced infrastructure, and well-developed extension services (World Intellectual Property Organization (WIPO) 2026). The primary objective in these contexts is optimization: enhancing input-use efficiency, reducing labor dependency, minimizing environmental externalities, and ensuring compliance with stringent sustainability and traceability standards (Chen 2025).

Conversely, in low- and middle-income countries, agricultural technology adoption is primarily influenced by affordability, accessibility, and the need to address structural constraints (Manzoor et al. 2025). These constraints include fragmented landholdings, weak extension networks, limited access to formal credit, and inadequate market connectivity. Consequently, adoption pathways are often platform-based and service-oriented, utilizing mobile technologies, digital advisory systems, electronic payment platforms, and e-voucher schemes (Manzoor et al. 2025). In these contexts, technology first addresses critical access gaps in information, inputs, finance, and markets and subsequently enables more advanced functions such as process optimization and automation (Ayim et al. 2022).

The diffusion of agricultural technologies is inherently context-specific, reflecting differences in agrarian systems and disparities in economic capacity. Identical technologies may produce distinct outcomes depending on the socioeconomic environment. In developed contexts, they primarily enhance efficiency, whereas in developing systems, they enable basic access and inclusion. This comparative perspective highlights the necessity for tailored innovation strategies that align technological design and deployment with local constraints, capacities, and development priorities.

16.1.1 Developed Nations: Precision, Automation, and Supply Chain Integration

In high-income, industrialized economies, the development of smart agriculture has proceeded incrementally, building upon established systems marked by advanced commercialization, mechanization, and technological sophistication. Adoption typically commences with precision agriculture tools, including GPS-guided machinery, automated steering, yield monitoring, remote sensing, and variable-rate input application, and advances toward fully integrated digital ecosystems (Barnes et al. 2018). These ecosystems facilitate seamless interaction among farm machinery, in-field sensors, meteorological data, satellite imagery, and decision-support platforms, which collectively enhance the speed, accuracy, and efficiency of farm-level decision-making (Inoue 2020). In these settings, digital agriculture functions as an integral component of a broader, data-driven agri-food system rather than as a standalone intervention.

A key characteristic of technology adoption in developed countries is its integration throughout the entire agricultural value chain, from primary production to processing, distribution, and retail (European Commission 2025). Technologies such as blockchain and the Internet of Things (IoT) are implemented within established supply chains where traceability, food safety, sustainability, and product differentiation are essential market requirements (Bekkouche and De-Magistris 2025). Blockchain systems offer immutable and transparent records of product origin, production practices, and transaction histories. IoT infrastructures support continuous, real-time monitoring of biophysical and logistical parameters, such as soil moisture, irrigation regimes, storage conditions, and cold-chain integrity (Nagaraja et al. 2019). For instance, Carrefour's blockchain-based traceability initiative in France (Commandré and Marty 2025) illustrates how digital technologies can enhance consumer trust and supply chain coordination by providing end-to-end product transparency. Likewise, Dutch horticultural systems demonstrate advanced IoT integration, with sensor-driven greenhouse environments that regulate microclimatic conditions, nutrient delivery, irrigation scheduling, and energy use to achieve optimized production outcomes in a globally competitive, export-oriented sector (Wageningen University and Research 2019).

Overall, the adoption of digital agriculture in developed economies is characterized by high capital intensity, extensive platform integration, and compliance with stringent regulatory and market standards (Wang et al. 2026). In these contexts, digital technologies primarily serve to optimize efficiency, precision, and sustainability within established agricultural systems, rather than to address fundamental constraints.

16.1.2 Developing Nations: Leapfrogging, Resilience, and Financial Inclusion

In developing countries, digital agriculture adoption follows a structurally distinct pattern, primarily motivated by the need to address systemic constraints rather than to optimize already efficient systems. Key limiting factors, such as fragmented landholdings, weak extension services, limited access to formal credit, poor market integration, and high exposure to climatic risks, drive a problem-oriented approach to technological uptake (Manzoor et al. 2025). As a result, adoption focuses on affordability, accessibility, and scalability. Mobile-enabled technologies, public digital infrastructure, and low-cost service delivery models form the foundation of digital transformation (Manzoor et al. 2025). In this context, technology serves not only as a productivity-enhancing input but also as a mechanism for improving access to information, finance, inputs, and markets, thereby reducing risk and strengthening resilience.

Technological leapfrogging is a defining feature of this pathway, as developing economies often bypass intermediate mechanization stages and directly adopt digital solutions such as mobile advisory platforms, digital financial services, crop registries, and service-based mechanization systems (Manzoor et al. 2025). India illustrates this approach through initiatives like the Digital Agriculture Mission, which emphasizes the creation of a national digital backbone that includes farmer databases, crop digitization, and geospatial mapping (Ministry of Agriculture and Farmers Welfare 2024). This infrastructure supports targeted subsidy delivery, improved crop insurance, real-time advisory services, and enhanced market access for smallholder farmers. In Kenya, widespread mobile connectivity and digital payment systems have enabled the development of integrated, mobile-first agricultural platforms that offer agronomic guidance, financial services, input procurement, and market information (KilimoKwanza 2025). This approach facilitates incremental, user-driven adoption without dependence on capital-intensive hardware.

The deployment of advanced technologies, such as IoT and blockchain, in developing countries further demonstrates this problem-driven orientation. IoT-based systems are mainly used for cost-effective irrigation management and field-level monitoring, which improve resource-use efficiency under conditions of scarcity (Dhanaraju et al. 2022). Blockchain applications are primarily implemented in supply chains where traceability, transparency, and fraud mitigation are essential, especially for high-value or export-oriented commodities (Demestichas et al. 2020). Overall, digital agriculture in developing contexts is shaped by the need to enhance inclusion, resilience, and coordination. Technological systems are designed to bridge infrastructural and institutional gaps rather than solely to improve production efficiency.

16.1.3 Comparative Case Studies of Blockchain and IoT in Developed and Developing Countries Contexts

Blockchain and the IoT illustrate how identical digital technologies can produce context-dependent outcomes across diverse economic environments. In high-income countries, blockchain is predominantly utilized to meet advanced supply chain requirements, such as food safety assurance, end-to-end traceability, sustainability certification, and fostering consumer trust (Saha et al. 2024). These applications operate within structured value chains in which data integrity and transparency are essential for regulatory compliance and market differentiation (Demestichas et al. 2020). Likewise, IoT technologies are implemented to optimize operations in large-scale or controlled-environment agricultural systems by integrating sensor networks, predictive analytics, and automated control mechanisms to improve efficiency, precision, and resource utilization (Saha et al. 2023).

Conversely, in developing countries, these technologies serve more foundational and problem-solving roles. Blockchain is frequently employed to strengthen traceability systems, mitigate transaction fraud, improve record-keeping, and facilitate access to formal financial services by creating verifiable digital histories (Demestichas et al. 2020). IoT applications primarily target practical challenges, including optimizing irrigation scheduling, enabling weather-informed decision-making, monitoring crop health, and minimizing postharvest losses in resource-limited settings (Nsoh et al. 2024). In these environments, the primary focus is on addressing critical gaps in information, coordination, and risk management rather than on system optimization (Kampan et al. 2022).

The effectiveness of blockchain and IoT in agriculture depends not only on their technical attributes but also on their integration with institutional frameworks, governance structures, value-chain organization, and economic viability at the user level (Elegbeleye et al. 2025). Their impact is greatest when implementation strategies are tailored to address specific systemic bottlenecks, rather than relying on uniform or technology-driven approaches. Table 16.1 presents the application of agricultural technologies in developed and developing countries, highlighting differences in adoption, scale, and functional focus.

Table 16.1 Application of technologies in developed and developing countries

Country/region	Income context	Lead actor	Technology focus	Primary objective	References
The United States	Developed	Mixed public–private ecosystem	Variable-rate application, autopilot systems, remote sensing, decision-support systems	Improve on-farm precision, optimize production, reduce environmental impact	Barbosa Júnior et al. (2024)
European Union (Belgium, Germany, Greece, the Netherlands, the United Kingdom)	Developed	Public-policy-influenced/farmer-adoption system	Machine guidance, variable-rate nitrogen application	Improve efficiency and support sustainable intensification across diverse farm conditions	Barnes et al. (2018)

(continued)

Table 16.1 (continued)

Country/region	Income context	Lead actor	Technology focus	Primary objective	References
The Netherlands	Developed	Private/ industry and data-governance innovation	IoT-enabled data governance, sensor-linked horticultural systems	Support data-driven horticulture, interoperability, and advanced greenhouse decision-making	Amir et al. (2024)
India	Developing	Government-led digital architecture	Digital agriculture architecture, farm management information systems, registries, digital advisory	Build a national digital backbone for agricultural services and future AgriStack integration	Balkrishna et al. (2023)
India	Developing	Private sector	ICT-enabled kiosks, market information, agricultural decision support	Strengthen agricultural market linkages and improve farmer decision-making across the supply chain	Ali and Kumar (2011)
Kenya	Developing	Private/ service-platform ecosystem	Mobile money, digital payments, advisory, digital services for agriculture	Improve finance, service access, and agricultural coordination for smallholders	Kieti et al. (2022)
France/ broader agri-food retail systems	Developed	Private sector/ retailer-led supply-chain innovation	Blockchain traceability, provenance, transaction veracity, food-safety transparency	Improve supply-chain transparency, trust, and traceability in formal agri-food chains	Eze et al. (2025)

16.2 Government-Led and Private Sector Initiatives

Institutional drivers of agricultural technology adoption differ markedly across national contexts due to variations in economic structure, governance capacity, and market maturity (Gill 2020). In certain systems, governments assume a central role by investing in digital public infrastructure, formulating enabling policies, strengthening extension systems, and establishing regulatory and data standards. In contrast, private-sector actors may lead innovation by developing digital platforms, analytics, farm machinery ecosystems, and service delivery models. Large-scale agricultural transformation typically results from the complementary interactions of public and private stakeholders within an integrated innovation ecosystem, rather than the efforts of a single actor (Ragasa et al. 2025).

In developed countries, government intervention generally focuses on creating enabling conditions, such as investing in rural broadband connectivity, establishing environmental compliance frameworks, funding research and innovation, and developing interoperable data infrastructure (World Intellectual Property Organization (WIPO) 2026). Within this environment, the private sector advances commercialization and scaling by providing advanced equipment, digital platforms, decision-support tools, and integrated logistics networks (McKinsey and Company 2020). In contrast, governments in developing countries often play a more direct and foundational role by establishing core digital systems, including farmer registries, land records, crop databases, subsidy delivery mechanisms, and climate-smart extension services. These public systems form the foundation upon which private firms develop user-oriented solutions, such as mobile advisory platforms, financial services, and market linkages (Porciello et al. 2022). This layered approach, combining public infrastructure with private innovation, is essential for fostering inclusive, scalable, and context-appropriate digital agriculture development (Ferroni and Castle 2011).

16.2.1 Government-Led Initiatives: Policy, Public Infrastructure, and Extension Systems

Government leadership in the digitalization of agriculture is primarily achieved through the provision of public goods, reduction of transaction costs, and enhancement of system-wide coordination. In developed economies, this leadership is evident in the establishment of enabling infrastructure and regulatory frameworks, such as rural broadband expansion, geospatial information systems, satellite-based monitoring, and targeted incentives, such as precision agriculture grants (Sanders et al. 2022). In the European Union, digital agriculture is integrated with climate policy objectives, nutrient management regulations, and conditional agricultural support mechanisms, with public institutions guiding the adoption of technology toward sustainability outcomes (European Commission 2025). In the United States, federal and state agencies promote the diffusion of digital technologies by investing in research, extension services, geospatial data provision, and connectivity policies, thereby fostering an environment conducive to private sector innovation (World Economic Forum 2023).

In contrast, developing countries often adopt a more contextual and foundational approach, with governments actively building digital infrastructure to support agricultural transformation (Manzoor et al. 2025). India serves as a prominent example through initiatives such as the Digital Agriculture Mission and AgriStack-type frameworks, which seek to establish interoperable systems including farmer registries, digital crop surveys, and geospatial databases. These platforms increase transparency in program delivery, enable more precise targeting of subsidies, improve crop insurance mechanisms, and provide real-time advisory services (Ministry of Agriculture and Farmers Welfare 2022). Furthermore, climate-smart extension systems, such as those implemented under NICRA and through Krishi Vigyan Kendras (KVKs), highlight the essential role of public institutions in disseminating localized, risk-responsive agricultural knowledge. By addressing informational and institutional gaps, these interventions promote broader inclusivity and support the adoption of resilient farming practices among smallholder farmers (Samuel et al. 2024).

16.2.2 Private Sector Initiatives: Platforms, Agribusiness Innovation, and Service Delivery

Private-sector innovation is a primary catalyst for agricultural technology adoption, integrating hardware, data, and services into comprehensive digital ecosystems. In addition to equipment manufacturing, private firms create value through advanced data analytics, telematics, interoperable farm management software, and subscription-based decision-support systems, which facilitate efficient digital workflows across farm operations. In high-income countries, these innovations extend into downstream value chains, where private actors are central to supply-chain traceability, product differentiation, and adherence to rigorous food-safety and sustainability standards (Charlebois et al. 2024). Retailer-led blockchain platforms, verification systems, and analytics-enabled reporting frameworks are increasingly incorporated into high-value agri-food markets, thereby strengthening consumer trust, regulatory compliance, and premium market positioning (Karkanias et al. 2026).

In developing countries, private sector innovation is typically oriented toward service-based, platform-driven models that address structural inefficiencies and access limitations (MarketsandMarkets 2023). Instead of prioritizing capital-intensive machinery, firms establish digital ecosystems that facilitate access to information, inputs, finance, and markets. Examples such as Safaricom's DigiFarm in Kenya (Safaricom 2018) and ITC's e-Choupal in India (APAARI 2014) offer integrated services, including agronomic advisory, digital payments, access to credit, and market linkages. These models aim to mitigate information asymmetries, reduce transaction costs, and enhance coordination among smallholder farmers and market participants (FAO 2022). Their effectiveness depends on affordability, mobile accessibility, and the delivery of immediate, tangible benefits such as weather forecasts, input recommendations, and assured procurement channels (Fernandez-Vidal and Alarcon 2025). Therefore, while private innovation in developed economies is primarily driven by efficiency, interoperability, and value-chain optimization, in developing contexts it is focused on improving inclusion, accessibility, and functional integration within fragmented agricultural systems (World Intellectual Property Organization (WIPO) 2026).

16.2.3 Public–Private Complementarity and Lessons from Comparative Implementation

Cross-country analyses reveal that public and private actors in agricultural digitalization play inherently complementary roles rather than substitutable ones (Kukk et al. 2022). Governments are optimally positioned to deliver foundational public goods, such as shared digital infrastructure, trusted registries, regulatory standards, integrated extension systems, and safeguards for public interests, including data governance and equity (OECD 2024). Conversely, private sector actors demonstrate strengths in user-centric innovation, rapid technological iteration, service delivery, logistics optimization, and the creation of scalable, market-oriented solutions (Benfica et al. 2023).

The effectiveness of national digital agriculture ecosystems depends on the extent of coordination between the public and private sectors (Doukas et al. 2022). Public digital infrastructure without private sector engagement risks underutilization (Tian et al. 2026). Conversely, private innovation lacking interoperable standards, regulatory clarity, and trust frameworks can result in fragmented and non-scalable digital systems (Adegoke et al. 2025). This interdependence is especially significant in agriculture, where data ownership, farmer trust, and institutional alignment critically affect adoption and scalability (Duncan et al. 2026). Synchronized public and private collaboration is therefore essential to achieve inclusive, efficient, and sustainable digital transformation in agriculture. Table 16.2 summarizes patterns of technology adoption across countries and the key agronomic, economic, and environmental outcomes associated with their use.

Table 16.2 Technology adoption and associated outcomes across different countries

Country/ region/context	Technology/ intervention	Main outcome dimension	Key findings	References
Global, multi-country meta-analysis	Technologies for precision agriculture	Profitability, environmental footprint, input-use efficiency	A meta-analysis of 85 real-world studies shows that precision agriculture, when properly implemented, increases ROI by 22.3% and net profit by 18.5%, while improving nitrogen-use efficiency by 15.1% and reducing pesticide use by 12.8% and greenhouse gas emissions by 9.4%.	Lan and Ban (2025)
Global review	Farm automation in field crops	Environmental footprint, cost	Automation can reduce environmental impacts by improving resource-use efficiency and enabling more precise input application, though the extent of benefits varies with crop type, region, and technology used.	Bahmutsky et al. (2024)

(continued)

Table 16.2 (continued)

Country/ region/context	Technology/ intervention	Main outcome dimension	Key findings	References
Brazil, sugarcane farming	Precision agriculture adoption	Productivity, technical efficiency	The adoption of precision agriculture improved technical efficiency and narrowed technology gaps, indicating that digital tools can enhance managerial performance even without significant yield gains.	Carrer et al. (2022)
Argentina, 5000 ha dryland farm	Zone-based precision agriculture	Productivity, profitability, risk reduction	Whole-farm precision management increases production and profitability while reducing risk, demonstrating that advanced precision systems deliver both agronomic and economic benefits at the commercial scale.	Monzon et al. (2018)
Ghana, smallholder context	Agricultural digitalization broadly defined	Productivity, livelihoods, inclusion	Digitalization in agriculture improves smallholder livelihoods by enhancing decision-making, market access, and overall welfare, extending benefits beyond yield gains in developing regions.	Addison et al. (2024)
Sub-Saharan Africa	Digital agriculture tools for smallholders	Productivity, resilience, adoption conditions	In Sub-Saharan Africa, digital technologies support marketing, extension services, and farmer advisory services, with the potential to enhance productivity and sustainability; however, their effectiveness depends strongly on infrastructure, affordability, and digital literacy.	Choruma et al. (2024)
Agri-food/ livestock supply chains	Blockchain traceability	Food safety, transparency, waste reduction	Blockchain enhances trust, efficiency, and security by enabling transparent, tamper-proof traceability from source to consumer, thereby reducing fraud, improving recall systems, and supporting supply-chain sustainability.	Patel et al. (2023)

(continued)

Table 16.2 (continued)

Country/region/context	Technology/intervention	Main outcome dimension	Key findings	References
Agri-food supply chains: review context	Blockchain in food supply chains	Traceability, sustainability, environmental management	Blockchain can enhance traceability, transparency, and sustainability management in food supply chains, with important implications for reducing waste and improving resource governance.	Ellahi et al. (2024)
Northwest India	Precision nutrient management in conservation agriculture	Profitability, nutrient-use efficiency, environmental footprint	Yield, nutrient-use efficiency, profitability, and GHG emissions are systematically assessed, demonstrating that precision nutrient management can enhance farm returns while reducing environmental impacts in developing-country contexts.	Sapkota et al. (2013)
US Midwest	Precision agriculture from farmer's perspective	Profitability, environmental performance	A survey of 1119 farmers identified profitability as the primary driver of precision agriculture adoption, with respondents also recognizing its benefits for site-specific management and environmental performance.	Wang et al. (2023)

16.3 Outcomes and Impact Metrics: Productivity, Profitability, and Environmental Footprint

The evaluation of digital agriculture technologies should prioritize measurable outcomes over technological novelty, with productivity, profitability, and environmental impact serving as the primary indicators. These outcomes are highly context-dependent and vary according to technology type, crop system, farm scale, institutional support, and national economic conditions. In developed countries, digital agriculture generally produces incremental but economically significant gains by improving timing, precision, reducing redundancy, lowering labor requirements, and enhancing regulatory compliance. Conversely, in developing countries, the impacts are often more transformative, especially when digital tools address fundamental constraints such as limited access to advisory services, credit, insurance, markets, and mechanization (Zhang and Zhu 2025). Therefore, impact assessment frameworks should be adapted to system-specific baselines rather than applying uniform expectations.

16.3.1 Productivity Outcomes: Yield, Efficiency, and Decision Quality

Productivity gains from smart agriculture are not always reflected solely in yield increases. In developed agricultural systems, improvements are primarily driven by enhanced decision-making, precise input application, reduced operational delays, and optimized labor allocation (Ye et al. 2024). Precision agriculture technologies facilitate spatially explicit management, aligning inputs with field variability to increase efficiency and minimize waste (Saha et al. 2025). In developing contexts, productivity improvements are often more substantial when technologies address previously unmet needs. Mobile-based advisory systems, weather forecasting, sensor-enabled irrigation, and pest surveillance tools can significantly increase yields by reducing uncertainty and improving the timeliness and quality of agronomic decisions (Kumar et al. 2025). As a result, productivity should be evaluated using multidimensional metrics, such as output per unit area, labor productivity, input-use efficiency, operational timeliness, and decision quality.

16.3.2 Profitability Outcomes: Cost Reduction, Market Access, and Risk Mitigation

Profitability impacts emerge through various pathways depending on the agricultural context (OECD 2017). In developed countries, gains are primarily achieved through cost optimization, including reduced input use, improved machinery efficiency, lower fuel consumption, and enhanced planning (Mühl et al. 2025). Digital tools also support market differentiation by enabling traceability and sustainability certification, thereby allowing products to command premium pricing (Villar et al. 2023). In developing countries, improvements in profitability are more closely associated with reduced transaction costs, improved price discovery, strengthened market linkages, and greater access to credit and insurance (Azevedo et al. 2023). Digital platforms and blockchain-based systems further enhance value addition by increasing trust, transparency, and traceability within specific supply chains (Pizzileo et al. 2025). Profitability should be understood as encompassing not only yield gains but also avoided losses, reduced uncertainty, improved bargaining power, and more stable integration into value chains.

16.3.3 Environmental Footprint: Resource Efficiency, Emissions, and Waste Reduction

Environmental performance is an increasingly important criterion for evaluating agricultural technologies. In developed systems, precision agriculture tools such as variable-rate application, site-specific irrigation, and sensor-based monitoring help reduce input overuse, improve water-use efficiency, and lower greenhouse gas emissions, supporting regulatory and sustainability objectives (Pizzileo et al. 2025). In developing countries, environmental benefits often result from improved crop planning, climate-smart advisory systems, optimized irrigation timing, reduced input misuse, and the promotion of sustainable practices such as agroforestry (FAO 2014). Technologies that reduce postharvest losses or enhance supply chain coordination can also indirectly decrease environmental impacts (Olaru et al. 2024). Blockchain technology primarily contributes by improving traceability and reducing waste in logistics systems, while the IoT provides direct environmental benefits through real-time monitoring and decision support for resource management (Marchese and Tomarchio 2022). Environmental impact assessment should be based on quantifiable indicators, including water- and input-use efficiency, emissions intensity, waste reduction, and pollution mitigation, to ensure that sustainability claims are both evidence-based and context-specific.

16.4 Barriers to Scalable Implementation and Key Lessons Learned

Although smart agriculture demonstrates significant potential, its widespread adoption is hindered by persistent and interconnected barriers. These barriers include affordability, limitations in connectivity and interoperability, constraints in human capabilities, and challenges related to data governance and institutional trust. The severity and nature of these obstacles vary between developed and developing regions, yet they collectively shape adoption patterns and scalability. High initial and ongoing costs for precision hardware, digital platforms, and analytics services often restrict adoption, especially among smallholder farmers (FAO 2022). Weak connectivity, fragmented digital ecosystems, and insufficient interoperability further diminish the practical value of these technologies. Additionally, successful adoption depends on farmer awareness, digital literacy, and access to advisory support, while unresolved concerns regarding data ownership, privacy, and trust complicate integration into agricultural systems.

16.4.1 Foundational Barriers: Economic Feasibility, Connectivity, and System Integration

Economic feasibility is a primary constraint in digital agriculture. Advanced technologies, including precision farming systems, IoT networks, and cloud-based analytics, require substantial capital investment and ongoing operational costs, which raises concerns about return on investment, particularly for smaller farms. In developing countries, affordability issues are even more pronounced, often restricting access to basic digital tools and services (Dibbern et al. 2024). Infrastructure deficiencies further intensify these challenges, as digital agriculture relies on dependable broadband connectivity, electricity supply, and data exchange systems. In many rural areas, limited Internet coverage, unstable power supply, and weak service ecosystems impede the effective deployment and maintenance of technologies (Guo et al. 2026). Interoperability is an additional barrier, as diverse platforms, devices, and software systems often lack compatibility, leading to fragmented workflows, vendor lock-in, and diminished user confidence. These factors highlight that scalability requires not only technological availability but also an enabling environment with affordable access, reliable infrastructure, and integrated digital systems.

16.4.2 Systemic Barriers: Human Capacity, Trust, Governance, and Institutional Alignment

In addition to infrastructure and cost, the adoption of digital agriculture is strongly influenced by human and institutional factors. Successful adoption requires not only access to technology but also the ability to understand, trust, and use it effectively. Limited digital literacy, absence of localized content, and inadequate advisory support can significantly hinder uptake, especially among smallholder farmers. Concerns related to trust, such as data privacy, security, ownership, and transparency in algorithmic decision-making, are increasingly significant as agriculture becomes more data-driven. Institutional coordination is also essential, as digital agriculture intersects with advisory services, market systems, financial services, and data governance. Fragmented or misaligned institutional frameworks can result in duplicated efforts, decreased efficiency, and restricted scalability.

Governance challenges are increasingly recognized as critical determinants of long-term adoption in both developed and developing contexts. Issues such as data ownership, benefit-sharing, platform dependency, and regulatory oversight directly affect farmer confidence and participation. Research indicates that scalable digital agriculture systems require not only technological innovation but also strong governance frameworks, participatory planning, and collaboration among multiple stakeholders (Bekee et al. 2024). Comparative analysis reveals that successful scaling occurs when technologies address specific problems, are economically viable, user-focused, and supported by interoperable infrastructure and reliable institutional systems (Yeo and Keske 2024).

16.5 Conclusion and Future Directions

The adoption of agricultural technology is inherently path-dependent, reflecting historical trajectories, institutional capacities, and economic structures across countries. High-income nations have advanced through mechanization, precision agriculture, and integrated data-driven farming systems. In contrast, developing countries have prioritized mobile-enabled services, public digital platforms, and low-cost innovations tailored to smallholder farming. Government support remains a critical driver in both contexts, with its effectiveness contingent upon the creation of enabling ecosystems. These ecosystems encompass interoperable systems, standardized data protocols, digital registries, robust extension networks, and trust frameworks. Concurrently, the private sector plays a vital role in translating these foundations into practical, user-centered solutions and scalable business models. The most impactful technologies are those that deliver multiple benefits, enhancing productivity and profitability while reducing environmental impacts. However, scaling these innovations cannot rely on replicating a singular model of smart agriculture, as local context remains a critical determinant. Future progress will require designing location-specific innovation ecosystems that align infrastructure, institutions, incentives, and value chain requirements. Strengthening farmer capacity and digital literacy is equally essential. A balanced approach will ensure that digital agriculture develops as an inclusive, efficient, and environmentally sustainable pathway for global food systems.

References

Addison M, Bonuedi I, Arhin AA et al (2024) Exploring the impact of agricultural digitalization on smallholder farmers' livelihoods in Ghana. Heliyon 10:e27541. https://doi.org/10.1016/j.heliyon.2024.e27541

Adegoke K, Adegoke A, Dawodu D et al (2025) Interoperability as a catalyst for digital health and therapeutics: a scoping review of emerging technologies and standards (2015–2025). Int J Environ Res Public Health 22:1535. https://doi.org/10.3390/ijerph22101535

Ali J, Kumar S (2011) Information and communication technologies (ICTs) and farmers' decision-making across the agricultural supply chain. Int J Inf Manag 31:149–159. https://doi.org/10.1016/j.ijinfomgt.2010.07.008

Amir A, Vinke C, Heijningen R (2024) Horti-IoT: a new data framework for Dutch horticulture. Int Food Agribus Manag Rev 27:1–24. https://doi.org/10.22434/ifamr1124

APAARI (2014) ITC e-Choupal—innovation for large-scale rural transformation: a success story. Asia-Pacific Association of Agricultural Research Institutions, Bangkok

Ayim C, Kassahun A, Addison C et al (2022) Adoption of ICT innovations in the agriculture sector in Africa: a review of the literature. Agric Food Secur 11:22. https://doi.org/10.1186/s40066-022-00364-7

Azevedo P, Gomes J, Romão M (2023) Supply chain traceability using blockchain. Oper Manag Res 16:1359–1381. https://doi.org/10.1007/s12063-023-00359-y

Bahmutsky S, Grassauer F, Arulnathan V et al (2024) A review of life cycle impacts and costs of precision agriculture for cultivation of field crops. Sustainable Prod Consumption 52:347–362. https://doi.org/10.1016/j.spc.2024.11.010

Balkrishna A, Pathak R, Kumar S et al (2023) A comprehensive analysis of the advances in Indian digital agricultural architecture. Smart Agric Technol 5:100318. https://doi.org/10.1016/j.atech.2023.100318
Barbosa Júnior MR, Moreira BRDA, Carreira VDS et al (2024) Precision agriculture in the United States: a comprehensive meta-review inspiring further research, innovation, and adoption. Comput Electron Agric 221:108993. https://doi.org/10.1016/j.compag.2024.108993
Barnes A, Soto I, Eory V et al (2018) Exploring the adoption of precision agricultural technologies: a cross regional study of EU farmers. Land Use Policy 80:163–174. https://doi.org/10.1016/j.landusepol.2018.10.004
Bekee B, Segovia MS, Valdivia C (2024) Adoption of smart farm networks: a translational process to inform digital agricultural technologies. Agric Hum Values 41:1573–1590. https://doi.org/10.1007/s10460-024-10566-3
Bekkouche S, De-Magistris T (2025) Digitalization in the European agri-food supply chain: a scoping review of traceability, transparency, and sustainability. Front Blockchain 8:1701872. https://doi.org/10.3389/fbloc.2025.1701872
Benfica R et al (2023) Food system innovations and digital technologies to foster productivity growth and rural transformation. In: von Braun J, Afsana K, Fresco LO et al (eds) Science and innovations for food systems transformation. Springer, Cham. https://doi.org/10.1007/978-3-031-15703-5_22
Carrer MJ, Filho HMDS, Vinholis MDMB et al (2022) Precision agriculture adoption and technical efficiency: an analysis of sugarcane farms in Brazil. Technol Forecast Soc Change 177:121510. https://doi.org/10.1016/j.techfore.2022.121510
Charlebois S, Latif N, Ilahi I et al (2024) Digital traceability in agri-food supply chains: a comparative analysis of OECD member countries. Foods 13(7). https://doi.org/10.3390/foods13071075
Chen X (2025) The role of modern agricultural technologies in improving agricultural productivity and land use efficiency. Front Plant Sci 16:1675657. https://doi.org/10.3389/fpls.2025.1675657
Choruma DJ, Dirwai TL, Mutenje MJ et al (2024) Digitalisation in agriculture: a scoping review of technologies in practice, challenges, and opportunities for smallholder farmers in sub-Saharan Africa. J Agric Food Res 18:101286. https://doi.org/10.1016/j.jafr.2024.101286
Commandré Y, Marty J (2025) Blockchain in the French agri-food sector: redefining transparency, standardization and power dynamics. J Sci Technol Policy Manag:1–24. https://doi.org/10.1108/JSTPM-12-2024-0473
Demestichas K, Peppes N, Alexakis T et al (2020) Blockchain in agriculture traceability systems: a review. Appl Sci 10(12). https://doi.org/10.3390/app10124113
Dhanaraju M, Chenniappan P, Ramalingam K et al (2022) Smart farming: internet of things (IoT)-based sustainable agriculture. Agriculture 12(10). https://doi.org/10.3390/agriculture12101745
Dibbern T, Romani LAS, Massruhá SMFS (2024) Main drivers and barriers to the adoption of digital agriculture technologies. Smart Agric Technol 8:100459. https://doi.org/10.1016/j.atech.2024.100459
Doukas YEL, Maravegias N, Chrysomallidis C (2022) Digitalization in the EU agricultural sector: seeking a European policy response. In: Mattas K, Baourakis G, Zopounidis C et al (eds) Food policy modelling. Springer, Cham. https://doi.org/10.1007/978-3-031-08317-4_6
Duncan E, Ruder S, Krishna Bahadur KC et al (2026) The political economy of agricultural data governance: farmer perspectives on trust in data sharing. J Rural Stud 123:104076. https://doi.org/10.1016/j.jrurstud.2026.104076
Elegbeleye F, Oguntona O, Elegbeleye I (2025) Blockchain and IoT for sustainable agriculture: innovations and impacts. J Inf Syst Inform 7:2038–2049. https://doi.org/10.51519/journalisi.v7i2.1136
Ellahi RM, Wood LC, Bekhit AED (2024) Blockchain-driven food supply chains: a systematic review for unexplored opportunities. Appl Sci 14(19). https://doi.org/10.3390/app14198944
European Commission (2025) Digitalisation of agriculture and rural areas in the EU. https://agriculture.ec.europa.eu/overview-vision-agriculture-food/digitalisation_en
Eze VHU, Eze EC, Alaneme GU et al (2025) Integrating IoT sensors and machine learning for sustainable precision agroecology: enhancing crop resilience and resource efficiency through data-

driven strategies, challenges, and future prospects. Discov Agric 3:83. https://doi.org/10.1007/s44279-025-00247-y
FAO (2014) Climate-smart agriculture: policies, practices and financing for food security, adaptation and mitigation. FAO
FAO (2022) The state of food and agriculture 2022: leveraging automation in agriculture for transforming agrifood systems. FAO, Rome. https://doi.org/10.4060/cb9479en
Fernandez-Vidal J, Alarcon S (2025) Financing agricultural innovation: challenges and alternatives to venture capital in the AgTech sector. Food Policy 136:102967. https://doi.org/10.1016/j.foodpol.2025.102967
Ferroni M, Castle P (2011) Public-private partnerships and sustainable agricultural development. Sustainability 3:1064–1073. https://doi.org/10.3390/su3071064
Finger R (2023) Digital innovations for sustainable and resilient agricultural systems. Eur Rev Agric Econ 50:1277–1309. https://doi.org/10.1093/erae/jbad021
Gill IS (2020) Technology adoption by firms in developing countries. World Bank
Guo X, Lv J, Zhu R (2026) Digital technology and sustainable agriculture: evidence from Henan Province, China. Sustainability 18(2). https://doi.org/10.3390/su18020780
Inoue Y (2020) Satellite- and drone-based remote sensing of crops and soils for smart farming – a review. Soil Sci Plant Nutr 66:798–810. https://doi.org/10.1080/00380768.2020.1738899
Kampan K, Tsusaka TW, Anal AK (2022) Adoption of blockchain technology for enhanced traceability of livestock-based products. Sustainability 14(20). https://doi.org/10.3390/su142013148
Karkanias C, Malamakis A, Banias GF (2026) Blockchain and megatrends in agri-food systems: a multi-source evidence approach. Foods 15(3). https://doi.org/10.3390/foods15030447
Kieti J, Waema TM, Baumüller H et al (2022) What really impedes the scaling out of digital services for agriculture? A Kenyan users' perspective. Smart Agric Technol 2:100034. https://doi.org/10.1016/j.atech.2022.100034
KilimoKwanza (2025) The value chain finance ecosystem in Kenya: a comprehensive market analysis and strategic map. Report, October 21, 2025
Kukk M, Põder A, Viira AH (2022) The role of public policies in the digitalisation of the agri-food sector: a systematic review. NJAS Impact Agric Life Sci 94:217–248. https://doi.org/10.1080/27685241.2022.2147870
Kumar SN, Suriyan K, Jacob AT et al (2025) Smart farming for a sustainable future: implementing IoT-based systems in precision agriculture. Bull Natl Res Cent 49:71. https://doi.org/10.1186/s42269-025-01366-8
Lan J, Ban Q (2025) The farm-level economic and environmental benefits of precision agriculture technology adoption: a meta-analysis of global evidence. Sustainability 17(24). https://doi.org/10.3390/su172411223
Manono BO, Khan S, Kithaka KM (2025) A review of the socio-economic, institutional, and biophysical factors influencing smallholder farmers' adoption of climate smart agricultural practices in Sub-Saharan Africa. Earth 6(2). https://doi.org/10.3390/earth6020048
Manzoor F, Wei L, Siraj M et al (2025) Digital agriculture technology adoption in low and middle-income countries—a review of contemporary literature. Front Sustain Food Syst 9:1621851. https://doi.org/10.3389/fsufs.2025.1621851
Marchese A, Tomarchio O (2022) A blockchain-based system for agri-food supply chain traceability management. SN Comput Sci 3:279. https://doi.org/10.1007/s42979-022-01148-3
MarketsandMarkets (2023) Smart agriculture market by offering (hardware, software, services), application, and region – global forecast to 2028
McKinsey & Company (2020) Agriculture's connected future: how technology can yield new growth
Ministry of Agriculture & Farmers Welfare (2022) India digital ecosystem of agriculture (IDEA) framework. Government of India
Ministry of Agriculture & Farmers Welfare (2024) Cabinet approves digital agriculture mission with an outlay of ₹2,817 crore. Press Information Bureau (PIB), Government of India

Monzon J, Calviño P, Sadras V et al (2018) Precision agriculture based on crop physiological principles improves whole-farm yield and profit: a case study. Eur J Agron 99:62–71. https://doi.org/10.1016/j.eja.2018.06.011

Mühl DD, Dos Santos Ramos F, De Oliveira L (2025) Price transparency and digital traceability: a framework for fairer, more sustainable food supply chains. Comput Electron Agric 238:110807. https://doi.org/10.1016/j.compag.2025.110807

Nagaraja GS, Soppimath AB, Soumya T et al (2019) IoT based smart agriculture management system. In: 2019 4th international conference on computational systems and information technology for sustainable solution (CSITSS). IEEE, pp 1–5. https://doi.org/10.1109/CSITSS47250.2019.9031025

Nsoh B, Katimbo A, Guo H et al (2024) Internet of things-based automated solutions utilizing machine learning for smart and real-time irrigation management: a review. Sensors 24(23). https://doi.org/10.3390/s24237480

OECD (2017) Improving energy efficiency in the agro-food chain. OECD green growth studies. OECD Publishing, Paris. https://doi.org/10.1787/9789264278530-en

OECD (2024) Digital public infrastructure for digital governments. OECD Public Governance Policy Papers No 68. OECD Publishing, Paris. https://doi.org/10.1787/ff525dc8-en

Olaru D, Smith B, Siddique KHM (2024) Decision-making in grain supply chains for sustainable food system. Sustainability 16(24). https://doi.org/10.3390/su162410816

Patel A, Brahmbhatt M, Bariya A et al (2023) Blockchain technology in food safety and traceability concern to livestock products. Heliyon 9:e16526. https://doi.org/10.1016/j.heliyon.2023.e16526

Pizzileo G, Colizzi L, Guerriero E et al (2025) Resource use efficiency and environmental sustainability in greenhouse agriculture through IoT-based irrigation and fertilization management. Smart Agric Technol 12:101180. https://doi.org/10.1016/j.atech.2025.101180

Porciello J, Coggins S, Mabaya E et al (2022) Digital agriculture services in low- and middle-income countries: a systematic scoping review. Glob Food Secur 34:100640. https://doi.org/10.1016/j.gfs.2022.100640

Ragasa C, Spielman D, Lynam J (2025) Agricultural innovation frames, policies, and instruments: evolution, lessons, and future research. Food Policy 137:103008. https://doi.org/10.1016/j.foodpol.2025.103008

Safaricom (2018) DigiFarm: transforming smallholder agriculture in Kenya. Safaricom Report/GSMA Case Study

Saha P, Kumar V, Kathuria S et al (2023) Precision agriculture using internet of things and wireless sensor networks. In: 2023 international conference on disruptive technologies (ICDT), Greater Noida, pp 519–522. https://doi.org/10.1109/ICDT57929.2023.10150678

Saha A, Raut RD, Kumar M (2024) Leveraging blockchain technology to combat food fraud in the agri-food supply chain. Int J Food Sci Technol 59:3469–3477. https://doi.org/10.1111/ijfs.16889

Saha S, Kucher OD, Utkina AO et al (2025) Precision agriculture for improving crop yield predictions: a literature review. Front Agron 7:1566201. https://doi.org/10.3389/fagro.2025.1566201

Samuel J, Rama Rao CA, Anshida Beevi CN et al (2024) Enhancing farm income resilience through climate smart agriculture in drought-prone regions of India. Front Water 6:1327651. https://doi.org/10.3389/frwa.2024.1327651

Sanders CE, Gibson KE, Lamm AJ (2022) Rural broadband and precision agriculture: a frame analysis of United States federal policy outreach under the Biden administration. Sustainability 14(1). https://doi.org/10.3390/su14010460

Sapkota TB, Majumdar K, Jat M et al (2013) Precision nutrient management in conservation agriculture-based wheat production of Northwest India: profitability, nutrient use efficiency and environmental footprint. Field Crop Res 155:233–244. https://doi.org/10.1016/j.fcr.2013.09.001

Tian X, Zhai Y, Sun Z (2026) Digital public infrastructure and agricultural modernization: causal evidence from the broadband China policy. Sustainability 18(5). https://doi.org/10.3390/su18052644

Villar PF, Kozakiewicz T, Bachina V et al (2023) The effects of agricultural output market access interventions on agricultural, socio-economic and food and nutrition security outcomes in low- and middle-income countries: a systematic review. Campbell Syst Rev 19:e1348. https://doi.org/10.1002/cl2.1348

Wageningen University & Research (2019) Greenhouse horticulture in the Netherlands: advanced technology and innovation. Wageningen University & Research Report

Wang T, Jin H, Sieverding H et al (2023) Understanding farmer views of precision agriculture profitability in the U.S. Midwest. Ecol Econ 213:107950. https://doi.org/10.1016/j.ecolecon.2023.107950

Wang X, Wu C, Wu Q et al (2026) Digital agriculture: past, present, and future. Adv Intell Discov 2:e202500193. https://doi.org/10.1002/aidi.202500193

World Economic Forum (2023) How digital public infrastructure could transform the agriculture landscape

World Intellectual Property Organization (WIPO) (2026) World intellectual property report 2026: technology on the move. WIPO, Geneva. https://doi.org/10.34667/tind.59025

Ye Z, Yin S, Cao Y et al (2024) AI-driven optimization of agricultural water management for enhanced sustainability. Sci Rep 14:25721. https://doi.org/10.1038/s41598-024-76915-8

Yeo ML, Keske CM (2024) From profitability to trust: factors shaping digital agriculture adoption. Front Sustain Food Syst 8:1456991. https://doi.org/10.3389/fsufs.2024.1456991

Zhang H, Zhu H (2025) The impact of agricultural digitization on land productivity: an empirical test based on micro panel data. Land 14(1). https://doi.org/10.3390/land14010187

FSC
www.fsc.org
MIX
Papier | Fördert
gute Waldnutzung
FSC® C083411

Kolibri 360 GmbH
Ferdinand-Jühlke-Straße 7
99095 Erfurt, Deutschland
produktsicherheit@kolibri360.de